Fundamentals of Ultra-Dense Wireless Networks

Discover the fundamental characteristics of ultra-dense networks with this comprehensive text. Featuring a consistent mathematical description of ultra-dense small cell networks while also covering real-world issues such as network deployment, operation and optimization, this book investigates performance metrics of coverage probability and area spectral efficiency (ASE) and addresses the aspects of ultra-dense networks that make them different from current networks. Insightful intuitions, which will assist decision-makers as they migrate their services, are explained and mathematically proven.

The book presents the latest review of research outcomes on ultra-dense networks, based on both theoretical analyses and network simulations, includes more than 200 sources from 3GPP, the Small Cell Forum, journals and conference proceedings, and covers all other related and prominent topics. This is an ideal reference text for professionals who are dealing with the development, deployment, operation and maintenance of ultra-dense small cell networks, as well as researchers and graduate students in communications.

David López-Pérez is a technical leader at Huawei Technologies France. He was recognized as a Bell Labs Distinguished Member of Staff in 2019 and is a co-author of *Heterogeneous Cellular Networks: Theory, Simulation and Deployment* (Cambridge, 2013).

Ming Ding is a senior research scientist at Data61, CSIRO, Australia. He served as one of the key members in the Sharp Company Best Team: LTE Standardization Patent Portfolio.

Fundamentals of Ultra-Dense Wireless Networks

DAVID LÓPEZ-PÉREZ

Advanced Wireless Technology Laboratory, Huawei Technologies

MING DING

Data61, CSIRO

CAMBRIDGE
UNIVERSITY PRESS

University Printing House, Cambridge CB2 8BS, United Kingdom

One Liberty Plaza, 20th Floor, New York, NY 10006, USA

477 Williamstown Road, Port Melbourne, VIC 3207, Australia

314–321, 3rd Floor, Plot 3, Splendor Forum, Jasola District Centre, New Delhi – 110025, India

103 Penang Road, #05–06/07, Visioncrest Commercial, Singapore 238467

Cambridge University Press is part of the University of Cambridge.

It furthers the University's mission by disseminating knowledge in the pursuit of education, learning, and research at the highest international levels of excellence.

www.cambridge.org
Information on this title: www.cambridge.org/9781108484695
DOI: 10.1017/9781108689274

© Cambridge University Press 2022

First published 2022

Printed in the United Kingdom by TJ Books Limited, Padstow Cornwall

A catalogue record for this publication is available from the British Library.

Library of Congress Cataloging-in-Publication Data
Names: López-Pérez, David, 1981– author. | Ding, Ming, 1981– author.
Title: Fundamentals of ultra-dense wireless networks / David López-Pérez, Ming Ding.
Description: Cambridge ; New York, NY : Cambridge University Press, 2022.
 | Includes bibliographical references and index.
Identifiers: LCCN 2021048856 (print) | LCCN 2021048857 (ebook)
 | ISBN 9781108484695 (hardback) | ISBN 9781108689274 (epub)
Subjects: LCSH: Wireless communication systems. | BISAC: TECHNOLOGY &
 ENGINEERING / Mobile & Wireless Communications
Classification: LCC TK5103.2 .L66 2021 (print) | LCC TK5103.2 (ebook) |
 DDC 621.384–dc23/eng/20211108
LC record available at https://lccn.loc.gov/2021048856
LC ebook record available at https://lccn.loc.gov/2021048857

ISBN 978-1-108-48469-5 Hardback

Importantly, the authors would also like to dedicate this book to anyone who has been affected, in any form or shape, by COVID-19 – a terrible disease, which has made the entire world suffer during the time this book was written. We pray for every soul on this planet that was tragically taken away by COVID-19.

Book Updates

The authors of this book would appreciate any potential feedback – of any form and kind – from the readers to further improve the content and the presentation of the subject matter of this book.

Contents

About the Authors

David López-Pérez is a telecommunications engineer who has devoted most of his career to the study of both cellular and Wi-Fi networks, where his main research interests are in network performance analysis, both theoretical and simulation-based, network planning and optimization, as well as technology and feature development. López-Pérez's main contributions are around the understanding of small cells and ultra-dense networks. He has also pioneered work on cellular and Wi-Fi interworking, and investigated both multi-antenna capabilities and ultra-reliable low latency features for future indoor networks. López-Pérez was recognized as Bell Labs Distinguished Member of Staff in 2019, has authored a book on small cells, and has published more than 150 research manuscripts on a variety of related topics. He has filed 52 patent applications with more than 25 granted as of today, and has received a number of prestigious awards. He is an editor of *IEEE Transactions on Wireless Communications*.

Ming Ding is a senior research scientist at Data61, CSIRO, in Sydney, NSW, Australia. His research interests include information technology, wireless communication networks, data privacy and security, and machine learning and artificial intelligence. He has authored more than 150 papers in IEEE journals and conferences, all in recognized venues, and around 20 3GPP standardization contributions, as well as a book titled *Multi-point Cooperative Communication Systems: Theory and Applications* (Springer, 2013). Moreover, he holds 21 US patents and has co-invented another 100+ patents on 4G/5G technologies in China, Japan, Korea, Europe and beyond. Currently, he is an editor of *IEEE Transactions on Wireless Communications* and *IEEE Communications Surveys and Tutorials*. Besides, he has served as guest editor, co-chair, co-tutor and technical program committee member for many IEEE top-tier journals/conferences and received several awards for his research work and professional services.

Foreword

In the wireless communications field and industry, a subtle "fact" long taken for granted is that the network can be densified to increase coverage and capacity. By densifying the network, we mean adding more wired access points and base stations, which enables denser spatial reuse of precious bandwidth. Twice as many access points or base stations means twice the capacity, in addition to improved coverage. Network densification has been arguably the single most important – yet underappreciated – contributor to the wireless revolution, for both Wi-Fi and cellular systems. With Wi-Fi, the densification has happened mostly organically one home or business at a time, whereas in cellular systems, it has happened more systematically, but also monotonically: once a new cell site is deployed, it stays deployed indefinitely.

But is this trend sustainable? What are the fundamentals – from an electromagnetic and information theory perspective – of network densification? Intuitively, it is easy to understand that adding more radiating infrastructure increases the interference, all else being equal. But as this book explains – all else is *not* usually equal. Historically, it has been an unqualified win to add more base stations, despite the nominal increase in interference. Nominal, because after all, closer base stations can reduce their transmit power considerably, while maintaining the same signal-to-noise ratio at the mobile devices. But, as networks become very dense, is there some "ultra-dense" threshold where they behave differently? Where and when does the new interference become more harmful to the network than the spatial reuse of bandwidth is beneficial? These often neglected but very important questions have major implications on the wireless networks for the coming decades, and even for 5G networks, some of which are close to or even perhaps past the ultra-dense threshold, where further densification gives rapidly diminishing returns unless specific countermeasures are taken.

In this new and needed book, López-Pérez and Ding provide a rigorous framework for understanding ultra-dense wireless networks. These two scholars have contributed significantly to our field's understanding of the ultra-dense phenomenon over the last several years, and they are uniquely placed to author this comprehensive treatment on dense cellular networks. This book provides a brief overview of the history of cellular network analysis along with an in-depth treatment of the new tools and models needed to understand ultra-dense networks in all their richness. I am confident that researchers and engineers can develop innovative technical approaches that allow one to continue the wireless evolution despite the limits to network densification; but for this to happen, it is necessary to be aware that those limits do in fact exist.

Jeffrey G. Andrews
Cockrell Family Endowed Chair in Engineering
University of Texas at Austin

Foreword

Fifth-generation (5G) and beyond fifth-generation (B5G) networks are playing – and will continue to play – a key role in the digitalization of modern society by delivering hyperconnectivity, higher data rates and ultra-low latency.

To provide such requirements, 5G/B5G networks will be much more heterogeneous in nature than any other preceding network technology, pioneering the use of higher-frequency bands, such as millimetre-wave, sub-terahertz, terahertz and visible light frequencies, and targeting new verticals with novel and specific latency-oriented networking features.

To make the most of the environment, 5G/B5G networks will also soon take advantage of city structures coated with intelligent meta-surfaces, commonly known as reconfigurable intelligent surfaces, to reflect incident signals in a customized manner, so as to optimize/recycle the signal propagation in future networks.

Moreover, unmanned aerial vehicle technology is becoming mature, and both *flying base stations* and *user terminals* will become commonplace in the near future in order to provide terrestrial coverage in isolated regions and enhance capacity in traffic hotspots, as well as to support new delivery or supervision services, respectively.

Importantly, end terminals will also gradually be equipped with computing and/or storage capabilities and immersed in a fog radio access architecture, thus shifting the design paradigm from ubiquitous connectivity to ubiquitous wireless intelligence and goal-oriented solutions.

The realization of these new network features and capabilities will require, without any doubt, a significant level network densification with a large number of diverse sets of networking nodes deployed per unit of area.

To support the optimal deployment and operation of these complex dense heterogenous networks of the future, large-scale network performance analysis will be crucial to formulate and translate relevant theoretical network insights into controllable network trade-offs, which can be used by network operators to govern such sophisticated systems.

Over the past decade, stochastic geometry has emerged as a powerful network performance analysis tool to evaluate the wide area network performance and reveal fundamental insights into the functioning, interactions and behaviour of dense heterogeneous networks.

With the imminent onset of this crucial new decade, where the global commercialization of 5G will occur, and essential research questions related to B5G are expected to emerge, stochastic geometry is expected to constitute an essential tool to unveil the fundamental properties of emerging ultra-dense radio access networks for 5G/B5G and quantify the benefits of key enabling technologies.

This book titled *Fundamentals of Ultra-Dense Wireless Networks* constitutes a unique guide for understanding, modelling, analyzing and optimizing ultra-dense heterogeneous networks by using the analytical tool of stochastic geometry. On the one hand, the book covers both the theoretical foundation and applications of sparse and dense cellular network deployments as well as the impact of practical communication aspects, such as the presence of line-of-sight links, the height of access points, the user density and the use of sleep modes. On the other hand, it derives and presents in an accessible manner fundamental scaling laws as well as design and engineering insights, which are essential to academic researchers and industrial practitioners in the telecommunications field. Remarkably, the authors identify the rich opportunities that stochastic geometry offers for understanding and optimizing future wireless networks. This book also provides a unified treatment to the analytical evaluation and optimization of ultra-dense heterogeneous networks, which makes it an essential guide for each researcher and engineer interested in the design and deployment of future wireless communications.

Marco Di Renzo
CNRS Research Director
Laboratory of Signals and Systems
CentraleSupelec, Paris-Saclay University

Preface

By reading this book, readers will embark on a thrilling journey to discover the fundamentals of ultra-dense wireless networks. This journey will particularly help explore the information capacity limits of sparse, dense and ultra-dense generic wireless networks, having its nodes communicating with the associated receivers via electromagnetic waves, under different channel and network characteristics. The questions answered in this book are of fundamental nature and intriguing, similar to those posed to introduce, for example, Newton's law of gravitation in our physics classes when we were teenagers. Why does the apple fall from the tree? The answers to such questions are thought provoking and equally exciting too.

First theoretical analyses proposed an information capacity law of a dense wireless network, which was compelling due to its simple, intuitive and beautiful nature. It was a linear law, stating that the network capacity would linearly scale with respect to the base station density in a wireless network. To put this in context, this linear law implies a golden rule, i.e. doubling the cell density would double the network capacity, thus providing a tool to indefinitely increase the performance of our communication systems. Importantly, the world has seen this law partially verified with the cell density increasing by a factor of around 3000× from 1950 to 2000, and the resulting network capacity growing accordingly.

Things, however, started to deviate from the main track when the need for significantly larger capacities appeared on the horizon. In 2014, in the quest to enhance end-user experience, Qualcomm, for instance, reported new real-world experimental results, indicating that a cell densification factor of 100× would only lead to a network capacity increase of 40×, and thus contradicting the aforementioned linear law. In the following years, the wireless research community was awakened by this alarming call and thoroughly re-examined its theoretical understanding. After years of work, it was concluded that the linear law could not characterize the network capacity when the base station density is high and that a different law should apply. In this book, we will present and discuss one such new law – step by step – and explain the intuitions behind the discovery.

Interestingly, the progress in the understanding of the capacity law of a dense wireless network resembles that in the understanding of the gravitation law of our universe, used earlier in this preface as an example. In the good old times, when Newton's law of gravitation ruled everything, it was empirically safe to state that the gravity would scale linearly with the product of the masses of two objects. However, it turned out that Newton's theory would break down when the considered masses are significantly high, say for black holes, where the general relativity should be used instead to characterize the gravity, via a completely different law.

Such a change of law when "the amount of things" significantly grows is no surprise any longer, and actually occurs in many areas of science and in many domains of our universe. Georg Wilhelm Friedrich Hegel referred to this change of law in his three laws of dialectics as the law of transformation of quantity into quality and vice versa. To provide a few examples, it is intriguing to see how such a meta-law works in physics, with the aforementioned example of gravity, in chemistry, with the change of the form of water as temperature varies, and in anthropology, with the transformation of the cognitive functions with the size of primate brains. Our journey, to be narrated by this book, serves as another instance of the quantity–quality transformation, full of surprises, as we will find out together.

As a kind reminder for our readers, note that the full picture of the quantity–quality transformation of the network capacity will not be revealed until Chapter 8. Please be patient, adventurers. Onwards!

Acknowledgements

The authors of this book would like to thank Dr. Xiaoli Chu, Dr. Adrian Garcia-Rodriguez, Dr. Giovanni Geraci, Dr. Amir Hossain Jafari and Dr. Nicola Piovesan for their help and reviews.

Further, the authors would like to thank Professor Youjia Chen, Dr. Holger Claussen, Dr. Lorenzo Galati-Giordano, Professor Jun Li, Prof. Zihuai Lin and Professor Guoqiang Mao for their valuable suggestions and expert guidance.

Dr. David López-Pérez would like to thank his parents, Juan Jose and Antonia, his sister, Charo, his nephew, Lucas, and the rest of his beloved family, together with his good friends and committed colleagues for their loving care, comprehension and support.

"You all have been very patient and always proudly stood by me. I love you all!"

Dr. López-Pérez would also like to dedicate this book to all those people, met during his journey, who were and still are passionately pursuing a dream.

"You all have enlightened and motivated me to carry on, abroad, far from my people, and do it, without any doubt, in an enjoyable manner. This book is for you!"

Dr. Ming Ding would like to dedicate this book to his wife, Minwen Zhou.

"Minwen, you have a brilliant mind and a caring soul. I'm so fortunate and happy to spend my lifetime with you in this universe. You are my best friend and soulmate, showing me and accompanying me on a journey of a wonderful life that surpasses my best imagination. Since only non-physical things have the potential to reach eternity, let me use these words to carry my deep love for you beyond the end of time and space."

Dr. Ding would also like to thank his parents and parents-in-law for their constant support and great understanding, especially when moving with his wife to their dream city, Sydney, almost 5,000 miles from all their family members. He would also like to express his gratitude to his friends, especially Dawei Wang and his wife, Xiaoyan Xia; Minlong Chen and his wife, Guoying Xia; Wuyang Jiang and Meng Zhang, for their kind help to his family while he and Minwen are living abroad.

"Thinking of you always brings back warm memories and joyful thoughts to me. I miss you all!"

List of Abbreviations

3GPP	3rd-Generation Partnership Project
4G	fourth generation
5G	fifth generation
ASE	area spectral efficiency
AWGN	additive white Gaussian noise
BLER	block error rate
BS	base station
CATV	community antenna television
CCDF	complementary cumulative distribution function
CDF	cumulative distribution function
CMF	cumulative mass function
CQI	channel quality indicator
CSI	channel state information
dB	decibel
dBi	isotropic-decibel
EIRP	effective isotropic radiated power
FD	frequency domain
FDD	frequency division duplexing
FDMA	frequency division multiple access
FDTD	finite-difference time-domain
GPS	global positioning system
HARQ	hybrid automatic repeat request
HD	high definition
HPBW	half power beam width
HPPP	homogeneous Poisson point process
HSPA	high speed packet access
IC	interference cancellation
IEEE	Institute of Electrical and Electronics Engineers
i.i.d.	independent and identical distributed
IMT	international mobile telecommunications
IoT	Internet of things
IP	internet protocol
ISD	inter-site distance
LIDAR	laser imaging detection and ranging

LoS	line-of-sight
LTE	long-term evolution
LUT	look up table
MAC	medium access control
MCS	modulation and coding scheme
MIMO	multiple-input multiple-output
MMSE	minimum mean square error
MOS	mean opinion score
MPC	multi-path component
MSE	mean square error
NLoS	non-line-of-sight
NR	new radio
OFDM	orthogonal frequency division multiplexing
PDCP	packet data convergence protocol
PDF	probability density function
PF	proportional fair
PGFL	probability generating functional
PHY	physical
PMI	precoding matrix indicator
PMF	probability mass function
PPP	Poisson point process
QoS	quality of service
RI	rank indicator
RR	round robin
RV	random variable
SAR	specific absorption rate
SIM	subscriber identity module
SINR	signal-to-interference-plus-noise ratio
SIR	signal-to-interference ratio
SLL	secondary lobe level
SNR	signal-to-noise ratio
SON	self-organizing network
SSS	secondary synchronization channel
TDD	time division duplexing
TDMA	time division multiple access
TRU	time resource utilization
UAV	unmanned aerial vehicle
UE	user equipment
UMTS	universal mobile telecommunication system
URLLC	ultra-reliable low latency communication
VoIP	voice over IP
WiMAX	wireless interoperability for microwave access

Part I

Getting Started

1 Introduction

As old as I have become, many developments my eyes have seen. When I was a
young man, changes used to take place from time to time, every now and then.
Nowadays, they occur so often, life itself seems to change every day.

*—A 97-year-old man commented, making reference to the increasing pace of the
developments in the communications field through the last century*

These comments, although exaggerated at first sight, may not be so, in light of the
advancements seen in the telecommunication industry since Graham Bell carried out
the first successful bidirectional telephone transmission in 1876 [1]. Since then, society has witnessed

- most long-distance communications having at least a wireless component, freed
  from wires and operated through air at the speed of light,
- wireless and mobile communications made available to over 8.6 billion
  connections across the globe [2],
- new types of communications and social interactions emerging through both the
  Internet [3] and social networking [4] and
- many other breakthroughs, which have certainly changed our everyday lives.

These developments, although of great importance to the 97-year-old man, are
probably just small steps towards a new era – the era of digital and pervasive communications – which will continue to change the world we are inhabiting in unpredictable
and fascinating manners.

As a matter of fact, today, we are on the brink of another significant societal
change. While the network has mainly served humans up to now, this capability will
increasingly be extended to machines in the near future too. By 2022, it is expected
that there will be not only 8.4 billion handheld or personal mobile-ready devices, but
also 3.9 billion machine-to-machine connections [2]. The emergence of this machine-
originated data traffic will drive further the demand for network capacity, but also
impose additional requirements on network performance, mainly in the areas of end-
to-end latency and reliability. These are currently the major challenges for many new
applications.

Nowadays, most of the data services reside on the Internet, far away from the user
equipment (UE), where the speed of light becomes one of the main factors limiting
end-to-end latency. To address this problem, processing will have to move closer to

the UE, e.g. into a cloud computing infrastructure, which will extend – and act as a ramification of – the network. In addition, an intelligent and adaptive network management and a well-designed congestion control can also help to significantly enhance reliability, thus enabling new real-time applications, such as augmented reality or efficient machine communication.

With these new requirements and changes, communication networks are evolving to become our main interface with the virtual world, and increasingly also with the physical one. This future network will simplify and automate many aspects of life, allowing one to effectively "create time," by improving the efficiency in everything we do [5].

Making this vision of the future network a reality will require from a technical perspective both

- ultra-broadband wireless access, providing orders of magnitude improved throughput, delay and reliability as well as quality of service (QoS) control and
- a highly adaptable and remotely programmable cloud computing infrastructure located close to the edge of the network.

Throughout this book, we argue that small cells, and more specifically, ultra-dense deployments are one of the answers to the technological challenges of creating an ultra-broadband wireless access that connects mobile UEs, machines and objects to a processing cloud engine. In more detail, this book serves as a tool to shed new light on the fundamental understanding of ultra-dense networks.

As an introduction to the content of the book, the remainder of this chapter first depicts the current industry capacity challenge and follows with an overview of the small cell technology and its history, from both an industry and an academic perspective. Then, the individual parts and chapters of the book are introduced, as well as their relationship to various aspects of deploying and operating small cell networks. To conclude, some of the key nomenclature used in this book is presented together with a list of the most relevant publications in the field of ultra-dense networks by the authors of this book.

As an important disclaimer, let us note that this book is going to focus on the study of single antenna UEs and base stations (BSs), thus using single-input single-output transmission modes. As a result, this book does not consider either multiple-input multiple-output (MIMO) or multi-cell coordinated transmissions/receptions, and all the statements within are done accordingly.

1.1   The Capacity Challenge

Voice-based services, such as voice over internet protocol (VoIP), were the killer applications at the beginning of this century, demanding an average of tens of kilobits per second per UE for this type of connection [6], while the streaming of high-definition (HD) video is probably the most popular service today, requiring tens of megabits per second per video feed [7]. Future services, however, such as three-dimensional visualization, augmented and virtual reality, online gaming using multiple displays and the robot-to-robot exchange of HD laser imaging detection and ranging (LIDAR) maps will require much more capacity, with expected average throughputs per UE exceeding 1 Gbps [8] – and who knows what else tomorrow will bring?

With this enormous challenge of improving the average throughput per UE by orders of magnitude, before making any decision on any technology investment, which is likely to be costly, it is advised that network operators and service providers understand well the different dimensions that they have to improve wireless capacity.

In a simplified form, the Shannon–Hartley theorem [9],

$$C = B \cdot \log_2 \left(1 + \frac{S}{N} \right), \tag{1.1}$$

provides an insight into which are the variables that influence the amount of information – capacity, C, in *bps* – that a transmitter can send to a receiver

- over a communication channel of a specified bandwidth, B, in Hertz
- with a received signal power, S, in Watts
- in the presence of an additive white Gaussian noise (AWGN) power, N, in Watts.

From this theorem, it can be inferred that the capacity, C, of a UE can be scaled up by increasing

- the bandwidth, B, per UE and/or
- the signal-to-noise ratio (SNR), $\frac{S}{N}$, of such UE, or more accurately, the signal-to-interference-plus-noise ratio (SINR), $\frac{S}{I+N}$, of the UE in a multi-cell multi-UE network, like the ones that will be studied in this book, where I stands for Gaussian interference in Watts.

Importantly, equation (1.1) also shows that to scale the capacity, C, of a UE, increasing the bandwidth, B, per UE is generally a more promising technique than increasing the SINR, $\frac{S}{I+N}$, of such a UE, since the former yields a linear scaling, while the latter only a logarithmic one.[1]

---

[1] Even though multi-antenna technology is not considered in this book, for completeness, it should be noted that multiple antennas can be used to either

- leverage spatial multiplexing through MIMO techniques or
- increase the SINR, $\frac{S}{I+N}$, of a UE via beamforming.

In particular, MIMO transmissions/receptions, can take advantage of the spatial resources, and linearly increase the capacity, C, of a UE with the number of spatial streams multiplexed. This can be treated as a "virtual" increase of the bandwidth, B, per UE. When taking a cell or a network perspective, it should also be noted that multi-user MIMO, coordinated beamforming and multi-cell coordinated transmissions/ receptions can be used to increase the spatial multiplexing in the cell or the network and/or the SINR, $\frac{S}{I+N}$, of a UE. Readers interested in related topics are referred to [10] and references therein.

$$C = B \; log_2 \left(1 + \frac{S}{I + N}\right)$$

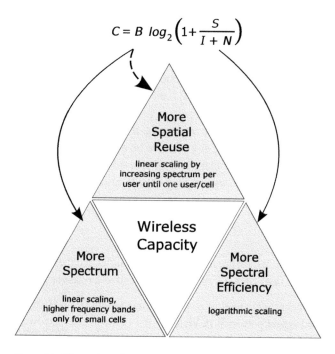

Figure 1.1 Dimensions for scaling the capacity of a UE in a wireless network [8].

With this in mind, a fundamental question arises:

How can we increase the bandwidth, B, per UE?

In a network with multiple UEs, the bandwidth, B, per UE can be scaled up by either increasing

- the amount of frequency resources invested into the network and/or
- the network densification, and in turn, its associated spatial frequency reuse.

For the sake of clarity, it should be noted that the reduced cell size in a denser network results in an improved spatial reuse of the frequency resources, since there are less UEs in each cell to share the available cell bandwidth. As a result, each UE has access to more of such frequency resources.

Overall, as depicted in Figure 1.1, this leaves one with three main approaches to enhance the capacity, C, of the UE, i.e. using a wider bandwidth, deploying a denser network and improving the signal quality – the SINR of the UE, where the two first ones may be more appealing due to their intrinsic linear scaling of capacity.

To put things in context, and show how each of these three degrees of freedom have historically contributed to the increase of the capacity of practical networks, Webb [11] put together an interesting analysis, indicating that, between 1950 and 2000, such network capacity has increased around

- 2700× from densifying the network with smaller cells,

- 15× by using more bandwidth in the sub-6 GHz bands (from 150 MHz to 3 GHz) and
- 10× by improving the spectral efficiency (waveforms and multiple access techniques, modulation, coding and medium access control (MAC) methods such as scheduling, hybrid automatic repeat request (HARQ), etc.).

From this study, it is clear that – by far – the majority of the capacity gains in the past were achieved by increasing the spatial frequency reuse through densifying the network with smaller cells. This leaves one with the following question:

How much further can we increase the spatial reuse by reducing the cell size?

Answering the above question from a theoretical perspective is one of the primary goals of this book.

For completeness, and before proceeding any further, it should be noted here at this point that the operation at higher carrier frequencies, e.g. millimetre wave bands, also offers the possibility of accessing large amounts of spectrum and the associated very wide bandwidths, thus enabling extreme data rates.

Higher carrier frequencies, however, are also associated with higher radio frequency attenuations, which limit their network coverage. Although this can be compensated to some extent by means of multi-antenna technologies – and more in detail through beamforming – a substantial coverage disadvantage will always remain for a network operating at higher carrier frequencies [12].

Another challenge with the operation at higher frequency bands is the regulatory aspects. For non-technical reasons, the rules defining the allowed radiation may change in these higher frequency bands, from a specific absorption rate (SAR)-based limitation to a more effective isotropic radiated power (EIRP)-like limitation. These restrictions may also impose further coverage constraints [13].

More importantly, the millimetre wave technology in general – and the beamforming one in particular – are not mature as of today, or at least, are not cost-effective for ultra-dense deployments. The features required to deal with

- the beam alignment and tracking, possibly needed at both communication ends and
- the related issues arising from unexpected device rotation, blockage and mobility

are still of high complexity, and lack of robust QoS provisioning. This together with the large energy consumption of current millimetre wave access points make this solution still too expensive.

As a result, the more mature and developed sub-6 GHz technology still remains a front runner for ultra-dense deployments, especially due to its ability to satisfy communication requirements in non-line-of-sight (NLoS) and outdoor-to-indoor propagation conditions. For these reasons, we focus on low carrier frequencies in this book, and analyze in detail the impact of network densification and the increase of spatial reuse on network performance, as posed earlier. However, due to its potential, we leave the door open to the analysis of millimetre wave deployments for next editions of this book, and in particularly to the inter-working of sub-6 GHz and millimetre wave technology.

1.2   Network Densification

In a multi-cell multi-UE network, UEs being served within a cell share the available bandwidth. Thus, reducing the cell size – while deploying more cells to maintain the same level of coverage – also reduces the number of UEs per cell, and in turn, increases the bandwidth per UE. Through this approach, the bandwidth per UE can be increased, until each cell only serves a single UE. When densifying further, beyond this point, only the signal power – and potentially the SINR – of the UE can be improved by reducing the distance between the UE and its serving small cell BS.

Overall, by increasing the bandwidth, B, per UE, the capacity, C, scales up linearly, until the one UE per cell limit is reached, after which the scaling becomes only logarithmic through improvements on the SINRs of the UEs, as indicated by equation (1.1).

Figure 1.2 illustrates this capacity scaling behaviour, showing that with increasing cell densities, the capacity

- initially increases quickly due to the spatial frequency reuse, but then
- slows down, when the one UE per cell limit is reached, and the gains are mainly dominated by improvements on the SINRs of the UEs, through proximity gains [8].

From Figure 1.2, it is also important to note that the results were obtained assuming an active UE density of 300 BSs/km^2, typical in some dense urban scenarios, and that the one UE per cell limit is reached for an inter-site distance (ISD) of around 30–40 m. This indicates that there is still plenty of room for network densification in major cities like Manhattan and London, where the average ISD is around 200 m.

A second aspect of densification is that the required transmit power may reduce to an extent where its contribution to the total energy consumption becomes insignificant, and the processing power of the small cell BS becomes the dominant factor. Moreover, with reduced cell sizes, the required number of small cells to provide coverage increases, and as a result, many of them may not serve any UE for most of the time. However, they still consume energy and transmit unnecessary pilot signals, which may cause inter-cell interference. This issue can be addressed by introducing idle mode capabilities, where small cell BSs are only woken up to actively serve UEs. With efficiently controlled idle modes, the network energy consumption reduces and the SINR of the UE significantly improves.

The main challenge of network densification, however, is the issue of increasing costs for equipment, deployment and operational expenses. In this light, it is important to note that the cost of a small cell BS, estimated in 2015, only accounts for approximately 20% of the total deployment costs associated with outdoor small cells.

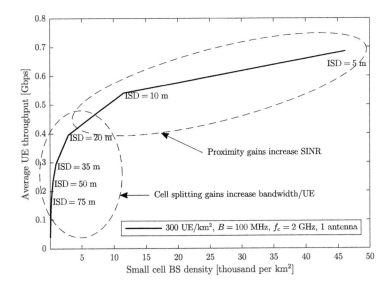

Figure 1.2 Capacity scaling with densification for different inter-site distances in a dense urban scenario considering a hexagonal BS deployment and a semi-clustered UE distribution. For more details on the scenario, models and results, the reader is referred to [8].

The majority of the costs are site leasing (26%), backhaul (26%), planning (12%) and installation (8%) [14]. The good piece of news is that this challenge can be addressed by changing the deployment model from an operator deployment to a "drop and forget" end-user deployment, and reusing the existing power and backhaul infrastructure. In this model, the end-user simply connects the small cell BS to the power and the backhaul, which then triggers a fully automatic configuration, and a continuous self-optimization process during operation. This end-user deployment model is feasible for both the residential and the enterprise markets.

Due to cost and performance reasons, it is also important to highlight that it becomes increasingly important to deploy the small cell BSs wherever the UEs are, since small cells cannot compensate for misplacement as well as larger cells do. If the small cell BSs are not deployed in an intelligent manner following the distribution of UEs, a larger number of small cell BSs will be required to achieve the one UE per cell limit. However, accurate UE demand distributions are hard to derive today, because of the limited accuracy of conventional localization techniques in cellular networks, such as triangulation. One may think of using more accurate techniques such as the global positioning system (GPS) for this purpose, but its performance is poor indoors, where 80% of the traffic demand is located [15]. Advanced planning tools for small cell deployment are still an open challenge.

In summary, densification continues to have a high potential to increase capacity until reaching the one UE per cell limit. To maintain high performance and energy efficiency, idle mode capabilities that switch off small cells when they are not serving

UEs are necessary. Transitioning to a "drop and forget" deployment by the end-user has a high potential to reduce the deployment and operational costs.

1.3 A Brief History of Small Cells

In this section, we provide an overview of the small cell technology and its history, first from an industry perspective, and subsequently from a theoretical one. This last part of the chapter serves as an introduction to the rest of the book, which will be formalized by the outline presented in the next section.

1.3.1 From an Idea to a Market Product

The idea of the small cell has been around for over three decades [16]. Initially, "small cell" was the term used to describe the cell size in a metropolitan area, where a macro-cell – with a cell diameter on the order of kilometres – would be split into a larger number of smaller cells with reduced transmit power, known today as metropolitan macrocells or microcells. These small cells had a cell diameter of a few hundreds of metres.

In the 1990s, picocells appeared with even a smaller cell size, between a hundred metres to around a few tens of metres [17]. These "more traditional" small cells were used for coverage and capacity infill, that is, where macrocell penetration was insufficient to give a good connection or where the macrocell was at its capacity limit. These types of small cell BSs were essentially a smaller version of the macrocell BSs, which also had to be planned, managed and interfaced with the network. This last point is probably the most important reason why small cells – other than metropolitan macrocells or microcells – have not gained much popularity for quite some time. Essentially, the costs associated with deploying and running a large number of small cells outweighed the performance advantage that this kind of cellular topology provided.

In the 2000s, new thinking on the deployment and configuration of cellular systems began to address the cost and the operational aspects of small cell deployments, which enabled the cost-effective deployment of even smaller cells [18]. Such thinking crystallized in the home BS concept first [19] and the femtocell one later [20]. A femtocell is a low-cost cellular BS with advanced auto-configuration and self-optimization capabilities, which allows the end-user – without any operator involvement – to deploy this small form factor BS in a plug-and-play manner within the home. Femtocells use a broadband internet connection as backhaul, and connect to the cellular network through dedicated gateways, which enables a better scaling to millions of femtocell BSs. Early results on the performance of 3G universal mobile telecommunication system (UMTS) femtocells were presented in [21–23], which were shortly afterward extended with a bulk of studies on self-optimization and offloading strategies, multiple antenna techniques and energy management methods [20, 24–30]. Soon after, results on fourth-generation (4G) wireless interoperability for microwave access (WiMAX) and long-term evolution (LTE) small cells followed too [31–36]. Femtocells also emerged as the first step towards a heterogeneous network deployment model [37–42].

Following this early research and development in the femtocell area, in 2007, several industry players advocating for small cell technology formed the Femto Forum – rebranded as the Small Cell Forum in 2012 – to create a venue for promotion, standardization and regulation. Moreover, governments started funding research projects on femtocells, e.g. the European Union ICT-4-248523 BeFEMTO project, which focused on the analysis and the development of 4G LTE-compliant femtocell technologies [43]. To highlight the success of the femtocell in the research fora, it is worth highlighting that the number of publications registered in the Institute of Electrical and Electronics Engineers (IEEE) Xplore digital library [44] including the word "femtocell" or "femtocells" increased from 3 in 2007 to 11 (2008), 52 (2009), 117 (2010), 1088 (2015) and 3178 (2019).

The first commercial deployments of residential femtocells started in 2008 when Sprint launched a nationwide service in the United States, followed in 2009 by Vodafone in Europe and Softbank in Japan. Since then, small cell technology has quickly proliferated. The number of deployed small cells for the first time exceeded those of macrocells in 2011 [45], and over 77 operators used small cells worldwide in 2015 [46]. The business impact of small cells took off with 4G LTE, with this small cell type being – by far – the most widely deployed today, where multimode cells with additional 3G UMTS and/or Wi-Fi capabilities are also widely available [47].

Due to its success, the scope of femtocells was then extended from residential deployments to public indoor spaces, which generate most of the cellular traffic. By 2015, 71 operators had deployed in-building small cells, known as pico- or microcells, in enterprise or public buildings [46], and from 2020 to 2025, it is expected that the number of such in-building small cells will grow from 10.7 to 208.6 million [48]. Their design is aimed at reducing planning and deployment costs, decreasing the need for large customer support teams and eliminating the need for massive reprovisioning. The generally good availability of internet protocol internet protocol (IP) backhaul, such as Ethernet, in enterprise and public buildings is an important deployment advantage. However, the overall system configuration and the overlap with outdoor macrocells and microcells must be monitored and well managed. To this end, in-building small cell deployments are equipped with per-call and QoS analytics, as well as self-organizing network (SON) features [49].

The expansion of femtocells to the outdoor space is more difficult due to challenges such as cost, site rental, backhaul availability, network provisioning and management as well as monetization issues. However, even in this area, small cells have proved to be a viable approach in the form of metrocells – smaller and more flexible versions of metropolitan macrocells or microcells, with whom they share many hardware and software features, most notably the support of a high number of simultaneous UEs. Metrocells, however, shine for their SON capabilities, providing self-configuration and self-optimization of, e.g. neighbour relation management, inter-cell interference mitigation and handover parameter configuration, among others. Outdoor small cells are mainly serving operator-deployed, public networks in urban, suburban or rural environments, although a number of outdoor small cell deployments are already dedicated to particular businesses and enterprises (e.g. oil drilling rigs or power stations). In the fifth-generation (5G) era, when more small cells will be related to industrial and internet of things (IoT) services, there will be more vertical-specific small cells deployed outdoors, managed by enterprise specialists. Recent surveys indicate that in the outdoor environment, the growth rate of small cell deployments will double that of indoor ones from 2020 to 2025, 41% versus 20% [48]. By 2025, outdoor small cell deployments will reach 2.76 million, with urban scenarios reaching a total installed base of 11.2 million small cells [48].

Overall, it is expected that the installed base of small cells will reach 70.2 million in 2025 as operators seek to densify their networks. This growth will likely be led by Asia-Pacific and North America, with Europe lagging [48]. While residential deployments will continue to rise, the non-residential market is expected to grow far more quickly, at an annual growth of 36%, accounting for 75% of annual deployments and 55% of the total installed base in 2025 led by the urban public and enterprise public sectors. The total installed base of 5G new radio (NR) or multimode small cells in 2025 is predicted to be 13.1 million, over one-third of the total in use [48]. From 2025, 5G NR small cells are expected to overtake 4G LTE-only as well as combined 4G LTE and 5G NR models [48].

1.3.2   The Evolution of the Small Cell Theoretical Understanding: A Brief Summary of This Book

As indicated in the previous sections, densification has a high potential to increase capacity, until reaching the one UE per cell limit, and the market is already heading down this path, deploying denser and denser networks in dense urban scenarios, as well as indoors, in enterprises and factories. However, there are fundamental questions around whether ultra-dense networks – which do not exist out there yet – will behave similarly as today's more sparse ones, or whether they follow different fundamentals, which may impact performance. For example:

Will the larger number of small cells create an inter-cell interference overload that will render any communication impossible?

Answering this and other fundamental questions related to ultra-dense networks is the objective of this book.

The Old Understanding

When it comes to small cell deployment and network performance analysis, the theoretical work of M. Haenggi, J. G. Andrews, F. Baccelli, O. Dousse and M. Franceschetti stands out. In their seminal work [50] and references therein, the authors created a mathematical framework based on stochastic geometry to analyze the performance of random networks in a tractable manner.

In a nutshell, this mathematical stochastic geometry framework allows one to theoretically calculate, sometimes even in a closed-form expression, the coverage probability of a typical UE, which is defined as the probability that the SINR, γ, of the typical UE is larger than an SINR threshold, γ_0, i.e. $\Pr\left[fl > \gamma_0\right]$. Based on this coverage probability – also known as success probability – the SINR-dependent area spectral efficiency (ASE) in bps/Hz/km^2 can also be investigated, among other metrics.

This framework has become the de facto tool for the theoretical performance analysis of small cell networks in the entire wireless community. Good tutorials and more references on the fundamentals of this mathematical tool can be found in [51–54] and the references therein. Chapter 2 of this book and in more detail Section 2.3 will also provide a more detailed introduction to the topic.

Many efforts have been made since 2009 to extend the capabilities of this stochastic geometry framework to improve the understanding of small cell networks. M. Haenggi et al. further developed the framework to account for different stochastic processes different that the basic homogeneous Poisson point process (HPPP) [55], as well as distinct performance metrics, such as the typicality of the typical UE [56] and the transmission delay [57], among others. T. D. Novlan et al. further extended the framework to study uplink transmissions, calculating the aggregated inter-cell interference, using the probability generating functional of the HPPP [58]. M. Di Renzo et al. also did a good number of extensions, by considering more detailed wireless channel characteristics in the modelling, such as other non-HPPP distributions, building obstructions, shadow fading and non-Rayleigh multi-path fast fading, of course, at the expense of tractability [59].

When it comes to the analysis of different wireless network technologies and features using stochastic geometry, it is worth highlighting the extensive work of J. G. Andrews, V. Chandrasekar, H. S. Dhillon et al., which touches on spectrum allocation [37], sectorization [38], power control [39], small cell-only networks [60], multi-tier heterogeneous networks [61], MIMO [62], load-balancing [63], device-to-device communications [64], content caching [65], IoT networks [66] and unmanned aerial vehicle (UAV) communications [67], to cite a few.

Regarding higher frequency bands and the massive use of antennas, the studies of R. W. Heath, T. Bai et al. stand out, for example, those on massive MIMO [68] and millimetre wave [69] performance analysis, random blockage [70], millimetre wave ad hoc networks [71] and secure communications [72], shared millimetre wave spectrum [73] as well as wireless power systems [74].

For further reference, and with regard to the analysis of other relevant network aspects, it is worth pointing out the research of G. Nigam et al. on coordinated multipoint joint transmission [75], H. Sun et al. on dynamic time division duplex [76] and Y. S. Soh et al. on energy efficiency [77]. Many other analyses studying different types of stochastic processes, performance metrics, wireless characteristics and network features can be found in the literature, which show both the generality and the strength of stochastic geometry framework.

Among all the mentioned results, one of the most important theoretical findings is that by J. G. Andrews and H. S. Dhillon et al., concluding that the fears of an inter-cell interference overload in small cell networks are not well-grounded, neither in a small cell-only network [60] nor in a heterogeneous one [61]. Instead, their results showed that the increase in the inter-cell interference power due to the larger number of interfering small cell BSs in a dense network is exactly counterbalanced by the increase in the signal power due to the closer proximity of transmitters and receivers. This conclusion is powerful, meaning that an operator can continually densify its network – no problem – and expect that

- the spectral efficiency in each cell stays roughly constant, and as a result that
- the network capacity – or in more technical words, the ASE – linearly grows with the number of deployed cells.

This behaviour – or capacity scaling law with the small cell BS density – is referred hereafter in this book as *the linear capacity scaling law*.

This exciting message created big hype in the community, and also in the industry, presenting the small cell BS as the ultimate mechanism in providing a superior broadband experience. Consequently, this raised a new fundamental question:

> Can we infinitely reduce the cell size to achieve an infinite spatial reuse, and in turn, an infinitely large capacity?

Unfortunately, a few important caveats to realize such a linear scale of capacity in an ultra-dense network were quickly found. Among them, it is worth highlighting

- the need for an open-access operation [78] and
- the impact of the transition of a large number of interfering links from NLoS to line-of-sight (LoS) [79],

which will be discussed in the following paragraphs.

The Effect of the Access Method

Closed-access operation provides an experience comparable to Wi-Fi access point, in which the owner of the small cell BS can select which UEs can associate to it. This is an appealing model for the small cell owner but prevents the UE to connect to the strongest cell. This degrades the SINR of the UE and breaks the linear scale of capacity observed in both [60] and [61]. This is because the inter-cell interference power can now grow much faster than the signal power. This is particularly true when the UE, let us say in a block of apartments, moves away from its closed-access small cell – to which it can connect – and gets closer to a neighbouring one – which it cannot access.

Open-access operation has been widely adopted in small cell BS products to address this issue and restore the linear capacity scaling law. Since the performance impact of closed- and open-access operation is intuitive and well understood [78], this topic is not theoretically treated in this book.

The Impact of the NLoS to LoS Inter-Cell Interference Transition

A more fundamental problem than the access method was found in [79], which showed that, even if open-access operation is adopted, the inter-cell interference power can grow faster than the signal power, with the consequent degradation of the SINR of the UE, when the small cell BS density grows ultra-dense.

To understand this phenomenon, it is important to note that the linear scale of capacity presented in both [60] and [61] was obtained with the assumption of a single-slope path loss model, meaning that both the inter-cell interference and the signal powers decay at the same pace, $d^{-\alpha}$, over a given distance, d, where α is the path loss exponent.

Although simplistic, when the path loss exponent is "fine-tuned," this single-slope path loss model is applicable to sparse networks, such as metropolitan macrocell and microcell ones. However, this model may be inaccurate for denser networks, where the small cell BSs are deployed below the clutter of man-made structures. This is because the probability of the received signal strength abruptly changing due to a change in the LoS condition of the interfering and/or serving links is much larger in a dense network, where the small cell BSs are deployed at street level.

To model this critical channel characteristic, whether the UE is in NLoS or LoS with a small cell BS, the use of

- a multi-slope path loss model considering NLoS and LoS transmissions and
- a probabilistic function governing the switch between them

was proposed in [79], and implemented over the theoretical analysis framework presented in [60] and [61].

Intuitively speaking, the key difference between the single-slope and this new multi-slope path loss model is that

- a UE always associates to the nearest small cell BS in the former, while
- a UE may be connected with a further but stronger small cell BS in the latter.

This probabilistic model introduces randomness and renders decreasing distances less useful. The results of this new analysis showed an important fact. There is a small cell BS density region where

- the strongest interfering links transit from NLoS to LoS, while
- the signal ones stay LoS dominated due to the close proximity between the UEs and their serving small cell BSs.

As a result, the SINR of the UE – and the spectral efficiency in the small cell – do not stay constant, and the network capacity does not grow linearly with the small cell BS density anymore.

In this book, we will refer to this important phenomenon – the loss of linearity in the scale of capacity due to the transition of a large number of interfering links from NLoS to LoS – as *the ASE Crawl.*

Importantly, it should be noted that the ASE Crawl is not the result of a mathematical artefact, and that its impact was shown by the real-world experiments in [80], in which a densification factor of $100\times$ led to a network capacity increase of $40\times$ – clearly not a linear increase.

This new theoretical finding showed that the small cell BS density matters, and that it should not be taken lightly during the planning of a small cell deployment, as the ASE Crawl has a counterproductive effect. For completeness, however, it is fair to note that once the network density is ultra-dense, and the strongest interfering links transit from NLoS to LoS, both the inter-cell interference and the signal powers will again grow at a similar pace, as they are all LoS dominated, and the path loss decays at a similar rate. This restores the linear scale of capacity with the small cell BS density, but at a lower rate, as the path loss exponent in LoS is generally smaller than that in NLoS.

At this point in our story telling, it is worth noting that

- the need for an open-access operation and
- the impact of the transition of a large number of interfering links from NLoS to LoS

served as a wake-up call to the theoretical research community, which began to realize the importance of an accurate network and channel modelling, and started to review their understanding of ultra-dense networks. Some asked themselves whether some other important details were overlooked. Details that could change the performance trends expected for ultra-dense networks until then. This brought back again the original question:

Will the network capacity linearly grow with the small cell BS density or not?

Two frameworks should be highlighted in this quest, which will be discussed in the following:

- the implications of considering the near-field transmissions [81]; and
- the effect of the antenna height difference between the UEs and the small cell BSs [82].

The Myth of the Near-Field Effect

While looking at a more accurate channel model that could reveal new findings, the research in [81] presented a reasonable conjecture, indicating that the path loss exponent should be an increasing function of the distance, and proposed to capture this in a multi-slope path loss model similar to that presented in [79]. To illustrate the thinking behind it, the authors provided the following argument: *"There could easily be three distinct regimes in a practical environment:*

- *a first distance-independent 'near-field', where $\alpha_1 = 0$,*
- *second, a free-space like regime where $\alpha_2 = 2$, and*
- *finally, some heavily-attenuated regime where $\alpha_3 > 3$,"*

and then posed the following question:

What happens if densification pushes many BSs into the near-field regime?

The mathematical results derived in [81] provided an answer to such a question and concluded that the inter-cell interference power can grow faster than the signal power when the network is ultra-dense, even if both the inter-cell interference and the signal powers are LoS dominated. The intuition behind this is that when the UE enters the near-field range,

- the signal power is bounded, as the path loss becomes independent of the distance between such UE and its serving BS, i.e. $\alpha_1 = 0$, while
- the inter-cell interference power continues to grow, since more and more interfering small cell BSs approach the UE from every direction, when the network marches into the ultra-dense regime.

As a result, once the signal power enters the near-field range, the SINR of the UE cannot be kept constant, and will monotonically decrease with the small cell BS density.

This finding raised the alarm again, as it indicates that the near-field effect could lead to a void of capacity in an extreme densification case due to the overwhelming inter-cell interference.

Subsequent results on the topic, based on measurements, however, have shown that this alarm was unfounded [83]. The measurements, shown in figure 1 of [83], indicate that the near-field effect only takes place at sub-metre distances in practical ultra-dense networks with a carrier frequency of around 2 GHz and an antenna aperture of a few wavelengths. This is in line with the near-field effect theory, which indicates that the near-field is that part of the radiated field, where the distance from the source, an antenna of aperture, D, is shorter than the Fraunhofer distance, $d_f = 2D^2/\lambda$ [84], where λ is the wavelength. As a result, a BS density of around 10^6 BSs/km^2 would be needed for this near-field effect to be an issue, and this is unlikely to be seen in practice – at least as of today – as it means having one small cell BS every square metre.

Such result renders the near-field effect issue negligible in practical deployments. For this reason, despite being interesting and relevant, this topic is not theoretically further treated in this book.

The Challenge of the Small Cell Base Station Antenna Height

In parallel with the work done to shed new light on the performance impact of the near-field effect, new investigations presented yet another reason why the inter-cell interference power could grow faster than the signal power in an ultra-dense network, i.e. the antenna height difference, L, between the UEs and the small cell BSs [82].

By considering the antenna heights of both the UEs and the small cell BSs in the multi-slope path loss model presented in [79], and carrying a theoretical performance analysis on a fully loaded network, it was shown that

- the distance between a UE and its interfering small cell BSs decreases faster than the distance between such UEs and its serving BS when densifying the network.

This is because the UE can never get closer than a distance, L, to its serving small cell BS, since the UE cannot climb up towards it and that

- the inter-cell interference power, as a consequence, increases faster than the signal power at such a UE in a dense network.

As a result, this faster increase of the inter-cell interference power results in

- a decline of the SINR of the UE, which can be fast with the increase of the small cell BSs density due to the sheer number of interfering small cell BSs in an ultra-dense network, and in turn,
- a potential total network outage in the ultra-dense regime,

thus putting again a question mark on the benefits of network densification.

Here, we should also highlight that the impact of this cap on the received signal power at the UEs due to the antenna height difference, L, between the UEs and the small cell BSs is not a mathematical construct. It was confirmed in [85] using the measurement data of [86], where an antenna height difference, $L = 4.5\,\mathrm{m}$, was considered. More importantly, and in contrast to the near-field effect, it should be also noted that this phenomenon occurs at more realistic and practical small cell BS densities of around 10^4 BSs/km^2, with small cell BS antenna heights of $10\,\mathrm{m}$, thus being a more realistic threat.

Hereafter in this book, we will refer to this phenomenon – the continuous decrease in network capacity due to the antenna height difference between the UEs and the small cell BSs – as *the ASE Crash*.

Exploiting a Surplus of Small Cell Base Stations

When analyzing the previous results and trying to understand their implications, it is important to note that they all follow a traditional, macrocell-centric modelling assumption, in which the network is fully loaded, i.e. the number of UEs is always much larger than the number of BSs, and thus it is always safe to assume that there is at least one UE in the coverage area of every BS considered in the study. This assumption fits with macrocell as well as sparse small cell scenarios. Moreover, it is quite handy, as it allows to derive the capacity of the network at full load, and also makes the theoretical analysis of the network more tractable. However, in a dense deployment with many relatively small cell BSs, things are different. The probability of a small cell not serving any UE at a given time can be significantly high in some scenarios, and thus the always-on control signals have two negative impacts [13]:

- they impose an upper limit on the achievable network energy performance; and
- they cause inter-cell interference, thereby reducing the achievable data rates.

To address this, a number of mechanisms have been introduced for switching on and off small cell BSs – or at least their always-on signals – in the last years. For example, the 3rd-Generation Partnership Project (3GPP) LTE Release 12 [87] introduced mechanisms for turning on and off individual small cell BSs as a function of the traffic load to reduce the power consumption and the inter-cell interference. Moreover, sophisticated "lean carrier" approaches to allow a more dynamic on and off operation have been developed in 3GPP NR [12].

Embracing these practical considerations, the work in [88] revisited the ultra-dense network system model, and made a leap in theoretical performance analysis, accounting for finite active UE densities and idle mode capabilities. This new modelling brought new viewpoints, more industry aligned, which have led to a significantly different understanding of ultra-dense networks. In a nutshell, this work theoretically demonstrated the true benefits of an ultra-dense network, in which the number of small cell BSs is much larger than the number of UEs. This allows to reach the one UE per cell limit, where every UE can simultaneously reuse the entire spectrum managed by its serving small cell BS, without sharing it with other UEs. More importantly, it also showed how UEs can benefit from an improved performance in an ultra-dense network of this nature, because every small cell BS can

- tune its transmit power to the lowest possible one, just to cover its small intended range,
- and switch off its wireless transmissions through its idle mode capability, if there is no UE in its coverage area.

This results in both energy savings and a mitigated inter-cell interference; the latter because the control signals usually transmitted by active – but not empty – small cell BSs are switched off now, and do not interfere neighbouring transmitting cells.

From a theoretical perspective, and since multi-cell coordination is not considered here, when all UEs are served in an ultra-dense network, and the number of active small cell BSs is equal to the number of UEs, it is important to note that the number of interfering small cell BSs is automatically bounded, and thus so is the inter-cell interference. Since every active small cell BS serves a UE, no more active small cell BSs are needed. As a result, this bounded inter-cell interference power leads to an increasing SINR of the UE when densifying the network beyond this point, as the signal power continues to grow due to the closer proximity between a UE and its serving small cell BS. This leads to an enhanced overall network performance.

In the sequel of this book, we will refer to this phenomenon – the continuous increase in network capacity due to the surplus of small cell BSs with respect to UE and their idle mode capabilities – as *the ASE Climb*.

Channel-Dependent Scheduling and Multi-User Diversity

While the much larger number of small cell BSs with respect to that of UEs allows one to reach the one UE per cell limit, and in turn, a larger bandwidth, B, per UE, it also brings about a disadvantage. The lower number of UEs per small cell BS leads to a reduced multi-user diversity. In other words, a small cell BS has less UEs to choose from during its scheduling process, and thus, it is increasingly harder to opportunistically take advantage of potential constructive multi-path fast fading gains. Not only that, when the network goes ultra-dense, due to the closer proximity between a UE and its serving small cell BS, and the resulting higher probability of LoS transmissions, the radio-channel variations on a given time-frequency resource also become smaller and smaller. This also leads to a reduced multi-user diversity, given that the scheduler finds it increasingly harder to opportunistically find large multi-path fast fading gains.

The research in [89] studied multi-user diversity, and showed how the widely used channel-dependent schedulers indeed lose their ability to select a better UE for each scheduled time-frequency resource in each scheduling decision period, with the resulting loss in small cell capacity. It should be noted that the one UE per cell limit is the extreme case, where channel-dependent scheduling does not have any degree of freedom to select UEs during the scheduling process. This fact advocates for the use of simpler schedulers at small cell BS in ultra-dense networks, such as a simple round robin (RR) policy, to save on hardware processing complexity.

A New Capacity Scaling Law

Considering the above introduced theoretical findings, i.e. the ASE Crawl, the ASE Crash and the ASE Climb, new fundamental are questions raised, which are summarized in the following:

> *Will the negative impact of the ASE Crawl and Crash overweight the positive one of the ASE Climb, or the other way around?*

> *Which is the resulting capacity scaling law that best characterizes an ultra-dense network when considering all these characteristics?*

A new theoretical performance analysis came to answer these questions, using stochastic geometry, and proposed a new capacity scaling law for ultra-dense networks [90]. This is probably the most complete model and comprehensive capacity scaling law in the literature, as of the time of writing this book and up to the authors' knowledge. In short, this new study considered for the first time a model able to capture the combined effects and interactions of

- the transition of a large number of interfering links from NLoS to LoS,
- the antenna height difference between the UEs and the small cell BSs,
- a finite UE density, and the surplus of small cell BSs with respect to UE, as well as
- the idle mode capability at the small cell BSs,

and derived a new capacity scaling law, indicating that both the coverage probability and the ASE will asymptotically reach a maximum constant value in the ultra-dense regime.

Theoretically speaking, this research showed that in a densifying network

- the signal power caps because of the antenna height difference between the UEs and the small cell BSs, while
- the inter-cell interference power becomes bounded due to the finite UE density as well as the idle mode capability at the small cell BSs.

This results in a constant SINR of the UE in a densifying ultra-dense network of the above characteristics, leading to the mentioned asymptotic behaviour with the increase of the small cell BS density – *a constant capacity scaling law*. From this new capacity scaling law, it can be concluded that, for a given UE density, the network densification should not be abused indefinitely, but should be instead stopped at a given level, because any network densification beyond such point is a waste of both invested money and energy consumption.

At this point, to recap, it is important to note that this new constant capacity scaling law in the ultra-dense regime is significantly different from

- the initial linear scale of capacity introduced in both [60] and [61],
- the pessimistic ASE Crawl and ASE Crash, leading to a disastrous network performance with the densification, presented in [79] and [82], respectively, and
- the optimistic ASE Climb, discussed in [88],

and shows how the theoretical performance analysis and the understanding of ultra-dense networks have both improved up to now.

Dynamic Time Division Duplex to Make the Most of Ultra-Dense Networks

As a by-product of the lower number of UEs per small cell BS in a densifying network, it is also important to note yet another impact. The per-small cell aggregated downlink and uplink traffic demands become highly dynamic. Sometimes a small cell BS may have much more downlink than uplink traffic, while the opposite may be true at a different time instant or for another small cell BS in a neighbouring location. To address such scenarios, dynamic time division duplexing (TDD), i.e. the possibility for dynamic assignment and reassignment of time resources between the downlink and uplink transmission directions, was developed in 3GPP LTE Release 12 [87], and is a key 3GPP NR technology component [12].

Contrary to a static or a semi-dynamic TDD system, where the number of time resources devoted to downlink and uplink transmissions are preconfigured, and may not match the instantaneous traffic demands of a small cell, when embracing dynamic TDD, each small cell BS can provide a tailored configuration of downlink and uplink time resources, e.g. subframes, to meet its instantaneous downlink and uplink traffic requests. In other words, the transmission direction can be dynamically changed on short time periods in each cell. However, it is important to note that such flexibility does not come free, but at the expense of introducing inter-cell interlink interference. For example, the downlink transmission of a small cell may interfere with the uplink reception in a neighbouring small cell (downlink-to-uplink inter-cell interference), and vice versa, the uplink transmission of a UE in a small cell may interfere with the downlink reception of another UE in a neighbouring small cell (uplink-to-downlink inter-cell interference).

As a consequence, dynamic TDD may present a trade-off that needs to be well understood for its proper operation in certain cases, as it may

- improve the efficiency in the usage of the time resources at the MAC layer, but
- degrade the performance of the physical (PHY) layer, introducing inter-cell interlink interference, which in turn, decreases the SINR of the UE, and as a result, the cell performance.

This performance degradation, if it takes place, may be particularly severe for the uplink reception due to strong downlink-to-uplink inter-cell interference. This is because the transmit power and the antenna gain of a small cell BS is usually larger than that of a UE. This calls for the implementation of downlink-to-uplink inter-cell interference mitigation techniques.

The work in [91] presents a system-level simulation analysis on these dynamic TDD trade-offs. Importantly, a new theoretical performance analysis based on the stochastic geometry framework presented in the previous sections of this *"brief history of small cells"* has been developed in [92] to study the dynamic TDD technology. Both MAC and PHY layer aspects are covered, exploring the mentioned trade-offs and the benefits of inter-cell interference cancellation.

1.4   Outline of This Book

In this section and following the description in the previous *"brief history of small cells,"* we present the outline of this book, aimed at answering fundamental questions about network densification, and shedding new light on ultra-dense deployments.

The structure of the book is designed to show the fundamental differences between traditional sparse or dense small cell networks and ultra-dense ones, while the content is meant to teach readers the basis of theoretical performance analysis and empower them with the knowledge to develop their own frameworks.

Chapter 2 introduces the main building blocks of any performance analysis tool and describes the basic concepts of the system-level simulation and the theoretical performance analysis frameworks used in this book, paying particular attention to stochastic geometry.

Chapter 3 summarizes the modelling, derivations and results of probably one of the most important works on small cell theoretical performance analysis, that of J. G. Andrews et al. [60], which concluded that the fears of an inter-cell interference overload in small cell networks were not well grounded, and that the network capacity – or in more technical words, the ASE – linearly grows with the number of deployed small cells.

Chapter 4 analyzes in detail – from a theoretical perspective – the first caveat towards such linear growth of capacity in the ultra-dense regime, that of the impact of the transition of a large number of interfering links from NLoS to LoS. This chapter shows that the theoretical tools used to analyze traditional sparse or dense small cell networks do not apply to ultra-dense ones, and neither do their conclusions. The modelling used, the derivations done and the results obtained are carefully presented and discussed in this book chapter for the better understanding of the reader.

Chapter 5 studies in detail – and also in a theoretical manner – yet another and more important caveat towards a satisfactory network performance in the ultra-dense regime, that of the impact of the antenna height difference between the UEs and the small cell BSs. Similarly, as in the previous chapter, the modelling used, the derivations done and the results obtained are carefully presented and discussed in this book chapter for the better understanding of the reader. Moreover, some deployment guidelines are provided to mitigate such fundamental issue.

Chapter 6 brings the attention to an important feature of ultra-dense networks, the surplus of small cell BSs with respect to the UEs. Building on this fact, the ability of next generation small cell BSs to go into idle mode, transmit no signalling meanwhile, and thus mitigate inter-cell interference is presented and shown in this chapter, as a key tool to enhance ultra-dense network performance, and combat the previous presented caveats. Special attention is paid to the modelling and analysis of the idle mode capability at the small cell BS.

Chapter 7 investigates the impact of ultra-dense networks on multi-user diversity. A denser network reduces the number of UEs per small cell in a significant manner, and thus can significantly reduce – and potentially neglect – the gains of channel-dependent scheduling techniques. These performance gain degradations are theoretically analyzed in this book chapter, and the performance of a proportional fair (PF) scheduler is compared to that of an RR one.

Chapter 8, standing on the shoulders of all previous chapters, presents a new capacity scaling law for ultra-dense networks. Interestingly, the signal and the inter-cell interference powers become bounded in the ultra-dense regime due to the antenna height difference between the UEs and the small cell BSs and the finite UE density as well as the idle mode capability at the small cell BSs, respectively. This leads to a constant SINR of the UE, and thus to an asymptotic capacity behaviour in such a regime. From this new capacity scaling law, it can be concluded that, for a given UE density, the network densification should not be abused indefinitely, and instead, should be stopped at a given level. Network densification beyond such point is a waste of both invested money and energy consumption.

Chapter 9, using the new capacity scaling law presented in the previous chapter, explores three relevant network optimization problems: (i) the small cell BS deployment/activation problem, (ii) the network-wide UE admission/scheduling problem and (iii) the spatial spectrum reuse problem. These problems are formally presented, and exemplary solutions provided, with the corresponding discussion on the intuition behind the solutions.

Chapter 10, in contrast to all previous chapters of this book, which focused on the performance of the downlink, analyzes the performance of the uplink of an ultra-dense network. Importantly, this book chapter shows that the phenomena presented in – and the conclusions derived from – all the previous chapters also apply to the uplink, despite its different features, e.g. uplink transmit power control, inter-cell interference source distribution. System-level simulations are used in this book chapter to conduct the study.

Chapter 11 shows the benefits of dynamic TDD with respect to a more static TDD assignment of time resources in an ultra-dense network. As mentioned before, the number of UEs per small cell reduces in a significant manner in a denser network. As a result, a dynamic assignment of time resources to the downlink and the uplink according to the load in each small cell can avoid resource waste, and significantly enhance its capacity. Dynamic TDD is modelled and analyzed through system-level simulations in this book chapter, and its performance carefully explained.

1.5   Definitions

In this section, we provide a number of definitions that may be handy for the reader to provide clarity and avoid confusion on some of the widely used concepts in this book. It is important to note that the definitions presented in the following are provided for guidance and are not exhaustive, and that a more complete definition of some of them – including modelling details – will be given in Chapter 2 of this book.

Subscriber The subscriber is the customer who has a contract with the service provider, in other words, the person named on the bill for the telephone line or internet connection, or the person who owns the subscriber identity module (SIM) card on a pay-as-you-go mobile contract. This may be an individual or an organization.

End-user The end-user is the individual actually using the phone or the internet connection. This will not always be the same person as the subscriber for example, he or she might be the subscriber's employee, a customer, a family member or a friend.

User equipment A UE is any device used directly by an end-user to communicate. It can be a handheld telephone, a laptop computer equipped with a mobile broadband adapter or any other device.

Base station A BS is a specialized radio transmitter/receiver which connects the UE to a central network hub, the core network, and allows the connection to a network.

Cell site A cell site is the geographical location where the equipment that conforms the BS is deployed.

Cell A cell is the physical geographical area covered by the BS.

Deployment A deployment refers to a collection of two or more BSs which provides access to the network to the UEs in a geographical area, which is generally larger than what a single BS can cover.

Orthogonal deployment Two BSs are said to be deployed in an orthogonal manner when they use a different carrier frequency to communicate. For example, a BS operating at 2 GHz and another one functioning at 3.5 GHz are orthogonally deployed, and do not interfere with each other.

Co-channel deployment Two BSs are said to be deployed in a co-channel manner when they use the same carrier frequency to communicate. For example, two BSs operating at 2 GHz are co-channel deployed, and interfere with each other.

Antenna A rod, wire or other device used to transmit or receive radio signals.

Antenna radiation pattern An antenna radiation pattern is a diagrammatical representation of the distribution of the antenna radiated energy into space, as a function of direction. The most energy is radiated through the main lobe. The other parts of the pattern where the radiation is distributed sideways are known as side lobes. These are the areas where the power is wasted. There is another lobe, which is exactly opposite to the direction of the main lobe. It is known as the back lobe.

Channel Broadly speaking, the channel is the physical or logical link that connects a data source, e.g. the UE, to a data sink, e.g. the small cell BS. The wireless channel is characterized by a large number of parameters, such as its carrier frequency and bandwidth, to cite a few. Understanding the variations of the wireless channel over frequency and time is of importance to analyze a system performance. Such variations can be roughly categorized in the following two groups, where some of the mentioned concepts are further developed in the posterior definitions [93]:

- Large-scale fading, due to (i) the path loss of the signal as a function of the distance and (ii) the shadowing by large obstructing objects, such as buildings and hills. This happens as the UE moves through distances of the order of such large obstructing objects and is typically frequency independent.
- Small-scale fading, due to the constructive and destructive interference of the multiple signal paths between the transmitter and receiver. This occurs at the spatial scale of the order of the wavelength of the carrier frequency and is frequency dependent.

Path loss The path loss is the attenuation in power density of an electromagnetic wave as it propagates through space. The path loss is influenced by the environment (dense urban, urban or rural, vegetation and foliage), the propagation medium (dry or moist air), the distance between the transmitter and the receiver, the height and the location of the antennas, as well as other phenomena such as refraction, diffraction and reflection.

Shadow fading A radio signal will typically experience obstructions caused by objects in its propagation path, thus generating random fluctuations of the received signal strength at the receiver, referred to as shadow fading. The number, locations, sizes and dielectric properties of the obstructing objects, as well as those of the reflecting surfaces and scattering obstacles are usually not known, or very hard to predict. Due to such unknown variables, statistical models are generally used to model shadow fading, where the received power due to the shadow fading may vary significantly (e.g. in tens of dB) over distances of the order of such obstructing objects, surfaces and obstacles.

Multi-path fast fading The obstructing objects in the propagation path from the transmitter to the receiver may also produce reflected, diffracted and scattered copies of the radio signal, resulting in multi-path components (MPCs). The MPCs may arrive at the receiver attenuated in power, delayed in time and shifted in frequency (and/or phase) with respect to the first and strongest MPC, usually the LoS component, thus adding up constructively or destructively. As a consequence, the received signal strength at the receiver may vary significantly over very small distances of the order of a few wavelengths.

Table 1.1. Author's research articles on ultra-dense wireless deployments – foundation of this book

Factors	DL	UL	LoS	antH	IMC	PF	dynTDD	Rician	Shadow	Energy	HetNet	D2D	Downtilt	MIMO	UAV	LAA	caching	CoMP	DNA-GA
[94]	✓		✓	✓	✓														
[95]	✓		✓	✓	✓														
[96]	✓		✓					✓											
[97]	✓		✓	✓	✓	✓		✓				✓							
[98]	✓	✓	✓	✓			✓						✓						
[92]	✓		✓	✓	✓	✓	✓												
[99]	✓		✓		✓	✓	✓				✓								
[100]	✓		✓	✓	✓			✓		✓	✓								
[101]	✓		✓	✓	✓			✓	✓	✓	✓								
[102]	✓		✓	✓	✓			✓	✓										
[103]	✓		✓		✓									✓					
[104]		✓	✓	✓	✓			✓											
[105]		✓	✓		✓														
[106]	✓		✓	✓	✓						✓								
[107]	✓		✓	✓	✓			✓						✓					
[108]	✓		✓		✓	✓	✓												
[109]	✓		✓	✓	✓														
[89]	✓		✓		✓														
[110]	✓		✓	✓	✓														
[88]	✓	✓	✓	✓	✓	✓	✓	✓	✓		✓		✓	✓					
[111]	✓		✓		✓			✓	✓										
[112]	✓	✓	✓		✓			✓											
[113]	✓		✓	✓															
[114]	✓		✓		✓														
[115]	✓		✓								✓								
[116]	✓	✓	✓																
[117]	✓		✓			✓													
[79]	✓		✓																
[118]	✓		✓																
[119]	✓		✓																

Table 1.1. (Cont)

[120]
[121]
[122]
[123]
[124]
[125]
[126]
[127]
[128]
[129]
[130]
[131]
[132]
[133]
[134]
[135]
[136]
[137]
[138]
[139]
[140]
[8]
[91]
[141]
[142]
[143]
[144]
[145]

Noise power The noise power is the measured total noise in a given bandwidth at the antenna of the receiving device when the signal is not present, i.e. the integral of noise spectral density over the bandwidth.

1.6   Related Publications

In this section, and through Table 1.1, we provide a comprehensive list of all the research papers published by the authors of this book on the fundamental understanding of ultra-dense networks. The content of this book builds upon such publications and references therein. The authors advise to read the papers to gain a deeper understanding of some of the concepts presented in this book.

2 Modelling and Analysis of Ultra-Dense Wireless Networks

Performance analysis tools that are able to evaluate the overall performance of large networks are required to aid vendors and operators in

- the development and deployment of wireless networks in general and small cells in particular and
- the refinement of existing network features, such as handover and radio resource management, to cite a few.

In this context, two families of such tools stand out, namely

- *network simulation, planning and optimization tools* as well as
- *theoretical performance analysis tools*.

Network simulation, planning and optimization tools – aka *system-level simulations* for simplicity – are widely used in the industry, while theoretical performance analysis is the de facto approach in the academic community. Having said that, nothing prevents the industry and academic community from making their own toolbox choices according to the problem they are tackling. Indeed, it is increasingly common to see, for example, academics developing and using system-level simulations.

System-level simulations model the elements and the operations of a network through a computer software and present a number of benefits. First, system-level simulations are cheaper and simpler to realize than real implementations and proofs of concept. Second, they are more accurate and reliable than theoretical performance analyses. The number of assumptions and simplifications made in system-level simulations depends on the level of detail of the computer software used, but the number of assumptions are generally much fewer than in theoretical performance analyses, where they are a must for mathematical tractability reasons. As a result, system-level simulations can model more complex networks. However, they also present some drawbacks. One of these drawbacks is that they usually require a significantly high computing power to obtain statistically representative results. This is particularly an issue if the number of network nodes involved in the study is large, in the order of thousands or more, which is typically the case in an ultra-dense network – the subject of this book. Another disadvantage of system-level simulations is related to the reproducibility of the results by different parties, e.g. vendors, operators, research labs, which usually require tedious calibration campaigns. This is because system-level simulations are complex tools mostly owned by these private entities and not shared in the public domain due to intellectual property issues.

Theoretical performance analysis, although less accurate and reliable compared to system-level simulations in terms of end results, is usually simpler to devise and can generally handle larger networks. This is due to the more compact representation of the elements and the operations of the network through advanced mathematical models. Such simplicity is one of the main advantages of theoretical performance analysis, which sometimes even allows for closed-form expression solutions, able to readily provide an understanding of the underlying processes that drive the network behaviour and performance. In essence, theoretical performance analysis permits one to understand why things fundamentally occur as well as the mathematical relationship between the different entities and parameters of the network. A good example of this is the Shannon–Hartley theorem [9], presented in equation (1.1), and shown here again for convenience:

$$C \text{ [bps]} = B \cdot \log_2 (1 + \gamma). \tag{2.1}$$

This closed-form expression provides a fundamental insight into what the variables are that influence the amount of information (capacity, C, in *bps*) that can be transmitted over a communication channel of a specified bandwidth, B, in Hertz with a received signal quality, γ. Remember that the received signal quality, γ, is the ratio of the received signal power, S, in Watts to the additive white Gaussian noise (AWGN) power, N, in Watts, i.e. $\gamma = \frac{S}{N}$.

As one can infer from the above discussion, system-level simulations and theoretical performance analysis are not orthogonal tools, in stark competition. Instead, they complement each other – indeed, quite well – and have different purposes. To illustrate this, we would like to share our modus operandi when it comes to research, i.e. our research methodology. First of all, we recommend paying special attention and spending some quality time on an observation phase, where one gets familiar with the true empirical problem and talks to experts on the subject about it. This is crucial to make sure that the problem one is trying to solve exists in reality and is relevant. Who wants to solve an irrelevant problem or a problem that only exists in a hypothetical universe? Once the problem is confirmed, and some conjectures have been made about it, one may want to develop – or use existing – system-level simulations in the next phase to corroborate the existence of the problem and the validity of the conjectures made. System-level simulations allow the researcher to reproduce the problem repeatedly in a controlled environment. Finally, once system-level simulations have confirmed one's understanding of the problem and its solutions, one can go ahead and develop solid theories, stripping down the problem with assumptions and simplifications for tractability reasons while keeping the essence of it. System-level simulations can be used to double-check that such assumptions and simplifications are acceptable and do not change performance trends, and thus that one's theories remain relevant with the right level of modelling.

With this in mind, in this chapter, we first introduce the main building blocks of any performance analysis tool, whether system-level simulation or theoretical performance analysis, and then describe their general concepts. The following points summarize each section in more detail.

- Section 2.1 introduces the models used in this book: the small cell base station (BS) layout and the user equipment (UE) deployment models, the traffic model, the UE-to-small cell BS association model, the antenna gain model, the path loss model, the shadow fading model, the multi-path fast fading model, the received signal strength model, the signal-to-interference-plus-noise ratio (SINR) model, and some key performance indicators.
- Section 2.2 presents the different families and types of simulation tools. It first differentiates between link- and system-level simulations, then dives into the latter, and introduces the differences between static and dynamic system-level simulations, and finally explains the choice of static system-level simulations and the implementation for its use in this book.
- Section 2.3 provides a literature review on – and discusses the state of the art of – theoretical performance analysis tools, and then presents the bases of the theoretical performance analysis framework selected in this book – stochastic geometry – together with its benefits and drawbacks. Its more important properties are also described.
- Section 2.4 introduces an intuitive network performance visualization technique used in this book to illustrate the performance of an ultra-dense network.
- Section 2.5, finally, summarizes the key takeaways of this chapter.

Before proceeding with the rest of this chapter, however, it is important to note here that the total channel gain, $G_{t,r}$, between the tth transmitter and the rth receiver, the pillar of any performance analysis tool, is modelled as the combination of individual channel gains and losses, namely,

- antenna gain, $\kappa_{t,r}$,
- path loss, $\zeta_{t,r}$,
- shadow fading gain, $s_{t,r}$ and
- multi-path fast fading gain, $h_{t,r}$.

Note that details on the precise operations and units will be given in the following sections, and that the notation of the tth transmitter and the rth receiver may be dropped for convenience in our discussion or interchanged for that of the bth small cell BS and the uth UE, when necessary. This is because a transmitter and a receiver can be either a small cell BS or a UE, and this generic description, although good to accommodate both the downlink and the uplink in most cases, does not fit in others.

Based on this notation, we now proceed with the rest of the chapter.

2.1   Performance Analysis Building Blocks

In this section, we describe the minimum collection of building blocks that have to be modelled to some degree in any performance analysis tool, whether it is a system-level simulation or a theoretical performance analysis. Specific parameters for each model will be given in the analyses of each chapter of this book.

2.1.1   Network Layout

The network layout models the positions of the small cell BSs within a geographical area. According to their nature, different deployments have different layouts. For example, unplanned deployments, driven by the end-customer like in the Wi-Fi case, will have a less structured layout when compared to carefully planned cellular deployments, carried out by network operators using advanced optimization tools.

In this section, we briefly discuss the two network layouts used in this book – the random and the hexagonal network layout – depicted in Figure 2.1.

Please refer to appendix A.4.1 of [146] for more details on the two network layout models presented in the following, and others.

Random Layout

Small cell BSs are usually deployed on demand to provide coverage or capacity where it is needed as it is needed. As a consequence, they have a more ad hoc deployment nature than macrocell ones, and their network layouts usually follow a random fashion. Therefore, unless otherwise stated, in most of our analyses in this book, we assume that small cell BSs are randomly deployed without much planning, following a homogeneous Poisson point process (HPPP) layout of a defined intensity, λ, as suggested in [50]. Figure 2.1b illustrates this type of network layout. Moreover, Section 2.3.1 will formally define the concept of a point process, in general, as well as that of an HPPP, in particular, and will introduce some of the key properties of this type of distribution. This deployment model is at the core of the network performance analysis presented in this book.

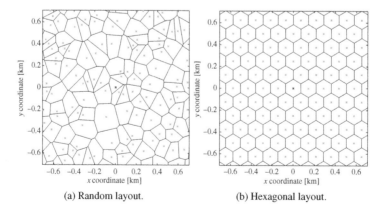

(a) Random layout.   (b) Hexagonal layout.

Figure 2.1 Illustration of two widely used types of small cell BS layouts, the random and the hexagonal layouts. Here, small cell BSs are represented by markers "x" and cell coverage areas for UE distribution are outlined by solid lines.

As a note, we would like to indicate that the set,[1] $x \subseteq \Phi$, generated by the process,[2] Φ, operating in the d-dimensional space, $S \subseteq \mathbb{R}^d$, will denote the set of small cell BSs – in more detail, the set of small cell BS locations – for study hereafter, and that the process, Φ, is an HPPP in this book, as indicated earlier. Section 2.3.1 will provide a more detailed definition and description of these sets, processes and spaces.

Hexagonal Layout

Small cell BSs may also be deployed in a more planned nature in some other scenarios, such as campuses, harbours, etc. To represent such planning, we adopt the general hexagonal layout used for the performance analysis of a multiplicity of scenarios in the 3rd Generation Partnership Project (3GPP) [147]. Figure 2.1a shows an example. In this model, BSs are deployed following a hexagonal grid, and the model is fully characterized by

- the inter-site distance (ISD), D_{ISD}, which defines the two-dimensional distance between any two BSs in the network and
- the number of tiers around the central hexagon or the bounds of the particular scenario of study.

It is important to note that this hexagonal layout provides optimistic results in terms of worst-case performance with respect to the random layout, as it maximizes the minimum distance among any two BSs, and in turn, minimizes the worst-case inter-cell interference. It also benefits from a lower system-level simulation computational complexity, particularly for dense networks, as multiple small cell BSs placements are not necessary to get statistically representative results.

2.1.2 User Equipment Position

Different UE deployment models can be found in the literature to define where the UEs are located within the scenario of study. In this book, due to its analytical tractability as well as its easy reproducibility, we adopt a uniform deployment, where UEs are dropped within the given network scenario according to another but independent HPPP of a defined intensity, ρ.

We would also like to indicate that the set, $y \subseteq \Phi^{UE}$, generated by the HPPP, Φ^{UE}, will denote the set of UEs of study hereafter.

Please refer to appendix A.4.10.1 of [146] for more details on more practical non-uniform deployment models.

---

[1] A set is a group or collection of objects or numbers, considered as an entity unto itself. Each object or number in a set is called a member or element of the set. Examples include the set of all computers in the world, the set of all oranges on a tree and the set of all prime numbers between 0 and 100.

[2] A stochastic process refers to a family of random variables indexed by some other variable or set of variables, e.g. time, realizations.

2.1.3   Traffic Model

The traffic model characterizes the amount of bits generated by the application layer. In this book, due to its analytical tractability as well as its easy reproducibility, we consider a full-buffer traffic model.

The full-buffer traffic model is a traffic model in which the information generated by the application layer of a UE generates at least as much data to transmit/receive as the air interface can support. This model is used to test and stress the air interface as much as possible. It represents a worst-case scenario and is useful to understand the maximum system load with which a network can cope. Very few real applications exhibit the behaviour of the full-buffer traffic model. However, despite its little applicability, it provides a reference to which other traffic models can be compared for evaluating the system performance.

Please refer to appendix A.4.10.2 of [146] for more details on this full-buffer model and others.

2.1.4   User Equipment-to-Base Station Association

Prior to any data transmission or reception, the UE needs to connect to the network. To start such network association procedure, the UE has first to

- find and acquire synchronization to a small cell BS in the network, and
- receive and decode the information needed to communicate and operate within such small cell BS, often referred to as system information.

Once the system information has been obtained, the UE can access the small cell BS by means of the random-access procedure [12].

The UE-to-small cell BS association model masks this procedure, and basically identifies the small cell BS to connect to for every UE, based on one of the following criteria.

Shortest Distance
Each UE associates with the small cell BS at the shortest Euclidean distance.

Strongest Received Signal Strength
Each UE associates with the small cell BS providing the strongest received signal strength.

Given a set of small cell BSs, $x \subseteq \Phi$, the UE-to-small cell BS association for the uth UE can be formulated as

$$b^* = \operatorname*{argmax}_{b \in x} \quad P^{\mathrm{RX}}_{b,u} \text{ [Watts]}$$

$$\text{subject to} \quad P^{\mathrm{RX}}_{b,u} \geq \chi_u,$$

where

- b^* is the serving small cell BS of the uth UE,
- $P_{b,u}^{RX}$ is the received signal strength of the pilot signal transmitted by the bth small cell BS at the receiver of the uth UE in Watts,[3] and
- χ_u is the sensitivity of the receiver of the uth UE in Watts.

In the following subsections, we will discuss how to calculate $P_{b,u}^{RX}$ considering multiple channel propagation factors.

Note that the UE association process is usually based on average received signal power, only accounting for slow fading – path loss and shadow fading – as the fast variations of the channel – multi-path fast fading – are averaged across frequency and time by the cell selection, the cell reselection and the handover procedures [12].

2.1.5   Antenna Gain

As described in the introduction of this chapter, the total channel gain between a transmitter and a receiver is modelled as the composition of different gains and losses. In this section, the concept of antenna gain is described.

The antenna gain is a relative measure of the ability of an antenna to concentrate radio frequency energy in a particular direction. Using a more formal definition, the antenna gain can be defined as the ratio of the power produced by the antenna from a far-field source in a particular direction to the power produced by a hypothetical lossless isotropic antenna in the same direction. Antenna gains are typically represented using antenna patterns and are usually expressed in isotropic-decibel (dBi). It is important to state that an antenna radiates energy in all directions, at least to some extent. The antenna pattern is thus three-dimensional. It is common, however, to describe this three-dimensional pattern with two planar patterns, called the principal plane patterns or polar diagrams. These principal plane patterns can be obtained by making two slices through the three-dimensional pattern by the maximum value of the pattern or by direct measurement. These principal plane patterns are also commonly referred to as the two-dimensional horizontal (azimuth) and vertical (elevation) antenna patterns. Importantly, it should be noted that characterizing the antenna gain with these two-dimensional horizontal and vertical antenna patterns works well for antennas that have well-behaved patterns. That is, not much information is lost when only two planes are shown [148].

When provided with the two-dimensional horizontal and vertical antenna patterns, the resulting antenna gain can be calculated, in its simplest form,[4] as

---

[3] Note that the pilot signals are signals transmitted by the small cell BS to drive the UE association procedure.

[4] This model applies when the antenna is formed by a single antenna element, and we are provided with the two-dimensional horizontal and vertical antenna patterns of the antenna element. If the antenna is comprised of a collection of antenna elements, arrayed in a given configuration, the array gains should also be considered. If we are still provided with the two-dimensional horizontal and vertical antenna patterns of the building antenna element, the array gains should be explicitly modelled according to the specific array configuration, e.g. linear array, planar array [149]. Alternatively, two-dimensional

$$\kappa\left(\varphi,\theta,\theta_{\text{tilt}}\right)[\text{dBi}] = \kappa_{\text{M}} + \kappa_{\text{H}}\left(\varphi\right) + \kappa_{\text{V}}\left(\theta,\theta_{\text{tilt}}\right), \qquad (2.2)$$

where

- $\kappa\left(\varphi,\theta,\theta_{\text{tilt}}\right)[\text{dBi}]$ is the antenna gain in dBi,
- φ and θ are the azimuth and the elevation angles of arrival/departure in the horizontal and the vertical planes in radians, respectively, which are defined according to a reference point and in accordance with that of the antenna pattern,
- θ_{tilt} is the down-tilt angle of the antenna in radians, which is also defined according to the mentioned reference point,
- κ_{M} is the maximum antenna gain in dBi,
- $\kappa_{\text{H}}\left(\varphi\right)$ is the horizontal attenuation offset in decibels (dB) and
- $\kappa_{\text{V}}\left(\theta,\theta_{\text{tilt}}\right)$ is the vertical attenuation offset in decibel (dB).

For illustration purposes, and to ease the understanding of the reader, Figure 2.2a illustrates the two-dimensional horizontal and vertical antenna patterns of a practical four-element half-wave dipole used in real small cell BSs, normalized with respect to the maximum antenna gain, κ_{M}, while Figure 2.2b shows the resulting spatial antenna gains in the horizontal plane. Note that, in Figure 2.2a, the antenna is supposed to be mounted in a vertical pole in the Y–Z plane, and the elevation angle, θ, in the vertical plane is measured against a reference angle – the 0 radian angle – located in the X–Y plane – the horizon. Moreover, in Figure 2.2b, even if the horizontal antenna pattern is omnidirectional, the spatial antenna gains in the horizontal plane are not uniform and vary with the distance. This is due to the effect of the directional vertical antenna pattern.

Building on this simple model, we briefly discuss the antenna models used in this book in the following.

Please refer to appendix A.4.3 of [146] for more details on the two antenna models presented in the following, and others.

Isotropic Antenna

An isotropic antenna is a theoretical point source of electromagnetic waves, which radiates with the same intensity in all directions. In other words, it has no preferred direction, and its uniform radiation resembles a sphere centred on the source. In reality, the practical implementation of a coherent isotropic antenna of linear polarization is impossible, since its radiation field would not be consistent with the Helmholtz wave equation – derived from the Maxwell's equations – in all directions simultaneously. However, similar to the role of the full-traffic model presented in Section 2.1.3, this isotropic antenna is used as a reference antenna, with which other antennas can be compared, for example, when determining their gains.

Note that the two-dimensional patterns of an isotropic antenna can be generally modelled as

---

horizontal and vertical arrayed antenna patterns may be implicitly provided, where the array gains are already embedded in these antenna patterns. This approach trades flexibility for simplicity.

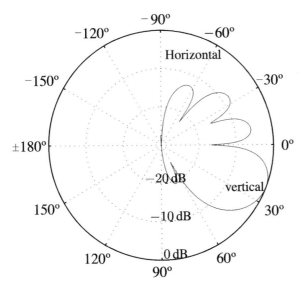

(a) Normalized four-element half-wave horizontal and vertical arrayed antenna patterns.

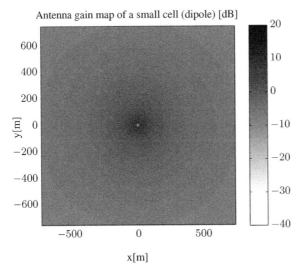

(b) Four-element half-wave spatial antenna gains.

Figure 2.2 Four-element half-wave antenna patterns and spatial antenna gains.

- Horizontal attenuation offset:

$$\kappa_{\mathrm{H}}\left(\varphi\right) [\mathrm{dB}] = 0, \tag{2.3}$$

    and
- Vertical attenuation offset:

$$\kappa_{\mathrm{V}}\left(\theta, \theta_{\mathrm{tilt}}\right) [\mathrm{dB}] = 0. \tag{2.4}$$

Four-Element Half-Wave Dipole Antenna

The dipole antenna is the simplest practical antenna from a theoretical perspective and is constructed based on two identical metal rods, oriented in parallel and displayed in line with respect to each other. The feeding current is applied in between the two rods.

The most common form of dipole is the half-wave dipole, in which each of the two rods are approximately $\frac{1}{4}$ wavelength long. Half-wave dipole antennas are typically used in femtocells for providing indoor residential coverage, while half-wave dipole-based antennas with multiple elements are more common in compact outdoor small cells.

To practically evaluate the impact of a more realistic small cell BS antenna than the isotropic one, we adopt the four-element half-wave dipole antenna presented in [150], whose two-dimensional arrayed antenna patterns are modelled as

- Horizontal attenuation offset:

$$\kappa_H \left(\varphi \right) [\text{dB}] = 0, \qquad (2.5)$$

and
- Arrayed vertical attenuation offset:

$$\kappa_V \left(\theta, \theta_{\text{tilt}} \right) [\text{dB}] = \max \left\{ 10 \log_{10} \left| \cos^n \left(\theta - \theta_{\text{tilt}} \right) \right|, F_V \right\}, \qquad (2.6)$$

where
☐ the antenna is supposed to be mounted in a vertical pole in the Y–Z plane,
☐ θ_{tilt} is the down-tilt angle, which in this case is defined as the angle between the horizon – X–Y plane – and the main beam direction in the vertical plane,
☐ θ is the elevation angle, which is also defined with respect to the horizon,
☐ n is a fitting parameter in linear units and
☐ F_V is the secondary lobe level (SLL) in dB.

2.1.6   Path Loss

An important contributor to the total channel gain between a transmitter and a receiver is the path loss.[5] The path loss models the average attenuation in received signal strength, which occurs when a radio signal travels from the antenna through space at a given distance at a given frequency.

A particular and important path loss model is the free-space path loss model, which indicates the path loss, ζ^{FS}, when a radio signal travels from the antenna through free space a distance, d, at a frequency, f_c, and can be formulated as

$$\zeta^{\text{FS}}(d)[\text{dB}] = -20 \log_{10}(d) - 20 \log_{10}(f_c) - 92.45, \qquad (2.7)$$

where

---

[5] Note that the path loss, although named as a loss, is defined as a gain in the mathematical sense, and thus the loss is expressed through its negative value.

- ζ^{FS} [dB] is the path loss in dB,
- d is the three-dimensional distance between the transmitter and the receiver in kilometres,[6] and
- f_c is the carrier frequency of the radio signal in gigahertz.

The free-space path loss model, however, does not hold for most environments due to its ideal conditions, e.g. the effects from the ground, the buildings, the vegetation and other features, are not considered.

To enhance the accuracy of the free-space path loss model and account for more realistic scenarios, deterministic or statistical path loss models are often adopted in network performance analysis [151]. Deterministic path loss models based on ray tracing, ray launching or finite-difference time-domain (FDTD) result in high accuracy at the expense of large computation complexity and the need for detailed input data, e.g. city maps, material properties. Statistical models, in contrast, are appealing due to their lower complexity. These models generally require a few input parameters and are based on empirical data – path losses measured and averaged in typical environments, such as dense urban, urban and rural scenarios as well as indoor venues – which is then processed through curve-fitting techniques.

A widely used statistical path loss model is the Okumura–Hata model [152], which was primarily built using measurements collected in Tokyo, Japan and enhanced later with different extensions. However, the Okumura–Hata model has some intrinsic drawbacks. For example, it neglects the terrain profile between the transmitter and the receiver, since the transmitters were located on hills well above the receivers, and only provides support for carrier frequencies up to 1.9 GHz.

To overcome such drawbacks and adapt to all types of conditions, while retaining a low complexity, a large number of statistical path loss models have been developed ever since, based on different measurement campaigns. In this book, the 3GPP path loss models are adopted, which can be applied for different antenna heights and to a large range of carrier frequencies. A key advantage of using the 3GPP models is that they are widely used within the industry and the academic community, and thus facilitate the reproducibility and comparison of the results from distinct parties.

In this section, we briefly discuss the basic component – or motivator – of the path loss models used in this book.

Please refer to appendix A.4.4 of [146] for more details on the path loss model presented in the following, and others.

Outdoor Small Cell Path Loss

The basis of the path loss models used in this book is the multi-slope path loss model for outdoor small cell deployments in urban areas defined by the 3GPP in [153], which was used to analyze "further enhancements to long-term evolution (LTE) time division duplex for downlink-uplink interference management and traffic adaptation" in Release 11.

---

[6] Note that three-dimensional distances will be denoted by variable, d, while two-dimensional ones will be denoted by variable, r.

This multi-slope path loss model presents a generic form, widely used in the literature, consisting of a line-of-sight (LoS) component, a non-line-of-sight (NLoS) component and a function dictating the probability of LoS, and thus governing the use of either the LoS or the NLoS component in a given link.

When using a carrier frequency, $f_c = 2\,\text{GHz}$, the LoS component, $\zeta^{L}(r)$, the NLoS component, $\zeta^{NL}(r)$, and the LoS probability function, $\text{Pr}^{L}(r)$, are characterized as

$$\zeta^{L}(d)[\text{dB}] = -103.8 - 20.9\log_{10}(d), \tag{2.8}$$

$$\zeta^{NL}(d)[\text{dB}] = -145.4 - 37.5\log_{10}(d) \tag{2.9}$$

and

$$\text{Pr}^{L}(r) = 0.5 - \min\left(0.5, 5 \cdot \exp\left(\frac{-0.156}{r}\right)\right) + \min\left(0.5, 5 \cdot \exp\left(\frac{-r}{0.03}\right)\right), \tag{2.10}$$

respectively, where

- ζ^{L} [dB] is the LoS path loss in dB,
- ζ^{NL} [dB] is the NLoS path loss in dB,
- Pr^{L} is the LoS probability in linear units and
- d is the three-dimensional distance between the transmitter and the receiver in kilometres.

To illustrate the behaviour of this outdoor small cell path loss model, Figure 2.3a shows how the LoS component, $\zeta^{L}(r)$, and the NLoS component, $\zeta^{NL}(r)$, vary as a function of the distance, d. Moreover, Figure 2.3b also shows the LoS probability function, $\text{Pr}^{L}(r)$, as a function of the distance, r.

2.1.7 Shadow Fading

The shadow fading models the random fluctuations of the average received signal strength at the receiver, created by the obstruction caused by the objects that a radio signal finds along its propagation path. The number, locations, sizes and dielectric properties of the obstructing objects, as well as those of the reflecting surfaces and scattering obstacles are usually not known, or very hard to predict. Due to such unknown variables, statistical models are generally used to model shadow fading.

A widely used model is the log-normal shadowing model, which has been shown able to model shadow fading with a good accuracy in both outdoor and indoor environments [154]. According to this model, the shadow fading gain, s, which occurs along the path between a transmitter and a receiver, can be a priori modelled using a log-normal random variable,

$$s[\text{dB}] \sim \mathcal{N}(\mu_s, \sigma_s^2), \tag{2.11}$$

where

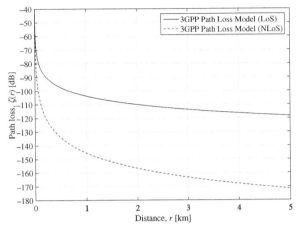

(a) 3GPP path loss model with the LoS and NLoS components.

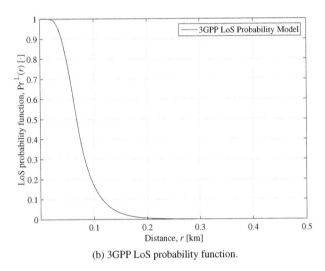

(b) 3GPP LoS probability function.

Figure 2.3 The 3GPP path loss model.

- s [dB] is the shadow fading gain in dB and
- μ_s and σ_s^2 are the mean and the variance of the log-normal random variable in dB, respectively.

When considering multiple transmitters (e.g. small cell BSs) and multiple receivers (e.g. UEs) in the scenario of study, the modelling of the shadow fading gain, s, however, is more intricate than that in equation (2.11). This is due to the spatial autocorrelation and cross-correlation properties of the shadow fading.

In this book, and according to [153], the cross-correlated shadow fading gain, $s_{b,u}$, which occurs along the path between the bth small cell BS and the uth UE, when considered, is modelled as

$$s_{b,u} \text{ [dB]} = \sqrt{\tau} \cdot s_u \text{ [dB]} + \sqrt{1-\tau} \cdot s_b \text{ [dB]}, \qquad (2.12)$$

where

- $s_{b,u}$ [dB] is the cross-correlated shadow fading gain between the bth small cell BS and the uth UE in dB,
- τ is the cross-correlation coefficient of the shadow fading in linear units and
- s_u [dB] $\sim \mathcal{N}(0, \sigma_s^2)$ and s_b [dB] $\sim \mathcal{N}(0, \sigma_s^2)$ are independent and identical distributed (i.i.d.) zero-mean Gaussian random variables in dB, attributable to the environments around the bth small cell BS and the uth UE, respectively, whose variances are the same and equal to σ_s^2.

Please refer to appendix A.4.6 of [146] for more details on this shadow fading model, and others, considering cross- and autocorrelation.

2.1.8 Multi-Path Fast Fading

The obstructing objects in the propagation path from the transmitter to the receiver may also produce reflected, diffracted and scattered copies of the radio signal, resulting in multi-path components (MPCs). The MPCs may arrive at the receiver attenuated in power, delayed in time and shifted in frequency (and/or phase) with respect to the first and strongest MPC, usually the LoS component, when present, thus adding up constructively or destructively. As a consequence, the received signal strength at the receiver may vary significantly over very small distances in the order of the wavelength, or fractions of it [154].

Since different MPCs travel over different paths of different lengths, a single impulse, sent from a transmitter, and suffering from multi-path, will result in multiple copies of such a single impulse being received at the receiver at different times. Thus, the channel impulse response, $z(t)$, of a multi-path fast fading channel can be modelled using a tapped delay line as

$$z(t) = \sum_{i=0}^{v-1} \xi(\tau_i) \cdot \delta(t - \tau_i), \qquad (2.13)$$

where

- v is the number of taps or resolvable MPCs,
- $\xi(\tau_i)$ is the normalized complex channel gain of the ith tap – no path loss or shadow fading is included [155] – and
- τ_i is the delay of the ith tap.

For a time-frequency resource of study,[7] calculating the multi-path fast fading channel gains in the frequency domain, h, using realistic and measured channel impulse

---

[7] Notice that the time-frequency resource of study is assumed to be "flat" in the time and the frequency domains, i.e. the channel gain does not change across such resource. This is the case of the orthogonal frequency division multiplexing (OFDM) narrow-band transmissions in 3GPP LTE and new radio (NR).

responses in the time domain, $z(t)$, requires an involved process, which is usually time consuming. In particular, for a flat time-frequency resource with a centre frequency, f, the multi-path fast fading channel gain in the frequency domain, h, should be calculated using the Fourier transform of the channel impulse response in the time domain, $z(t)$, as follows:

$$h = \left| \int_{-\infty}^{+\infty} z(t) \exp(-j2\pi ft) \, dt \right|^2 . \tag{2.14}$$

In system-level simulations and theoretical performance analysis, multi-path fast fading needs to be calculated for the carrier signal and all the interfering signals in every time-frequency resource for each UE involved in the analysis. For a state of the art computer, it may take hours to generate an adequate number of these multi-path fast fading realizations.

To ease this computation, in this section, we briefly discuss the two lightweight multi-path fast fading models used in this book. For the following two models, we will focus on the distributions of the multi-path fast fading channel gains in the frequency domain, h, as it is the variable that will be later used to analyze the SINR of the UE in a time-frequency resource of study. Having said that, note that these two models are named after the distributions of the multi-path fast fading channel amplitude, $\sqrt{h}$, and not the gain, h. This may lead to confusion to those who are unfamiliar with the skill of channel modelling. We will make this clear, when necessary, during our discussion.

Please refer to appendix A.4.7 of [146] for more details on the two multi-path fast fading models presented in the following, and others, considering measured channel impulse responses.

Rayleigh Multi-Path Fast Fading

The Rayleigh fading assumes that the magnitude or envelope of a signal that passes through the wireless channel will vary randomly, or fade, according to a Rayleigh distribution – the radial component of the sum of two uncorrelated Gaussian random variables. Rayleigh fading is viewed as a reasonable model for signal propagation in the presence of heavily built-up dense urban environments. In other words, Rayleigh fading is most applicable when there is no LoS propagation between the transmitter and the receiver, and the signal is expected to arrive at the receiver from a multitude of directions of space. If there is LoS propagation, dominant or not, Rician – and not Rayleigh – multi-path fast fading may be more applicable.

The multi-path fast fading channel gain in the frequency domain, h, of a random variable whose amplitude follows a Rayleigh distribution can be obtained through an exponential distribution, whose probability density function (PDF), $f(h)$, is given by [156]

$$f(h)[\cdot] = \exp(-h), \tag{2.15}$$

where

- h is the multi-path fast fading channel gain in the frequency domain in linear units and
- $f(h)$ is the PDF of the multi-path fast fading channel gain, h, in linear units.

As described above, remember that the Rayleigh fading distribution is meant for the amplitude, $\sqrt{h}$, and not for the multi-path fast fading channel gain, h, itself, which follows an exponential one.

Rician Multi-Path Fast Fading

The Rician fading occurs when there is a statistically invariant – and usually strong – LoS path. Similar to the case of Rayleigh fading, the in-phase and quadrature-phase component of the received signal are i.i.d. jointly Gaussian random variables. However, with Rician fading, the mean value of (at least) one component is non-zero due to the deterministic strong LoS path.

The multi-path fast fading channel gain in the frequency domain, h, of a random variable whose amplitude follows a Rician distribution can be obtained through a non-central chi-squared distribution, whose PDF, $f(h)$, is given by [156]

$$f(h)[\cdot] = (K + 1) \exp(-K - (K + 1)h)$$
$$\times I_0 \left(2\sqrt{K(K+1)h} \right), \tag{2.16}$$

where

- h is the multi-path fast fading channel gain in the frequency domain in linear units,
- $f(h)$ is the PDF of the multi-path fast fading channel gain, h, in linear units,
- K is the K-factor in linear units, which is defined as the ratio between the power in the direct path and to the total power in the other diffused paths, and modelled in this book as K [dB] $= 13 - (0.03 \cdot 1000 \cdot d)$, where
  - ☐ K [dB] is the K-factor in dB and
  - ☐ d is the three-dimensional distance between the transmitter and the receiver in kilometres [157]

  and
- $I_0(\cdot)$ is the 0th order modified Bessel function of the first kind [156].

As above, remember that the Rician fading distribution is meant for the amplitude, $\sqrt{h}$, and not for the multi-path fast fading channel gain, h, itself, which follows a non-central chi-squared one. It is also worth noting that when the K-factor equals to 0, i.e. $K = 0$, the Rician distribution equals to a Rayleigh one.

To ease the understanding of the reader, Figure 2.4a and b illustrate the PDF, $f(h)$, and the cumulative distribution function (CDF), $F_H(h)$, of the multi-path fast fading channel gain, h, respectively, when considering the Rayleigh multi-path fast fading model as well as the Rician one. This last one with different K-factors corresponding to different link distances.

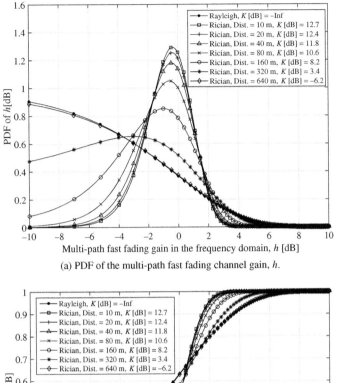

(a) PDF of the multi-path fast fading channel gain, h.

(b) CDF of the multi-path fast fading channel gain, h.

Figure 2.4 Distribution of the multi-path fast fading channel gain in the frequency domain, h.

2.1.9 Received Signal Strength

Having defined in the previous sections,

- the antenna gain, κ,
- the path loss, ζ,
- the shadow fading gain, s and
- the multi-path fast fading gain in the frequency domain, h,

the received signal strength, $P_{t,r}^{\mathrm{RX}}$, of the signal transmitted from the tth transmitter at the rth receiver in the time-frequency resource of study can be calculated as

$$P_{t,r}^{\mathrm{RX}} \text{ [Watts]} = P_t \cdot \kappa_{t,r} \cdot \zeta_{t,r} \cdot s_{t,r} \cdot h_{t,r}, \qquad (2.17)$$

where

- $P_{t,r}^{\mathrm{RX}}$ is in Watts,
- P_t is the transmit power applied by the tth transmitter in the time-frequency resource of study in Watts,
- $\kappa_{t,r}$ is the antenna gain from the tth transmitter to the rth receiver in linear units,
- $\zeta_{t,r}$ is the path loss from the tth transmitter to the rth receiver in linear units,
- $s_{t,r}$ is the shadow fading gain from the tth transmitter to the rth receiver in linear units and
- $h_{t,r}$ is the multi-path fast fading gain from the tth transmitter to the rth receiver in the time-frequency resource of study in linear units.

2.1.10 Noise Power

The noise power models the amount of noise that the specific radio receiver introduces when operating at a given bandwidth. If the radio receiver were perfect, no noise would be added to the signal when it passed through it, and the signal-to-noise ratio would be the same at both the output and the input of the radio receiver.

A widely used noise power model is the following, in which the noise power, P_r^{N}, at the rth receiver due to thermal noise in the time-frequency resource of study is modelled as

$$P_r^{\mathrm{N}} \text{ [Watts]} = 1000 \cdot k^{\mathrm{B}} \cdot T_r \cdot B \cdot \psi, \qquad (2.18)$$

where

- P_r^{N} is in Watts,
- k^{B} is Boltzmann's constant in Joules per Kelvin,
- T_r is the temperature at the rth receiver in Kelvin,
- B is the bandwidth of the time-frequency resource of study in Hertz and
- ψ is the noise figure of the rth receiver in linear units.

Note that $10 \log_{10}(1000\, k^{\mathrm{B}}\, T_r) \approx -174 \frac{\mathrm{dBm}}{\mathrm{Hz}}$ for a temperature, $T_r = 300$ Kelvin.

2.1.11 Signal Quality

Having defined in the previous sections,

- the received signal strength, $P_{t,r}^{\mathrm{RX}}$, of the signal transmitted by the tth transmitter at the rth receiver in the time-frequency resource of study and
- the noise power, P_r^{N}, at such rth receiver in such frequency time-frequency of study,

the SINR, $\gamma_{t,r}$, of the signal transmitted between them – in a single-input single-output mode – can be modelled as

$$\gamma_{t,r}[\cdot] = \frac{P_{t,r}^{RX}}{\sum_{\substack{t=1 \\ t' \neq t}}^{|\mathcal{T}|} P_{t',r}^{RX} + P_r^{N}}, \qquad (2.19)$$

where

- $\gamma_{t,r}$ and $P_{t,r}^{RX}$ is in linear units, and
- $|\mathcal{T}|$ is the cardinality of the set of transmitters, $\mathcal{T}$, where the set of transmitters, $\mathcal{T}$, in the downlink is the set of active small cell BSs, $x \subseteq \Phi$, while in the uplink is the set of active UEs, $y \subseteq \Phi^{UE}$.

The SINR as a Measurement Event, a Random Variable and a Real Number

Having defined the SINR, $\gamma_{t,r}$, of the link between the tth transmitter and the rth receiver at a particular time-frequency resource in equation (2.19), let us go deeper here around some concepts that will be useful in this book.

First of all, let us note that the SINR, $\gamma_{t,r}$, defined in equation (2.19) is a real number – the result of a performance measurement. However, such real number is not enough to meaningfully characterize the performance of a network or even a link, since such SINR, $\gamma_{t,r}$, is a local measure, and can quickly vary in the frequency, the time and/or the spatial domains due to the many reasons already introduced in this chapter. Let us understand this better hereafter.

A network is composed of a plurality of small cell BSs, $x \subseteq \Phi$, and a plurality of UEs, $y \subseteq \Phi^{UE}$, distributed across space, and the quality of their signals changes according to a large number of factors, for example, the small cell BS density, λ, the small cell BS locations, the UE locations, the probabilistic LoS and NLoS transmissions, the antenna gains, the shadow fading and the multi-path fast fading, as described earlier in this chapter. As a result of the randomness introduced by these factors, the SINRs, $\gamma_{t,r}$, of the links between all UEs and their serving small cell BSs greatly vary through the network.

Embracing this variability, let us define in the following a number of concepts that allow one to go beyond the SINR, $\gamma_{t,r}$, as a local measure.

DEFINITION 2.1.1   The set of SINR measurement events, Ω

Let the set of SINR measurement events, Ω, define all possible – or a subset of all possible – SINR measurement events that can take place in a network in the frequency, the time and the spatial domains. Imagine an SINR measurement event as an object, where this object may be characterized by multiple features, e.g. frequency, time, location, quantitative measure, and that every UE in the network periodically takes these SINR measurements over multiple time-frequency resources resulting in such objects.

DEFINITION 2.1.2   The SINR random variable, Γ
Let the SINR random variable, Γ, map the set of SINR measurement events, Ω, to the set of real numbers, $\mathbb{R}$, formally

$$\Gamma : \Omega \to \mathbb{R}, \tag{2.20}$$

where, following our previous example, such random variable maps a particular SINR measurement event, ω, to a particular real number, γ.

DEFINITION 2.1.3   The SINR realization, $\gamma = \Gamma(\omega)$
Let the real number, $\gamma = \Gamma(\omega)$, associated to an SINR measurement event, ω, in the set of SINR measurement events, Ω – referred to as an SINR realization of the SINR random variable, Γ – be the corresponding SINR, γ, defined in equation (2.19). For completeness, let us refer to the set, R_Γ, of all possible realizations, γ, as the support.

Importantly in this book, it should be noted that the set of SINR measurement events, Ω, the SINR random variable, Γ, and the SINR realization, $\gamma = \Gamma(\omega)$, are all function of many parameters as mentioned earlier, among others, the small cell BS density, λ. Such dependency is not expressed in the notation of such concepts but should be understood and remembered.

The Distributions of the SINR Random Variable

Having clarified the difference between these three concepts, let us also define in the following the PDF, the CDF and the complementary cumulative distribution function (CCDF) of the SINR random variable, Γ, which will be of use in this book.

DEFINITION 2.1.4   The PDF of SINR random variable, Γ
Let the distribution, $f_\Gamma(\gamma)$, be the PDF of the SINR random variable, Γ, which provides a relative likelihood of the value returned by such random variable, Γ, being equal to that of an SINR realization, γ.

DEFINITION 2.1.5   The CDF of SINR random variable, $F_\Gamma(\gamma_0)$
Let the distribution, $F_\Gamma(\gamma_0)$, be the CDF of the SINR random variable, Γ, which represents the probability that an SINR realization, γ, takes a value less than or equal to a certain value, γ_0, i.e. $F_\Gamma(\gamma_0) = \Pr[\Gamma \leq \gamma_0]$.

DEFINITION 2.1.6   The CCDF of SINR random variable, $\bar{F}_\Gamma(\gamma_0)$
Let the distribution, $\bar{F}_\Gamma(\gamma_0)$, be the CCDF of the SINR random variable, Γ, which represents the probability that an SINR realization, γ, takes a value greater than a certain value, γ_0, i.e. $\bar{F}_\Gamma(\gamma_0) = \Pr[\Gamma > \gamma_0]$.

Formally, it should be noted that

$$\bar{F}_\Gamma(\gamma_0) = \Pr[\Gamma > \gamma_0] = 1 - F_\Gamma(\gamma_0). \tag{2.21}$$

2.1.12   Key Performance Indicators

As indicated earlier, the Shannon–Hartley theorem, presented in equation (2.1) allows one to compute the maximum capacity, C, of a wireless channel – the link between a UE and its serving small cell BS – with a given bandwidth, B, and a given signal quality, γ, in the presence of Gaussian noise. This theorem can also be extended to interfered systems, under the assumption that the interference can be modelled as AWGN, where γ is thus an SINR realization. Moreover, a penalty of several dBs can be added to the SINR realization, γ, of a UE to account for physical (PHY) layer imperfections, finite block lengths and realistic modulation and coding schemes (MCSs). A cap on such SINR realization, γ, can also be included to consider the maximum efficiency of the highest MCS, and never go beyond it [158].

It is important to note, however, that this type of performance metric – the maximum capacity of a wireless channel – is local and specific to a link. As a result, it is not meaningful enough on its own to characterize the performance of a network, in a wider sense.

Two network-wide key performance indicators often used in network performance analysis are the coverage probability and the area spectral efficiency (ASE). Building on the Shannon–Hartley theorem, these two metrics are used to go beyond the link point of view and assess the performance of large networks in a simple and intuitive manner. Before proceeding with their formal definitions, however, let us introduce the important concept of the *typical UE*.

The Typical UE

According to the properties of an HPPP, i.e. the network layout model presented in Section 2.1.1, and more in detail, according to the Slivnyak's theorem [52, 159], the performance of a wireless HPPP-distributed network can be characterized by the performance of a typical UE located at the origin, o. Due to the stationarity and the independence of all the points in an HPPP, conditioning on a typical UE located at the origin, o, does not change the distribution of the underlying HPPP. The properties seen from a point in the HPPP are the same whether we condition on having a point located at the origin, o, or not. As a result, the SINR of – and rate experienced by – such typical UE have the same characteristics as those of any other UE in an HPPP-distributed network.

Importantly in this book, it follows from the previous formal definition that the set of SINR measurement events, Ω, across the network introduced in Section 2.1.11 is equivalent to the set of the SINRs measurement events of the typical UE in an HPPP-distributed network, as the typical UE can be assumed to be located anywhere on the plane. As a consequence, we can safely reuse some of the defined concepts and notations in the previous section, that is

- the SINR random variable, Γ, can also be used to map the SINR measurement events, Ω, of the typical UE to the set of real numbers, $\mathbb{R}$ and
- the real number, $\gamma = \Gamma(\omega)$, can also be used to represent an SINR realization of the SINR random variable, Γ, of the typical UE.

Section 2.3.1 will provide the formal definition of a point process in general, and an HPPP in particular, and then introduce the key properties of this type of distribution in a structured form.

Coverage Probability

With the previous definitions related to the typical UE in Section 2.1.12, and considering a network with a small cell BS density, λ, we can unequivocally define the coverage probability, $p^{\mathrm{cov}}(\lambda, \gamma_0)$, of such small cell BS network as

- the probability that the SINR, Γ, of the typical UE is above the SINR threshold, γ_0, which is equivalent to
- the value of the CCDF, $\bar{F}_{\Gamma}(\gamma_0)$, of the SINR random variable, Γ, sampled at the SINR threshold, γ_0,

i.e.

$$p^{\mathrm{cov}}(\lambda, \gamma_0) = \Pr[\Gamma > \gamma_0] = \bar{F}_{\Gamma}(\gamma_0), \tag{2.22}$$

where

- γ_0 is the minimum working SINR of the network in linear units, i.e. the minimum necessary SINR to successfully transfer information form the transmitter to the receiver using the minimum MCS defined by the particular technology in use.

Area Spectral Efficiency

The ASE, A^{ASE}, of the above small cell BS network is defined as the number of bits per unit of time, spectrum and area that such a network can successfully transfer, and for the sake of clarity, let us mention that it is usually expressed in bps/Hz/km^2.

According to [52], the ASE, $A^{\mathrm{ASE}}(\lambda, \gamma_0)$, can be formally defined as

$$A^{\mathrm{ASE}}(\lambda, \gamma_0) = \lambda \int_{\gamma_0}^{\infty} \log_2 (1 + \gamma) f_{\Gamma}(\gamma) d\gamma, \tag{2.23}$$

where

- $f_{\Gamma}(\gamma)$ is the PDF of the SINR, Γ, of the typical UE.

Importantly, given that the coverage probability, $p^{\mathrm{cov}}(\lambda, \gamma_0)$, in equation (2.22) can also be defined as the value of the CCDF, $\bar{F}_{\Gamma}(\gamma_0)$, of the SINR random variable, Γ, sampled at the SINR threshold, γ_0, according to equation (2.21), it follows that the PDF, $f_{\Gamma}(\gamma)$, in equation (2.23) can be calculated as

$$f_{\Gamma}(\gamma) = \frac{\partial \left(1 - p^{\mathrm{cov}}(\lambda, \gamma)\right)}{\partial \gamma}. \tag{2.24}$$

Based on the partial integration theorem in [160] and the definition of the coverage probability, $p^{\mathrm{cov}}(\lambda, \gamma_0)$, in equation (2.22), the ASE, $A^{\mathrm{ASE}}(\lambda, \gamma_0)$, in equation (2.23) can be computed by substituting equation (2.24) in equation (2.23), and solving the integral by parts. The resulting ASE, $A^{\mathrm{ASE}}(\lambda, \gamma_0)$, can be obtained as

$$A^{\text{ASE}}(\lambda, \gamma_0) = \frac{\lambda}{\ln 2} \int_{\gamma_0}^{+\infty} \frac{p^{\text{cov}}(\lambda, \gamma)}{1 + \gamma} d\gamma$$

$$+ \lambda \log_2 (1 + \gamma_0) \, p^{\text{cov}}(\lambda, \gamma_0). \tag{2.25}$$

As it is important in this book, let us stress that the small cell BS density, λ, plays a role as

- a direct multiplier, linearly increasing the ASE, $A^{\text{ASE}}(\lambda, \gamma_0)$, but also
- affecting the coverage probability, $p^{\text{cov}}(\lambda, \gamma_0)$, where more concretely, the SINR, Γ, of the typical UE depends on it.

2.2   System-Level Simulation

In this section, to give the reader some general notions on the network simulation field, we introduce and classify different types of network simulation approaches, explaining the differences between a link- and a system-level simulator as well as those between a static and a dynamic system-level simulator. Then, we motivate the use of a static system-level simulator for

- the performance analysis of ultra-dense networks and
- the validation of the theoretical models developed in this book.

Finally, we also draw some suggestions on how the interested reader can implement a static snapshot-based system-level simulation.

2.2.1   Link- and System-Level Simulations

Simulating the transmission of every single bit between all UEs and all small cell BSs in a dense small cell network is prohibitive because of the large computational costs. This is mostly due to the transmission of very large amounts of data, gigabits or even terabits. As a result, and to ease complexity, the simulation of a network is usually divided into two distinct levels, as shown in Figure 2.5:

- *link-level simulations* and
- *system-level simulations.*

At the link-level simulation [161], the behaviour of a radio link between a transmitter and a receiver – or a reduced number of them – is analyzed, considering

- all important characteristics of the radio propagation at small temporal scales [154], e.g. with an OFDM symbol resolution, as well as
- all relevant features of the PHY layer [162, 163].

In the PHY layer, functions such as scrambling, modulation, coding, rate matching, layer mapping, precoding as well as antenna port and resource block mapping should be carefully modelled according to the corresponding standard. Some features of the medium access control (MAC) layer, whose behaviour must be analyzed at the bit

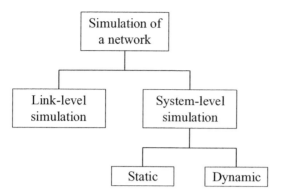

Figure 2.5 Diagram of network simulations.

level, such as hybrid automatic repeat request (HARQ), are also included in some – and studied through – link-level simulations [164]. In OFDM systems, link-level simulations have been widely used to investigate, for example, the performance of adaptive MCS methods and subcarrier mapping schemes [165].

At the system-level simulation [161], on the contrary, the performance of a dense small cell network as a set of UEs and a set of small cell BSs is studied, considering MAC and higher-layer aspects, such as mobility and radio resource management, to cite a few [41, 166–168]. In this case, the target is not to analyze the performance of a specific radio link, bit by bit, over small temporal scales, but to evaluate the performance of an entire network in terms of coverage, capacity, delay, etc. over long periods of time, e.g. minutes, hours and even days, depending on the amount of network detail considered.

Link-level and system-level simulations are usually independently executed due to their different time scales and computational costs. However, to achieve an accurate and a reliable network evaluation, they usually interact through static interfaces, referred to as look up tables (LUTs) [158, 169]. LUTs are generated by link-level simulations, and present link performance results in a simplified manner. For example, the radio link performance in terms of block error rate (BLER) can be mapped through an LUT to a radio link quality indicator, e.g. the SINR of the UE. Then, these LUTs and others can be queried during system-level simulations to assess the radio link performance and encapsulate specific features and details of the PHY layer, which may not be natively captured in the specific system-level simulation.

Due to the focus of this book on the network-wide performance analysis of ultra-dense networks, and not on the specific characteristics of a single link, in the rest of this chapter and the book, we will focus on and use system-level simulations to validate our theoretical performance analyses.

2.2.2   Static and Dynamic System-Level Simulations

Although the general target of system-level simulations is to characterize the overall performance of a network, they are used to address a very wide range of studies. For example, they can be used to analyze the impact of a new self-organization technique on network performance, or to evaluate the benefits and the drawbacks of deploying a large number of small cell BSs over an existing macrocell network, as is the case in this book. As a result, the specific and the right system-level simulation tool should be carefully selected for and tailored to each problem. In this chapter, we classify system-level simulations into

- *static system-level simulations* and
- *dynamic system-level simulations.*

Note that these are general concepts, and combinations between these two different techniques are also possible and commonly used.

Practically, and broadly speaking, any system-level simulation is a computer software, where every entity of the network is represented as a class and realized through an object, which can be seen, for those not familiar with object-oriented programming,[8] as a data structure with a number of methods associated to it. A UE, a small cell BS, an antenna, a path loss model, all the entities involved in the system-level simulation, even the simulation itself, are realized through objects.

Before describing the differences between static and dynamic system-level simulations, let us define the concepts of simulation, simulation campaign and drop. These definitions will be helpful to avoid confusion, and it is important to note that they generally apply to any type of system-level simulation. To illustrate these definitions, let us assume that one wants to analyze the performance of a given key performance indicator, e.g. the coverage probability or the ASE, as a function of a given range of models and/or parameters, e.g. the small cell BS density.

DEFINITION 2.2.1   Simulation
A simulation takes a specific set of models and parameters, for example, some of those presented in Section 2.1 and produces some results after performing a number of operations. These results are thus intrinsically related to the selected models and parameters and should not be extrapolated to others. In our example, a simulation would allow the analysis of the coverage probability or the ASE for one particular small cell BS density.

DEFINITION 2.2.2   Simulation campaign
A simulation campaign is comprised of multiple simulations, each one of them with a specific but different set of models and parameters. The results of the simulation campaign allow one to study given key performance indicators as a function of a given range of models and/or parameters. A simulation campaign thus allows for the study of scaling laws, i.e. how a magnitude changes with respect to another. In our example, a simulation campaign would permit the study of the coverage probability or the ASE with respect to a range of small cell BS densities.

DEFINITION 2.2.3   Drop
As mentioned earlier, to get statistically representative results going over the randomness present in all the models used in a simulation, multiple snapshots have to be run

---

[8] Note that a system-level simulator does not need to be developed through object-oriented programming, and that we use the terminology of "class" and "object" for clarity and convenience.

within a simulation. All the snapshots of a simulation use the same set of models and parameters, but the seeds of the random numbers are i.i.d. in each snapshot. A snapshot could be represented in the previous "programming" nomenclature by a simulation object. Thus, one can deploy and run multiple objects of the simulation class with a specific set of models and parameters to realize a simulation. One should note the possibility of parallelizing a simulation by running different objects of the simulation class in different cores of the computing engine. As for the mentioned randomness, it should be noted that it specifically originates in this particular static system-level simulation model from the random deployment of small cell BSs and UEs, the LoS probability function in the path loss model as well as the shadow fading and multi-path fast fading realizations. The name "drop" was originally given after the idea of dropping BSs and UEs in different locations.

Static Snapshot-Based Approaches

Static snapshot-based system-level simulations are aimed at analysis of the average performance of a network over large areas and for long periods of time. In this case, simulations are typically based on Monte Carlo approaches, where a very large number of independent drops, also referred to as snapshots, are used to average out time and frequency dependent fluctuations and evaluate the network performance in a statistical manner. Monte Carlo simulations have been widely used in network planning and optimization.

Importantly, in each snapshot, the time domain is neglected. There is no correlation between any two snapshots of a static system-level simulation. The random variables that define the network characteristics and its behaviour are i.i.d. in each drop, and they do not evolve through time. The performance of the overall network is then analyzed in terms of mean coverage and capacity, by averaging the results of the different snapshots. The coverage probability, the ASE, as well as the cell and the UE throughputs are widely used key performance indicators.

Since the time domain is neglected, static system-level simulations are easier to implement, and much faster to run than dynamic ones. For example, static system-level simulations do not require to constantly and coherently update multi-path fast fading gains for every link in the network – a process that may take a significant amount of time. Of course, this reduced complexity comes at the expense of reduced accuracy.

When talking about complexity, however, one should not underestimate the running time of a static system-level simulation. To obtain a statistically representative result of the average network performance, a simulation still requires a very large number of snapshots, in the other of 10,000 snapshots or more.[9]

Dynamic Event-Driven Approaches

In a dynamic event-driven system-level simulation, the target of the simulation is to accurately model the functioning of the network with a high degree of detail. In

---

[9] This is the number of drops carried out per simulation in results presented in the rest of this book.

this case, the evolution of the network over time is taken into account, and thus the simulation software must allow the network to live as a function of time or a series of events.

To capture such a high degree of detail, both the end-to-end behaviour of a network together with its dynamic characteristics should be simulated. Among such dynamic features, we should highlight the need for mobility and traffic models, as well as for models capturing the fluctuations of the radio channel over time and frequency, e.g. correlated shadow fading and correlated multi-path fast fading. In this way, the behaviour of different techniques, such as mobility and radio resource management can be analyzed over time in terms of optimality, convergence, delay, etc.

Accordingly, the running time of a dynamic system-level simulation significantly increases with the size of the scenario of study considered, and thus smaller areas and shorter times with respect to static system-level simulations are usually analyzed. In dynamic system-level simulations, network performance is typically assessed by measures such as call block and drop rates, cell and UE throughputs, end-to-end delay and jitter as well as packet losses and retransmission ratios.

It is important to note that since some random variables involved in a dynamic system-level simulation evolve – and are explored – throughout the simulated timeline, a much lower number of drops is needed in this type of simulation to go over their randomness. This is, for example, the case for both the shadow fading and the multi-path fast fading. As a rule of thumb, the number of drops in this case is inversely proportional to the simulated time.

2.2.3   Our Static Snapshot-Based System-Level Simulation

As mentioned earlier, a system-level simulation approach should be chosen according to the objective of the analysis. Since we are interested in the fundamental understanding of ultra-dense networks through theoretical performance analyses in this book, and the validation of theoretical models, a static system-level simulation is selected. Dynamic system-level simulations are disregarded, as we are not planning to study the detailed functioning of any network feature through time, even in the scheduling section. Instead, we concentrate always on the performance of a specific and representative time-frequency resource, and the focus is on the understanding of the network performance and the accuracy of our theoretical performance analyses for such resource, when an increasing number of small cell BSs are added to the network. Of course, the right level of detail and feature modelling should be ensured in this simpler static system-level simulation to guarantee that the true behaviour of a practical and realistic network is captured in general terms. This is the art of modelling! In other words, a useful model should be neither too simple to ignore important factors nor too complex to render simulation computationally impossible.

With this in mind, and with the intention to help new students, researchers and engineers in the art of system-level simulation, in this section, we sketch the main steps towards the implementation of a static system-level simulation, similar to the one

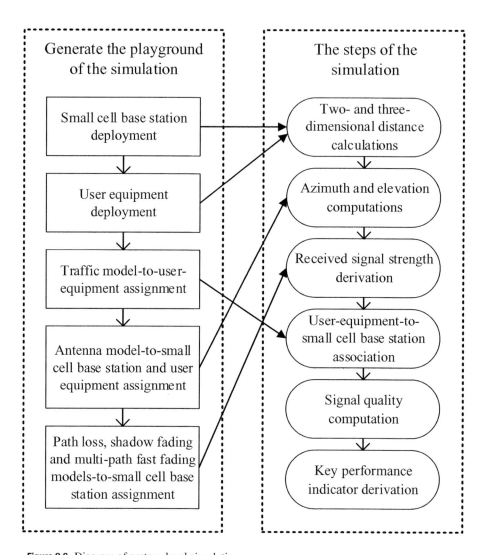

Figure 2.6 Diagram of system-level simulations.

used by the authors of this book to evaluate the accuracy of the theoretical performance analyses, which will be presented in each of the subsequent chapters.

The simulation object – the main, the driver of the code – has multiple tasks as illustrated in Figure 2.6. The first one is to generate the playground of the simulation, in other words, defining the scenario and the laws within the simulation. A number of steps are generally followed in the simulation object to carry out this task, which are mostly designed to generate all the other objects that will play a role in the simulation. Those steps are:

- *Small cell BS deployment:* The small cell BS deployment generates the small cell BS objects. The number and parameters of such small cell BS objects are generated according to one of the network layout models presented in Section 2.1.1. For example, each small cell BS object contains information such as the particular small cell BS location, its height, etc. This information is filled according to the model.

- *UE deployment:* The UE deployment is realized through one of the UE deployment models presented in Section 2.1.2 and generates the UE objects. A UE object may contain information such as the UE location, its height, etc.
- *Traffic model-to-UE assignment:* This step associates a traffic object, generated according to the model presented in Section 2.1.3, with each one of the UEs. A traffic object may contain information such as the link direction, the packet size, the packet arrival rate, etc.
- *Antenna model-to-small cell BS and UE assignment:* This assignment pairs the corresponding antenna object, generated according to one of the models presented in Section 2.1.5, with each one of the small cell BS and UE objects. An antenna object may contain information such as the maximum antenna gain, the antenna patterns, etc.
- *Path loss, shadow fading and multi-path fast fading models-to-small cell BS assignment:* In this step, the corresponding path loss, shadow fading and multi-path fast fading objects, generated according to one of the models presented in Sections 2.1.6, 2.1.7 and 2.1.8, respectively, are paired with each small cell BSs. A path loss object may contain information such as the path loss exponent, probability of LoS function, etc. Similar examples would apply to the shadow fading and the multi-path fading.

Once the playground has been set up, and all the objects of the simulation are in place, the steps that the simulation per se should follow are – broadly speaking – the following:

- *Two- and three-dimensional distance calculations:* Using the small cell BSs and UEs locations, such distances are calculated and stored in, e.g. the small cell BS, the UE objects or the simulation object. This is a design choice, which will not be discussed hereafter.
- *Azimuth and elevation computations:* Using the small cell BSs and UEs locations, such angles can also be computed, and thereafter stored in the selected object(s).
- *Received signal strength derivation:* Once the distances and angles are known, antenna gains, path losses as well as shadow fading and multi-path fading gains between all small cell BSs and UEs can be derived, as well as their received signal strengths, following the formulation in Section 2.1.9. All this information should also be stored in memory, as it will be needed in the following computations of the simulation.
- *UE-to-small cell BS association:* The serving small cell BS for each UE is identified according to a performance metric, for example, the minimum distance or the maximum received signal strength (see Section 2.1.4 for more details).
- *Signal quality computation:* Once the serving small cell BS of each UE is known, the SINR of each UE can be obtained, using the formulation in Section 2.1.11.

Note that the computation of inter-cell interference becomes more involved when a mixture of downlink and uplink transmissions are considered, e.g. in a dynamic time division duplexing (TDD) scenario. The noise power should be priorly computed according to the model in Section 2.1.10.

- *Key performance indicator derivation:* Finally, key performance indicators, such as coverage probability, ASE, as well as cell and UE throughputs, can be calculated taking some assumptions on network operation, e.g. scheduling mechanism and others, which will be discussed throughout this book. Definitions for such key performance indicators are presented in Section 2.1.12 and will also be discussed in the correspondent chapters of this book.

2.3   Theoretical Performance Analysis

Having described the working methodology of a static system-level simulation, it is fair to acknowledge that an increasingly higher computational complexity is required to obtain statistically representative results as the small cell BSs density, λ, and the UE density, grow. The number of links in the network exponentially grows with them, and more computational power is needed accordingly to generate the respective channel gains and compute the resulting SINRs. Therefore, fast static system-level simulations, being able to support thousands of snapshots within a limited period of time, are generally needed to statistically assess the overall network performance. As one can imagine, the feasibility of performing statistically representative simulation campaigns closely relates to the computing capabilities that one may have at hand.

Theoretical performance analysis, although less accurate and reliable compared to these system-level simulations in terms of end results due to the number of assumptions taken for mathematical tractability, is a valid alternative to cope with the complexity of system-level simulations and ease the computational power. As mentioned earlier, theoretical performance analyses are generally simpler to devise, and can handle larger networks.

We should recall, however, that system-level simulations and theoretical performance analysis are not competing tools, and that they can complement each other well. Theoretical performance analysis can provide a better understanding of why things fundamentally occur as well as of the mathematical relationship between the different entities and parameters of the network.

Let us now briefly discuss the different existing approaches to theoretical performance analysis, and then focus on stochastic geometry – the state of the art tool – for the performance analysis of ultra-dense networks.

A very common theoretical performance analysis assumption has been – and still is – to consider a single cell. This helps tractability, accounts for the intra-cell interference, but completely negates the inter-cell interference; one of the most important characteristics of a network. To solve this issue, while keeping complexity low, other widely used assumptions in the literature have been

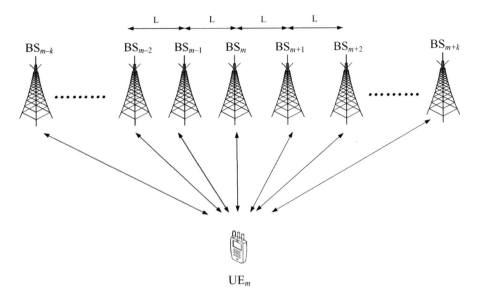

Figure 2.7 Illustration of the Wyner model.

- to consider a single interfering cell [170] or
- to distil the inter-cell interference to a fixed but adjustable value of interfering cells [171, 172],

where the inter-cell interference is modelled as a constant factor of the total interference.

In the two-cell case, at least the inter-cell interference and the SINRs of the UEs vary depending on the positions of the UEs, and possibly fading, but naturally such an approach still neglects most sources of inter-cell interference in the network and is highly idealized. A discussion on such models for inter-cell cooperation to reduce inter-cell interference is given in [173].

Among the models considering a fixed but adjustable number of interfering cells, the most popular is the multi-cellular one originally proposed by Wyner [174]. Compared to a real network, the most popular linear version of the Wyner model makes three major simplifications (see Figure 2.7):

- only the inter-cell interference from two adjacent cells is considered,
- the random locations of the UEs, and thus the corresponding path loss variations, are ignored and
- the inter-cell interference intensity from each neighbouring BS is oversimplified and characterized by a single fixed parameter, which varies from 0 to 1.

Despite its wide adoption, this model, unfortunately, has been shown to be a highly inaccurate model, unless there is a very large amount of inter-cell interference sources averaged over space, such as in the uplink of heavily loaded wireless systems [175]. For cellular systems using orthogonal multiple access techniques, such as the 3GPP

LTE and NR, the Wyner model and the related mean-value approaches are particularly inaccurate, since the SINRs of the UEs over a cell vary dramatically, and the model is not able to capture these fluctuations. Regardless of its shortcomings, this model has been commonly used to evaluate the "capacity" of multi-cell systems under various types of multi-cell cooperation [176–178].

Practicing systems engineers and researchers in need of more realistic models typically model a two-dimensional network of BSs on a regular hexagonal lattice (see Section 2.1.1 and Figure 2.1a for more details), or slightly more simply, on a square lattice. Tractable analysis can sometimes be achieved for a fixed UE with a small number of interfering BSs, for example, by considering the "worst-case" location of the UE – the cell-edge – and finding the SINR of such a UE at such a position [179]. The resulting SINR is still a random variable in the case of shadow fading and/or multi-path fast fading, from which performance metrics like worst-case average rate and outage probability relative to some target rate can be determined. Naturally, such an approach gives very pessimistic results, which do not provide much guidance to the performance of most of the UEs in the system. It is also important to realize that, although widely accepted, these grid-based models are themselves highly idealized and may become increasingly inaccurate for the analysis of large and random networks, where, for example, the cell radii vary considerably due to the non-grid based deployment, and other factors.

It is important to note that all these simplified approaches to inter-cell interference modelling were still considered state of the art in network performance analysis at the time when the small cell BS appeared, back in 2008, and shows the large effort that the research community has made to develop the more realistic and tractable approaches existing today. One such approach – the cornerstone – is stochastic geometry.

2.3.1   Stochastic Geometry and Point Processes

In mathematics, stochastic geometry [180] is a rich branch of applied probability, which allows the analysis of random spatial patterns on the plane or in higher dimensions. In more detail, stochastic geometry studies the relations between geometry and probability theory, and it is intimately related to the theory of the point processes.

Based on [50, 52], in the following, we provide a brief description of the main tools – definitions and properties – within the stochastic geometry framework that will be used in this book. In more detail, we will go over the concepts of point process, Poisson point process (PPP) and HPPP, stating and explaining some of its key properties. Please note that given the richness and the complexity of this mathematical discipline, the descriptions in this chapter are not exhaustive, and they are provided for guidance. For more details on formal definitions and elaborated proofs, the reader is referred to [181, 182] as well as [52] and the references therein.

Point Process

The most basic objects studied in stochastic geometry are the point processes. In a nutshell, a point process is a method for randomly allocating points to intervals of a real line or to rectangles or other higher-dimensional enclosures. Spatial point processes govern, for example, the locations of objects in a d-dimensional space, $S \subseteq \mathbb{R}^d$, where a dimension, $d = 2$, is of interest in many applications. Spatial point processes are used for modelling in a variety of scientific disciplines, including agriculture, bacteriology, climatology, epidemiology, geography and seismology, among others. They are also used in wireless communications to characterize, e.g. the locations of small cell BSs, as introduced in Section 2.1.1.

With this in mind, and for the sake of completeness and understanding, in the following, we will provide a formal – but by no means exhaustive – definition of point process, concentrating on the point process, Φ, whose realizations, $x \subseteq \Phi$, are locally finite subsets of the space, S.[10]

Let the operator, $n(x)$, denote the cardinality of the subset, $x \subseteq S$. Moreover, let the subset, $x_B \subseteq S$, be the intersection of the subset, $x \subseteq S$, and the subset, $B \subseteq S$, where the subset, $B \subseteq S$, is a bounded subset of the space, S, i.e. $x_B = x \cap B$.[11] Then, the subset, $x \subseteq S$, is said to be a locally finite subset, if the cardinality of the subset, $x_B \subseteq S$ is finite, i.e. $n(x_B) < \infty$. Otherwise, it is not finite.

Because we are concentrating on the point process, Φ, whose realizations, $x \subseteq \Phi$, are locally finite subsets of the space, S, we can then formally define the space, N_{lf}, from which such realizations, $x \subseteq \Phi$, are taken, as follows

$$N_{lf} = \{x \subseteq S : n(x_B) < \infty, \forall B \subseteq S\}, \tag{2.26}$$

where

- B is a bounded subset, as indicated earlier.

Note that a realization, $x \subseteq \Phi$, on the space, N_{lf}, is also referred to as a locally finite point configuration of the point process, Φ.

To be formally correct in our subsequent definitions, let us also assume that the space, S, is a complete, separable metric space with metric, d, and that both the space, S, and the space, N_{lf}, are equipped with the Borel sigma algebra, $\mathcal{B}$, where $\mathcal{B}_0$ is the class of bounded Borel sets,[12] and $\mathcal{N}_{lf}$ is the sigma algebra on the space, N_{lf}, i.e.

$$\mathcal{N}_{lf} = \sigma \left(\{x \in N_{lf} : n(x_B) = m\} : B \in \mathcal{B}_0, m \in \mathbb{N}_0 \right), \tag{2.27}$$

where

- $\mathbb{N}_0$ is the set on natural numbers, including 0, i.e. $\mathbb{N}_0 = \mathbb{N} \cup \{0\}$.

---

[10] Remember that the process, Φ, was used to denote the process in charge of generating HPPP-based small cell BS layouts in Section 2.1.1, and for consistency, we use the same notation in this section.

[11] Let $M = (A, d)$ be a metric space. Let $M' = (B, d_B)$ be a subset of the metric space, M. The subset, M', is bounded if and only if there exits at least an element of the set, A, within a finite distance of all elements of the subset, B. Otherwise, the subset, M', is said to be unbounded [183].

[12] In a nutshell, a Borel set is any set in a topological space that can be formed from open sets – or equivalently, from closed sets – through the operations of countable union, countable intersection and relative complement. The collection of all Borel sets on a topological space forms a sigma algebra. The Borel algebra on such topological space is the smallest sigma algebra, containing all open sets – or equivalently, all closed sets [184].

With these assumptions, definitions and notation, a point process, Φ, which can be depicted as a random collection of points in the space, S, can be formally defined as:

DEFINITION 2.3.1   Point process

A point process, Φ, defined on a space, $S \subseteq \mathbb{R}^d$, is a measurable mapping between a probability space,[13] $(\Omega, \mathcal{F}, \mathcal{P})$, and a measurable space,[14] $(N_{lf}, \mathcal{N}_{lf})$, i.e.

$$\Phi : (\Omega, \mathcal{F}, \mathcal{P}) \rightarrow (N_{lf}, \mathcal{N}_{lf}), \tag{2.28}$$

where this mapping induces a distribution, P_Φ, in the points of the point process, Φ, given by

$$P_\Phi(F) = P(\{\omega \in \Omega : \Phi(\omega) \in F\}), \tag{2.29}$$

which is defined by the function, $F \in \mathcal{N}_{lf}$.

It is a mathematically fruitful idea to define point processes as random counting measures. That is, a point process, Φ, can be defined as a random distribution of points in the space, S, with a given distribution, P_Φ, where $N(B)$ is the number of points in a bounded subset, $B \in \mathcal{B}_0$ – a random variable.

For completeness, let us mention that a point process, Φ, is completely character-ized by its finite-dimensional distributions, its void probabilities and its generating functional, whose details are omitted in this book, unless otherwise stated. The inter-ested reader in such characteristics is referred to [181, 182].

Importantly, there are different types of point processes, all with different proper-ties. A point process can be simple, Poisson, stationary, isotropic, and/or marked. To ease the discussion, in the following, we will focus on the point process of interest in this book, the PPP, and its properties.

Poisson Point Process

When a point process is a PPP, it possesses the property of no interaction between points or complete spatial randomness. In other words, there are no correlations between the locations of the points in a realization or between realizations themselves. As a result, there is a handy computational framework for different network quantities of interest.

Before proceeding with the formal definition of PPP, however, let us explain a few concepts, including that of the intensity function and the intensity measure.

Let us consider a point process, Φ, defined on a space, $S \subseteq \mathbb{R}^d$, and a realization, $x \subseteq S$. With this, let us define

---

[13] A probability space, $(\Omega, \mathcal{F}, \mathcal{P})$, is a mathematical construct, which provides a formal definition of a random process or experiment [185], where
  - the sample space, Ω, is a non-empty set, representing all possible outcomes of the experiment.
  - The event space, $\mathcal{F}$, is a collection of subsets of the sample space Ω. If the sample space, Ω, is discrete, the event space, $\mathcal{F}$, is usually given by the power set – the set of all subsets – of the sample space, Ω, i.e. $\mathcal{F} = \text{pow}(\Omega)$.
  - The probability function, $\mathcal{P}$, is a function, $\mathcal{P} : \mathcal{F} \rightarrow \mathbb{R}$, assigning probabilities to the events in the event space, $\mathcal{F}$. This is sometimes referred to as the probability distribution over the sample space, Ω.
[14] A measurable space is a set with a distinguished sigma algebra for all its subsets [185].

- an intensity function, $\Lambda(x)$, mapping the space, S, into a real number, $[0, \infty)$, which is locally integrable, i.e.

$$\Lambda : S \to [0, \infty), \qquad (2.30)$$

where

□ $\int_B \Lambda(x)\, dx < \infty, \ \forall B \subseteq S$

and

- an intensity measure, $\lambda(B) = \int_B \Lambda(x)\, dx$, where this measure is locally finite, $\lambda(B) < \infty \ \forall B \subseteq S$, and diffuse,[15] $\lambda(\{x\}) = 0 \ \forall x \in S$ [187]. Note that the term, $\{x\}$, refers to a singleton, i.e. a set with exactly one element.

With these prior notions, formally, a PPP can be defined as:

DEFINITION 2.3.2   PPP

A point process, Φ, defined on a space, $S \subseteq \mathbb{R}^d$, is a PPP with an intensity function, Λ, and a resulting intensity measure, λ, if the following two properties hold:

- For all subsets, $B \subseteq S$, the random variable, $N(B)$, measuring the number of points in a given subset, B, follows a Poisson distribution with a finite intensity measure, $\lambda(B)$, i.e. $N(B) \sim \text{Pois}\,(\lambda(B))$.
- For all subsets, $B \subseteq S$, the random variable, $N(B)$, measuring the number of points in a subset, B, is i.i.d.

For completeness, let us represent this PPP, Φ, as

$$\Phi \sim \text{Pois}\,(S, \Lambda)\,. \qquad (2.31)$$

Importantly, the intensity measure, $\lambda(B)$, determines the expected number of points for any subset, $B \subseteq S$, i.e.

$$\mathbb{E}\,\{N(B)\} = \lambda(B). \qquad (2.32)$$

As a result, if the space, S, is also bounded, this gives one a simple way to simulate a PPP, Φ, on such space, S.

- First draw the random variable,

$$N(B) \sim \text{Pois}\,(\lambda(B))\,, \qquad (2.33)$$

and then

- deploy $N(B)$ independent points uniformly on the space, S.

Using this previous nomenclature, let us now define an important subclass of PPPs – the HPPP, which will be the focus of this book.

---

[15] An atom is a measurable set, which has a positive intensity measure, and contains no other subset of smaller positive intensity measure. A measure, which has no atoms is called non-atomic or diffuse [186].

DEFINITION 2.3.3   Homogeneous Poisson point process
A PPP, $\Phi \sim \text{Pois}(S, \Lambda)$, is an HPPP, if the intensity function, Λ, is constant. Otherwise, it is said to be inhomogeneous.

Importantly, note that an HPPP is stationary and simple,[16] where formally, a stationary point process can be defined as:

DEFINITION 2.3.4   Stationary point process
A point process, Φ, defined on a space, $S \subseteq \mathbb{R}^d$, is stationary, if its distribution is invariant under translations.[17]

Campbell's Theorem
One important property of a stationary point processes, and thus of an HPPP, is the tractable expression of Campbell's theorem [188], which in general can be defined either as a particular equation – or a set of results – relating the expectation of a function summed over a point process to an integral involving the intensity measure of such a point process. This allows for the calculation of the mean and the variance of a random sum on the points of a process.

Let us elaborate on Campbell's theorem in the following. To this end, let us define the local configuration of neighbouring points of point, x_i, in a realization, $x \subseteq S$, generated by the stationary point process, Φ, as a mark of this point, where such a mark is defined by the collection of points in a ball of radius, R, centred at the point, x_i. Note that if the radius, R, is infinity, this mark is referred to as the universal mark of point, x_i, i.e. the point process seen from the point, x_i. Then, the Palm probability,[18] P^o, of this stationary point process, Φ, is the law of this universal mark, and importantly, it can be shown to be the same for all points, x_i, in a realization, $x \subseteq S$, generated by such a stationary point process, Φ. In other words, the distribution seen by all points is the same. This formalizes the notion of a "typical" point of the process, presented in Section 2.1.12, and suggests, as proved by the Slivnyak's theorem [159], that the Palm distribution of a stationary point process is identical to the distribution of the original point process with a point added at the origin, o, and as a result, conditioning on the origin, $o \in \Phi$, is the same as adding a point at the origin, o, to the stationary point process, Φ, as explained in [52].

A direct consequence of this definition is that Campbell's theorem for a stationary point process, Φ, with constant intensity measure, $\lambda > 0$, on a space, $S \subseteq \mathbb{R}^d$, reduces to a volume integral. In more detail, for a bounded non-negative function, $f(x)$, the mean and the variance values of the sums on the points of such stationary point process, Φ, can "simply" be computed as

---

[16] A point process is a simple point process when the probability of any two points coinciding in the same position, on the underlying space, is 0 [187].

[17] Translation is a point process operation, in which points of a point process are randomly moved from some locations to other locations on the underlying space [187].

[18] The Palm probability, often denoted $P^o(\cdot)$, is the probability or expectation conditioned on a specified event, for example, occurring at position, o, or time, 0. In this case, it is the probability of an event given that the point process contains a point at some position [52].

$$\mathbb{E}_{[\Phi]}\left\{\sum_{x\in\Phi}f(x)\right\} = \lambda\int_S f(x)\,dx \qquad (2.34)$$

and

$$\text{var}_{[\Phi]}\left\{\sum_{x\in\Phi}f(x)\right\} = \lambda\int_S f^2(x)\,dx, \qquad (2.35)$$

respectively, where

- $f(x)$ is formally a measurable function, mapping the space, S, into a bounded non-negative value in the range $[0,1]$, i.e. $f(x)\colon S\to[0,1]$ with $\{x\subseteq S\colon 0\le f(x)\le 1\}$,

Let us emphasize again that these equations naturally hold for the point process of interest in this book – the HPPP – as it is a stationary point process of the mentioned characteristics, i.e. with a constant intensity measure, $\lambda > 0$. Moreover, let us also note that more complex expressions than those presented above to deal with more general point processes exist.

The Probability Generating Functional

The probability generating functional (PGFL) of a discrete random variable is a power series representation of the probability mass function of such a random variable. Probability generating functions are often used for their succinct description of the sequence of probabilities in the probability mass function for a random variable, and to make available the well-developed theory of power series with non-negative coefficients.

For a PPP, Φ, with constant intensity measure, $\lambda > 0$, on a space, $S\subseteq\mathbb{R}^d$, the expression of its PGFL is also a direct consequence of the above presented Campbell's theorem.

Following from the previous definitions and explanations, a key property of a PPP – and thus of an HPPP – is that conditioning on the exact number, m, of points in a given subset, B, i.e. $N(B) = m$, such that m points are all independently, and in the case of an HPPP, homogeneously distributed in the subset, $B\subseteq S$.

This property leads to an explicit representation of the PGFL of the PPP, which takes the form

$$\mathbb{E}_{[\Phi]}\left\{\prod_{x\in\Phi}f(x)\right\} = \exp\left(-\int_S (1-f(x))\,\Lambda(x)\,dx\right), \qquad (2.36)$$

where

- $f(x)$ is the previously presented measurable mapping function in equations (2.34) and (2.35), and
- $\Lambda(x)$ is the intensity function at x.

If the PPP is an HPPP, the PGFL in equation (2.36) simplifies to a more tractable form, as the intensity function, Λ, is constant, which simplifies the integration.

Importantly, the PGFL formulation in equation (2.36) will be key in this book for the computation of the aggregated inter-cell interference in an HPPP-based ultra-dense small cell network. This will be further elaborated in Chapter 3.

Before concluding this section on point processes, let us indicate that another appealing feature of the PPP is its invariance to a large number of key operations. For example, the independent thinning of a PPP is again a PPP. This property has been used in network performance analysis to study the impact of inter-cell interference coordination techniques, where thinning is used, e.g. for "muting" transmitters and thus controlling the inter-cell interference [38]. The superposition of two or more independent PPPs is also a PPP. This property has been useful to study the performance of a heterogeneous network with multiple network tiers, e.g. a macrocell network with small cells overlaid on it [61].

2.3.2   Our Stochastic Geometry-Based Theoretical Performance Analysis

With the intention to help new students, researchers and engineers in the art of theoretical performance analysis, in this section, we sketch the main steps towards the derivation of the two network-wide key performance indicators presented in Section 2.1.12, i.e. the coverage probability and the ASE. Note that the method presented hereafter will be used in each of the subsequent chapters, with each of them

- embracing different network features and thus
- requiring different mathematical expressions and derivations to obtain our two key performance indicators of interest.

As mentioned earlier, we will assume in this book that the set, $x \subseteq \Phi$, generated by the process, Φ, operating in the d-dimensional space, $S \subseteq \mathbb{R}^d$, denotes the set of small cell BS locations of study, and that the process, Φ, is an HPPP. As a result, there is a convenient number of properties and tools, which can be used to derive magnitudes of interest related to the process, Φ, which in turn, will help us to derive the coverage probability and the ASE.

In a nutshell, our final objective is to calculate the ASE, $A^{\mathrm{ASE}}(\lambda, \gamma_0)$, of an ultra-dense network, defined in equation (2.25), which is a function of the coverage probability, $p^{\mathrm{cov}}(\lambda, \gamma_0)$.

As a result, our main problem is to compute the coverage probability, $p^{\mathrm{cov}}(\lambda, \gamma_0)$, formulated in equation (2.22), which in turn, is a function of the SINR, Γ, of the typical UE and the SINR threshold, γ_0. Remember that due to Slivnyak's theorem, sketched earlier, the coverage probability, $p^{\mathrm{cov}}(\lambda, \gamma_0)$, does not depend on the given location where it is measured. In more detail, it does not matter whether the measurement location is part of the underlying HPPP, Φ, or not, as long as its contribution is not considered. This allows one to condition – for convenience – on a typical UE, located at the origin, o.

With this in mind, the coverage probability, $p^{\mathrm{cov}}(\lambda, \gamma_0)$, presented in equation (2.22), can be further developed as

$$p^{\text{cov}}(\lambda, \gamma_0) = \Pr[\Gamma > \gamma_0] = \int\limits_{0}^{+\infty} \Pr\left[\Gamma > \gamma_0 \middle| r\right] f_R(r) dr, \qquad (2.37)$$

where

- $f_R(r)$ is the PDF of the random variable, R, characterizing the distance between the typical UE and its serving small cell BS (assuming a given type of UE-to-small cell BS association criteria, for example, that the typical UE is associated to the nearest small cell BS) and
- $\Pr[\Gamma > \gamma_0 | r]$ is the conditional probability of the SINR, Γ, of the typical UE being larger than the SINR threshold, γ_0, for a particular distance realization, $r = R(\omega)$.[19]

The question now is how to solve equation (2.37). Intuitively, we basically need proper expressions for both the PDF, $f_R(r)$, and the conditional probability, $\Pr[\Gamma > \gamma_0 | r]$, which are functions of the distance, r. Once we have them, we can perform the corresponding integrations, and solve the puzzle.

In the following, let us first elaborate in more detail in the derivation of the PDF, $f_R(r)$, and then provide some pointers around the calculation of the conditional probability, $\Pr[\Gamma > \gamma_0 | r]$.

The Probability Density Function, $f_R(r)$

Let us denote by

- Φ an HPPP with intensity measure, λ, and by
- $r_i = R(\omega)$, $\forall i = 1, 2, \dots, n$ the set of distances from a randomly chosen location up to the nth point in the HPPP, Φ.

In this subsection, we are interested in the PDF, $f_R(r)$, i.e. the PDF of the distance, r_i, $\forall i = 1, 2, \dots, n$.

The literature elaborating on this derivation is rich. According to [189], it was Hertz who first solved this problem in 1909 for the nearest neighbour problem, $n = 1$, and for an arbitrary dimension [190]. Extensions for more neighbours, $n > 1$, have followed [191–193]. In [194], Thompson further extended all these previous results, obtaining the joint distribution of distances up to the nth neighbour. In the following, we provide an explanation on the intuition behind the derivations, using the discussion in [195].

Remember that the probability of having n points in a region of the assumed HPPP, Φ, follows a Poisson distribution, and let this region be a circle of area, πr^2. Assume also that the centre of the circle is located at a random location, which does not necessarily coincide with a point of the HPPP, Φ. For convenience, let that location be that of the typical user, i.e. the origin, o. Importantly, it should be noted that

- if a circle of a radius, r, contains exactly $(n - 1)$ points and that
- if the nth point is located on the perimeter of such a circle,

---

[19] Note that the distance, $r = R(\omega)$, is a random variable of a two-dimensional distance.

the radius, r, is – as a result – the distance from the origin, o, to such an nth neighbour.

With this in mind, let us note that the probability of a disk of a radius, r, containing exactly n points is

$$\Pr(n, r) = \frac{(\lambda \pi r^2)^n}{n!} \exp(-\lambda \pi r^2), \tag{2.38}$$

and that the probability of finding at least n points within such circle of a radius, r, is given by

$$\Pr(\geq n, r) = 1 - \sum_{i=0}^{n-1} \Pr(n, r) = 1 - \sum_{i=0}^{n-1} \frac{(\lambda \pi r^2)^i}{i!} \exp(-\lambda \pi r^2). \tag{2.39}$$

From these equations, it follows that the probability of the nth nearest point being found in the interval, $[r, r + \Delta r]$, is equal to the probability of this point being located in an annulus with inner radius, r, and outer radius, $r + \Delta r$, i.e.

$$\Pr(n, [r, r + \Delta r]) = \Pr(\geq n, r + \Delta r) - \Pr(\geq n, r), \tag{2.40}$$

where

- Δr is an incremental distance.

Using this mathematical construct, letting the incremental distance, Δr, tend to 0, i.e. $\Delta r \to 0$, and differentiating with respect to the radius, r, we can derive the PDF, $f_R(r)$, of such radius, r. Remember again that due to Slivnyak's theorem, the radius, r, is equivalent to the distance from an arbitrarily chosen point, which may or may not coincide with a point of the HPPP, Φ, to its nth nearest point.

After a number of operations omitted here for brevity, we can find that the PDF, $f_R(r)$, has the following general expression:

$$f_R(r) = \frac{2(\pi \lambda)^n}{(n-1)!} r^{2n-1} \exp(-\pi \lambda r^2), \forall r > 0, n \in N. \tag{2.41}$$

Finally, let us note that when looking for the nearest neighbour, $n = 1$, the PDF, $f_R(r)$, simplifies to

$$f_R(r) = 2\pi \lambda r \exp(-\pi \lambda r^2). \tag{2.42}$$

This expression will be of importance when considering a UE to small cell BS association paradigm based on the shortest distance.

The Conditional Probability, $\Pr[\Gamma > \gamma_0 \mid r]$

First of all, it should be noted that the SINR, Γ, of the typical UE – and in turn, the conditional probability, $\Pr[\Gamma > \gamma_0 \mid r]$ – strongly depend on the network features considered. As a result, this section will only give pointers – and each of the following chapters will provide a more detailed discussion – about the calculation of the conditional probability, $\Pr[\Gamma > \gamma_0 \mid r]$, in accordance to the specific network features considered in each one.

Having said that, the characterization of the aggregate inter-cell interference, I_{agg}, at the typical UE is one of the cornerstones for the derivation of the conditional probability, $\Pr[\Gamma > \gamma_0 \,|\, r]$. Such aggregate inter-cell interference, I_{agg}, is a function of a number of variables, in more detail,

- the transmit power, P,
- the antenna gain, κ,
- the path loss, ζ,
- the shadow fading gain, s and
- the multi-path fast fading gain, h,

as can be derived from Section 2.1.

Assuming that the function, $f(x)$, calculates the inter-cell interference from a single link to the typical UE – see Section 2.1 for such inter-cell interference strength calculation – stochastic geometry provides an elegant set of tools to characterize the aggregate inter-cell interference, I_{agg}, resulting from the sum of a large number of inter-cell interfering links in an HPPP-based ultra-dense network.

In more detail, making some specific – but sensible – assumptions in our theoretical performance analysis on all the above-mentioned variables, which will be specified in each chapter of this book, we can find that such aggregate inter-cell interference, $I_{agg} = \left[\sum_{X \in \Phi} f(x)\right]$, at the typical UE can be characterized in a tractable form on an ultra-dense network generated from the HPPP, Φ.

Particularly convenient is the case where shadow fading is neglected, and both omnidirectional antennas and Rayleigh i.i.d. multi-path fast fading gains are considered – the case in Chapter 3. In this case, due to the exponential distribution of the CCDF of the Rayleigh i.i.d. multi-path fast fading gains, i.e. $\bar{F}_H(h) = \exp(-h)$, the aggregate inter-cell interference, $I_{agg} = \left[\sum_{X \in \Phi} f(x)\right]$, can be characterized by its Laplace transform,

$$\mathcal{L}_{[I_{agg}]}(s) = \mathcal{L}_{[\sum_{X \in \Phi} f(x)]}(s) = \mathbb{E}_{[\Phi]} \left\{ \exp\left(-s \sum_{X \in \Phi} f(x)\right) \right\}, \qquad (2.43)$$

evaluated at the variable value, s, which could represent any variable that is independent of the random variable, X.

To clarify the precedence of the above Laplace transform, $\mathcal{L}_{[I_{agg}]}(s)$, let us make the following annotations on a general case:

- By definition, for any random variable, X, the expectation, $\mathbb{E}_{[X]}\{\exp(-sX)\}$, can be calculated as

$$\mathbb{E}_{[X]}\{\exp(-sX)\} = \int\limits_{-\infty}^{+\infty} \exp(-sX) f_X(x) dx, \qquad (2.44)$$

where
  □ $f_X(x)$ is the PDF of such random variable, X.

- On the other hand, for any PDF, $f_X(x)$, its Laplace transform is defined as [156]

$$\mathcal{L}_X(s) = \int_{-\infty}^{+\infty} \exp\left(-sX\right) f_X(x)\, dx. \tag{2.45}$$

- As a consequence, we can say that the following equality is true,

$$\mathcal{L}_X(s) = \mathbb{E}_{[X]}\left\{\exp\left(-sX\right)\right\}. \tag{2.46}$$

Substituting the random variable, X, by our random variable of study, $\sum_{X \in \Phi} f(x)$, in equation (2.46) we can obtain equation (2.43).

Importantly, for this particular case, the Laplace transform, $\mathcal{L}_{I_{agg}}(s)$, can be solved using the PGFL of the HPPP, Φ, presented in equation (2.36) as follows:

$$\mathcal{L}_{\left[\sum_{X \in \Phi} f(x)\right]}(s) = \mathbb{E}_{[\Phi]}\left\{\exp\left(-s\sum_{X \in \Phi} f(x)\right)\right\}$$

$$= \mathbb{E}_{[\Phi]}\left\{\prod_{x \in \Phi} \exp\left(-sf(x)\right)\right\}$$

$$\overset{(a)}{=} \exp\left(-\int_{\mathbb{R}^d} \left(1 - \exp\left(-sf(x)\right)\right) \Lambda(x)\, dx\right), \tag{2.47}$$

where

- step (a) follows from the PGFL presented in equation (2.36) with the change, $f(x) = \exp\left(-sf(x)\right)$.

This expression will be key in all chapters to derive the conditional probability, $\Pr[\Gamma > \gamma_0 \,|\, r]$.

2.4   NetVisual

In this book, in addition to the presented static snapshot-based system-level simulation and theoretical performance analysis, we also introduce an intuitive network performance visualization technique to illustrate the performance of an ultra-dense network. This tool allows one to produce striking two-dimensional maps to represent the coverage probability and open the understanding of the results to a wider audience. This approach is referred to as *NetVisual*, and its methodology – based on a particular static snapshot-based system-level simulation with specific UE locations – is explained in the following.

First, we take the hexagonal layout presented in Section 2.1.1, and generate several random small cell BS deployments over it. In this book, and when using NetVisual, we use three different small cell BS densities, λ, as a reference, $\lambda = \{50, 250, 2500\}$ BSs/km^2, which represent a sparse, a dense and an ultra-dense small

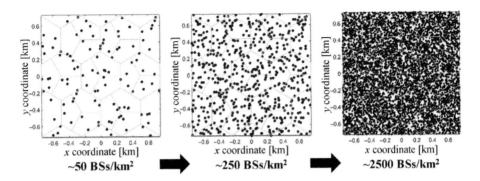

Figure 2.8 Random small cell BS deployments, where small cell BSs are shown as dots.

cell BS deployment, respectively. Figure 2.8 illustrates an example of such three small cell BS deployments, and shows the basic canvas used in NetVisual.[20] Note that small cell BSs are represented by dots in this figure, and are uniformly distributed over an area of 1.5×1.5 km. This figure also shows how the hexagonal layout can guide the random deployment of small cell BSs in the network, allowing one to roll out an exact number of small cell BSs per sector, if needed.

Second, and in contrast to the static snapshot-based system-level simulation presented in Section 2.2.3, the network scenario is divided in a square grid fashion now, and we place a probe UE at a given position, x, within such a grid, referred to as position zero. Based on the explicit modelling of the signal and the inter-cell interference power levels as well as that of the noise power presented in Sections 2.1.9 and 2.1.10, we then calculate the SINR, γ_x, of this probe UE at such a location, x, by using the formulation in Section 2.1.11. Thereafter, we derive its coverage probability, p_x^{cov}, by using equation (2.22). This process is repeated until all points of the grid are visited. Note that *10,000* static system-level simulation experiments are performed per location to go over the randomness of the random variables that are involved in the calculation of the SINR, e.g. shadow fading and multi-path fast fading.

Finally, since the coverage probability, p_x^{cov}, at location, x, ranges between 0 and 1, we show the network performance using per-location coverage probability heat maps, with bright and dark areas indicating a high or a low coverage probability, p_x^{cov}, respectively. With such per-location coverage probability heat maps, the significant performance impacts of the fundamental characteristics of an ultra-dense network can be translated into striking visual plots, intuitively showing the changes in performance as the network densifies.

Clearly, NetVisual is a static snapshot-based system-level simulation. However, let us comment in the following on its direct relationship with the theoretical performance analysis presented in Section 2.3.2. Most of the theoretical work based on stochastic geometry study the network performance in terms of the "average" coverage probability, p^{cov}, and it is important to stress that this is the average performance, that

---

[20] The specific parameters, under which the graphs of this example have been generated, will be introduced later in Chapter 3.

of the typical UE. In a nutshell, such performance metric averages the per-location coverage probability, p_x^{cov} $\forall x$, in the spatial domain of NetVisual, and gives a single performance measure at a *macroscopic* level, i.e. the coverage probability, p^{cov}. In other words, the average colour of a NetVisual heat map in a given scenario is equal to the average coverage probability, p^{cov}, derived through our stochastic geometry-based theoretical performance analysis. Importantly, NetVisual – in addition – provides *microscopic* information on the network performance, e.g. the variance of the coverage probability and the locations of the coverage holes, which may be useful in academic studies and for business proposals with clients. For example, the proposed technique can provide visual answers to the staff of network operators and service providers, who has a less technical background, but are still willing to understand the performance impact of, for example, incrementing their small cell BS density, λ, by a given factor.

2.5   Conclusions

In this chapter, we have presented the main building blocks needed in any performance analysis tool, either system-level simulations or theoretical performance analysis. These blocks are the small cell BS layout and UE deployment models, the UE-to-small cell BS association model, the traffic model, the antenna gain model, the path loss model, the shadow fading model, the multi-path fast fading model, the received signal strength model, the SINR model as well as the capacity one.

Moreover, we have introduced the different types of network simulation approaches, for completeness, explaining the differences between a link- and a system-level simulator as well as between a static and a dynamic system-level simulator, and thereafter, have motivated the choice of static system-level simulations in this book. When it comes to the network-wide performance analysis of ultra-dense networks, with tens – and even hundreds – of thousands of small cell BSs per square kilometre, and the validation of theoretical models, static snapshot-based system-level simulations are the best – and perhaps the only possible – choice with the most appealing complexity-insight trade-off. We have also drawn some suggestions on how the interested reader can implement a static snapshot-based system-level simulation.

Finally, we have discussed the basics of our stochastic geometry-based theoretical performance analysis, which mainly involves the definition of the PGFL of a PPP, and the calculation of the Laplace transform of the aggregate inter-cell interference.

3 Performance Analysis of Traditional Sparse and Dense Wireless Networks

3.1 Introduction

Back in 2008, when the first commercial deployments of residential small cells started – with Sprint launching a nationwide service in the United States, followed in 2009 by Vodafone in Europe and Softbank in Japan – the small cell technology presented a significant challenge to network deployment, management and optimization. This was a new technology, which although known and understood through some existing – but still limited – performance analyzes, was never deployed in larger numbers before [33]. There were many open questions about its deployment, functioning, performance, etc. Cost, site acquisition, backhaul and inter-cell interference were identified as some of the most important challenges that small cells would need to overcome to become a successful business model, and time has shown indeed that those were – and still are – the critical aspects of this technology [40].

As one can imagine, the cost to deploy a small cell was – and still is – one of the most pressing challenges, particularly outdoors. For every small cell base station (BS), an operator may need to

- gain site and equipment approvals,
- negotiate fees with the city or the landlord,
- deploy, provision and maintain the small cell BS,
- ensure it has appropriate backhaul and power and
- conform to the city's aesthetic and environmental regulations [196].

Making sure that all these aspects are addressed for each small cell BS deployed on the field can take up quite some time, and the more the small cell BSs that needed to be deployed, the more it is not economically viable to follow the traditional macrocell deployment procedure for small cells, e.g. negotiating a different set of approvals, fees and processes for every site.

To solve – or at least mitigate – these issues, operators drove significant efforts to negotiate with governments standardized rules and fees that could be applied across a whole country or region. Moreover, the mobile industry developed technology to make it easier to allay the concerns of cities in areas like aesthetics and environment, e.g. designing a wide range of new form factor small cell BSs, which could be easily hidden in existing street furniture or even pavements or trees. The mobile industry also performed a lot of enhancements on self-organizing techniques, to ease

Figure 3.1 Coverage map of an indoor small cell BS transmitting with an EIRP of 10 dBm [33].

the deployment processes, making it quicker to roll out small cell BSs at scale, while requiring minimum human involvement and minimizing disruption of community activities/services. Many improvements have been done – and continue to be done – in this regard up to now [196].

While such practical questions – and many others – were asked and answered, however, a more fundamental technical question raised important concerns:

> *Will the deployment of a large number of small cells BSs in an unplanned manner create an inter-cell interference overload, which will significantly affect the performance of each small cell, and thus render their deployment counterproductive?*

As can be seen in Figure 3.1, a small cell BS deployed indoors, inside a household in this case, can leak a significant amount of power outdoors, to the street, thus raising the inter-cell interference that an end-user walking down the street – and connected to a different outdoor cell – would suffer in the area in front of the house. This inter-cell interference increase would lower the signal quality of such communication, and in some cases, even prevent the user equipment (UE) from staying synchronized with the network. This would be a dramatic outcome for any small cell deployment, since as a result of it, a UE, which could have carried a service before the small cell deployment, cannot do so now. Imagine the frustration, since the owner of such a UE has probably had to pay a premium to finance such counterproductive small cell deployment [33]. Importantly, and more concerning, this inter-cell interference will for sure increase

- if such small cell BSs were deployed outdoors or
- if a larger number of such interfering small cell BSs would exist [40].

As a result, some feared that a large deployment of small cell BSs could result, following this example, in an inter-cell interference overload that could significantly and negatively affect network performance.

This fundamental question needed an urgent answer. If not addressed in a timely fashion, such inter-cell interference overload threatened to derail the business case for network densification, particularly outdoors, or at least force operators to delay or scale back their plans, until proper inter-cell interference mitigation countermeasures were effectively put in place.

Given the importance of the problem, and the fairly remarkable size of the telecommunications industry, one would think that this problem – the one of the inter-cell interference overload – would have been already thoroughly analyzed with a wide range of tools at that time, and the answer to such a question would be common knowledge in the telecommunications community. However, while a large number of system-level simulations may have been carried out, a theoretical answer with solid mathematical roots was certainly missing. Tractable models to accurately characterize the inter-cell interference and its impact on a large network were still and surprisingly unavailable.

The pioneering work of J. G. Andrews et al. in [60] was one giant leap for tackling this problem in a successful manner. As a matter of fact, the framework developed by J. G. Andrews et al. in such work became soon after the de facto tool for the theoretical performance analysis of small cell networks, and it is still used today. This seminal work addressed the problem in question by introducing an additional source of randomness in the analysis: the positions of the small cell BSs. Instead of assuming that small cell BSs were deterministically placed on a regular grid, as traditionally done until then, their location was modelled as a homogeneous Poisson point process (HPPP) of intensity, λ. The main advantage of this new approach to the modelling of small cell BS deployment is that the small cell BS positions can be assumed to be all independent, which allowed substantial tools to be brought to bear from stochastic geometry (see Section 2.3.1 for more details).

It should be mentioned, however, that such an approach to the modelling of a BS deployment – the one in [60] – was not new. Indeed, the use of HPPPs – and other processes – was considered for this purpose as early as 1996 in, e.g. [197–199]. However, the new breakthrough by J. G. Andrews et al. was the derivation of two key metrics, i.e. the coverage probability of the typical UE and the mean data rate achievable over a cell (closely related to the area spectral efficiency (ASE) of the network), which were not known until then. See Section 2.1.12 for the definition of both the coverage probability and the ASE, but let us remind the reader here that the coverage probability is the probability that the signal-to-interference-plus-noise ratio (SINR), Γ, of the typical UE located at the origin, o, is able to surpass some SINR threshold, γ_0. These two metrics were then further derived in closed-form expressions for a number of special cases, namely combinations of

- single-slope path loss with a path loss exponent of four,
- exponentially distributed interference power, i.e. Rayleigh multi-path fast fading and
- interference-limited networks, i.e. thermal noise power was ignored.

These special cases had increasing tractability, and in the case that all three simplifications were taken, a remarkably simple expression for the coverage probability that only depends on the SINR threshold, γ_0, could be found. Because of its simplicity, this expression allowed the drawing of remarkably intuitive – and equally powerful – conclusions, which similarly to the Shannon–Hartley theorem, could be explained with a simple formula.

Due to the importance of the framework in [60] and that of the results and conclusions obtained with it, we devote this chapter to its introduction and its analysis. In the rest of the book, we will use this framework as the reference model and the benchmark, and further enhance it to shed new light on the fundamentals of ultra-dense networks. In what follows,

- Section 3.2 introduces the system model and the assumptions taken in this theoretical performance analysis framework.
- Section 3.3 presents the theoretical expressions for the coverage probability and the ASE.
- Section 3.4 provides results for a number of small cell BS deployments with different densities and characteristics, and presents the conclusions drawn by this theoretical performance analysis. Importantly, and for completeness, this section also studies via system-level simulations the impact of (i) a hexagonal deployment layout – instead of a random one – as wells as (ii) a lognormal shadow fading and a Rician multi-path fast fading – instead of a Rayleigh one – on the derived results and conclusions.
- Section 3.5, finally, summarizes the key takeaways of this chapter.

3.2   Reference System Model

In this section, we introduce the system model used in [60], as said before, probably the most widely used in the literature for the theoretical performance analysis of sparse and/or dense networks. Importantly, it should be noted that this system model will serve as the reference system model – and the benchmark – throughout this book.

Before starting, however, it is important to note that in this particular chapter, and for the majority of this book, we will focus on the analysis of the downlink, i.e. transmissions from the small cell BSs to the UEs. Chapter 10, instead, will present the modelling assumptions for the analysis of the uplink, i.e. transmissions from the UEs to the small cell BSs, together with the related performance analysis and the discussion of the results.

3.2.1   Deployment

In this book, an orthogonal deployment of outdoor small cells and macrocells is considered, where the small cell and the macrocell BSs operate in a different frequency band [158]. In this way, small cells can complement the macrocells, adding targeted capacity, while the macrocells provide a blanket coverage for mobile UEs. The main benefit of this deployment type is the little interaction between the small cells and the macrocells, which do not have any effect on each other in terms of inter-cell interference. Interactions among them only occur at the upper layers, to orchestrate the network connectivity and the use of radio resources. Carrier aggregation at the medium access control (MAC) layer or dual connectivity at the packet data convergence protocol (PDCP) layer are examples of features through which such interactions and coordination can be realized [12].

Importantly to this system model, due to the non-existent inter-cell interference between the small cell and the macrocell tiers, the performance of each tier can be analyzed independently, thus allowing us in this book to focus on the performance analysis of small cell-only networks.

3.2.2   Small Cell Base Station

In this model, following the description in Section 2.1.1, it is assumed that the small cell BSs were deployed on-demand to provide coverage or capacity where it was needed as it was needed. As a consequence of this ad hoc deployment, the small cell BS layout follows a random fashion, modelled as a stationary HPPP. In more detail, it is considered that

- the small cell BSs follow a stationary HPPP, Φ, of intensity, λ, in BSs/km^2 on a two-dimensional plane and that
- all the small cell BSs have the same constant transmit power, P, in Watts over time.

Since an orthogonal deployment of outdoor small cells and macrocells is considered in this case, and because we only focus the analysis on the small cell BS tier, it is important to note that the impact of the transmit power – whether constant or jointly and equally varied per small cell BS – on the coverage probability and the ASE is mostly negligible in an interference-limited scenario. This is mainly due to

- the transmit power of the useful signal and that of the inter-cell interference signals cancel each other out in the SINR expression, see equation (1.1), and because
- the aggregated inter-cell interference power is much larger than the noise power.

Note that the above would not apply if each small cell BS would independently set its transmit power.

3.2.3   User Equipment

With regard to the UE locations, following the description in Section 2.1.2, it is assumed that the active UEs also follow a stationary HPPP, Φ^{UE}, of intensity, ρ, in UEs/km^2 on the same two-dimensional plane as the BSs. The HPPP, Φ^{UE}, is independent from the small cell BS one, Φ. Remember that all UEs are considered to be deployed in an outdoor scenario.

For tractability reasons, it was also assumed in [60] that the network is fully loaded, thus we also adopt such a consideration here, meaning that the active UE density, ρ, is sufficiently larger than the small cell BS one, λ, and thus every small cell BS has at least one active UE in its coverage area. As a consequence, it follows that every small cell BS is active, and thus all small cell BSs should be considered in the theoretical performance analysis.

With regard to the concept of active UE, it is important to remember that an active UE is that which has traffic to receive or transmit at a particular time instant. In general, only active UEs are considered in a theoretical performance analysis from a capacity standpoint, since non-active UEs are not involved in any data transmission, and thus can be "safely" ignored. In this system model, however, all UEs follow the full-buffer traffic model described in Section 2.1.3, and thus, all UEs are active UEs.

As a reminder, it is also worth mentioning that, while the total number of UEs in a typical cellular network could be very large, the active UEs with data traffic demands on a certain frequency band and at a given time slot may not be that many, when considering a more realistic traffic model. A typical active UE density in a dense-urban scenario is around 300–600 UEs/km^2, and those are the numbers that we will use later in this chapter.

Finally, recall that the typical UE is defined as a statistically representative active UE, and that for convenience is located at the origin, o, as described in Section 2.1.12.

3.2.4   Antenna Radiation Pattern

Following the description in Section 2.1.5, the typical UE and all small cell BSs are equipped with a single isotropic antenna – or at least one with an omni-directional antenna pattern in the horizontal plane – both with a maximum antenna gain, $\kappa_M = 0$.

Multi-antenna transmissions are neither considered in this system model nor treated in this book. This is mainly because the multi-antenna transmission technology is a topic on its own, especially considering the upcoming adoption of massive multiple-input multiple-output (MIMO) in future wireless communication networks.

3.2.5   Single-Slope Path Loss Model

Contrary to the more complex but practical path loss model presented in Section 2.1.6, in [60], the path loss, $\zeta(r)$, between the typical UE and an arbitrary small cell BS at a distance, r, was modelled through a simpler single-slope path loss model, of the following form,

$$\zeta(r) = Ar^{-\alpha},\tag{3.1}$$

where

- ζ is the path loss in linear units,
- r is the two-dimensional distance between the typical UE and the arbitrary small cell BS in kilometres,[1]
- A is the path loss at a reference distance, $r = 1$ km, for a given carrier in linear units and
- α is the path loss exponent in linear units.

Without loss of generality, in this book, for this single-slope path loss model, we adopt the parameters of the non-line-of-sight (NLoS) component of the multi-slope path loss model presented in Section 2.1.6. The specific values for these parameters will be provided in Section 3.4.1, when constructing the specific use cases to analyze.

3.2.6   Shadow Fading

For tractability reasons, shadow fading is usually not modelled in the traditional stochastic geometry reference model, and this is also the case in this chapter. However, note this assumption is justified later in Section 3.4.5, by analyzing the error introduced when neglecting the shadow fading through system-level simulations.

3.2.7   Multi-Path Fading

As proposed in [60], the multi-path fast fading gain, h, between the typical UE and an arbitrary small cell BS is modelled as a normalized Rayleigh random variable, thus following an exponential distribution with unitary mean. Moreover, it should be noted that the multi-path fast fading random variables of any two pairs of links are independent and identical distributed (i.i.d.).

It is important to recall that in theoretical performance analysis – with stochastic geometry being no exception – the multi-path fast fading is usually modelled with such Rayleigh fading for tractability reasons. However, a more accurate and practical model is that based on a generalized Rician fading, which accounts for the different degrees of scattering in a scenario. To have a better understanding of the impact of the Rician fading versus the Rayleigh one on network performance, Section 3.4.5 presents simulation results comparing the network performance obtained with both multi-path fast-fading models.

Please refer to Section 2.1.8 for more details on both Rayleigh and Rician fading.

---

[1] Note that three-dimensional distances will be denoted by variable, d, while two-dimensional ones will be denoted by variable, r.

Table 3.1. System model

Model	Description	Reference
Transmission link		
Link direction	Downlink only	Transmissions from the small cell BSs to the UEs
Deployment		
Small cell BS deployment*	HPPP with a finite density, $\lambda < +\infty$	Section 3.2.2
UE deployment	HPPP with full load, resulting in at least one UE per cell	Section 3.2.3
UE to small cell BS association		
Strongest small cell BS	UEs connect to the small cell BS at the shortest distance	Section 3.2.8 and references therein
Path loss		
Traditional	Single-slope path loss	Section 3.2.5, equation (3.1)
Multi-path fast fading		
Rayleigh**	Highly scattered scenario	Section 3.2.7 and references therein equation (2.15)
Shadow fading		
Not modelled**	For tractability reasons, shadowing is not modelled since it has no qualitative impact on the results, see Section 3.4.5	Section 3.2.6 references therein
Antenna		
Small cell BS antenna	Isotropic single-antenna element with 0 dBi gain	Section 3.2.4
UE antenna	Isotropic single-antenna element with 0 dBi gain	Section 3.2.4
Small cell BS antenna height	Not considered	–
UE antenna height	Not considered	–
Idle mode capability at small cell BSs		
Always on	Small cell BSs always transmit control signals	–
Scheduler at small cell BSs		
Round robin	UEs take turns in accessing the radio channel	Section 7.1

*   Section 3.4.4 presents simulated results with a hexagonal deployment layout of small cell BSs to demonstrate its impact on the obtained results.

** Section 3.4.5 presents simulated results with correlated lognormal shadowing and Rician multi-path fast fading to demonstrate their impact on the obtained results.

3.2.8 User Association Strategy

Each UE associates with the small cell BS at the shortest distance in [60], which considering the presented simplified channel model with a single-slope path loss model and no shadowing is equivalent to a maximum received signal strength association (see Section 2.1.4).

To finalize this section, Table 3.1 presents a concise summary of the system model presented here in this section.

3.3 Theoretical Performance Analysis and Main Results

In this section, the coverage probability of the typical UE and the ASE of the network are derived and formally presented, while embracing the above presented reference system model. Detailed step-by-step explanations around the derivations are provided in the appendices to help the understanding of the reader.

Note that these two metrics – the coverage probability and the ASE – were formally introduced in Chapter 2, and they are summarized for convenience of the reader in Table 3.2.

Before presenting the derivations and the results, let us also highlight that the scaling law of the ASE presented in this section was the first widely accepted capacity scaling law derived for small cell networks, and became the de facto understanding for many years.

3.3.1 Coverage Probability

In this subsection, we present the theoretical results on the coverage probability, p^{cov}, considering the above introduced system model, where

- the single-slope path loss model and
- the shortest distance UE association strategy

should be highlighted.

Formally, the coverage probability, p^{cov}, of a network of density, λ, is defined as the probability that the SINR, Γ, of the typical UE located at the origin, o, in such a network is larger than a given SINR threshold, γ_0, i.e.

Table 3.2. Key performance indicators

Metric	Formulation	Reference
Coverage probability, p^{cov}	$p^{cov}(\lambda, \gamma_0) = \Pr[\Gamma > \gamma_0] = \bar{F}_\Gamma(\gamma_0)$	Section 2.1.12, equation (2.22)
ASE, A^{ASE}	$A^{ASE}(\lambda, \gamma_0) = \frac{\tilde{\lambda}}{\ln 2} \int_{\gamma_0}^{+\infty} \frac{p^{cov}(\lambda, \gamma)}{1+\gamma} d\gamma$ $+ \tilde{\lambda} \log_2(1 + \gamma_0) p^{cov}(\lambda, \gamma_0)$	Section 2.1.12, equation (2.25)

$$p^{\text{cov}}(\lambda, \gamma_0) = \Pr\left[\Gamma > \gamma_0\right], \tag{3.2}$$

where

- γ_0 is the minimum working SINR of the network in linear units, i.e. the SINR necessary to successfully use the minimum modulation and coding scheme of the particular technology in use and
- $\gamma = \Gamma(\omega)$ is an SINR realization of the typical UE, and can be calculated as

$$\gamma = \frac{P\zeta(r)h}{I_{\text{agg}} + P^{\text{N}}}, \tag{3.3}$$

where

- ☐ P is the transmit power of an arbitrary small cell BS,
- ☐ h is the multi-path fast fading gain between the typical UE and its serving small cell BS, modelled as Rayleigh fading,
- ☐ P^{N} is the received additive white Gaussian noise (AWGN) power at the typical UE and
- ☐ I_{agg} is the aggregated inter-cell interference given by

$$I_{\text{agg}} = \sum_{i:\, b_i \in \Phi \backslash b_o} P\beta_i g_i, \tag{3.4}$$

where

- ○ b_o is the small cell BS serving the typical UE,
- ○ b_i is the ith interfering small cell BS and
- ○ β_i and g_i are the path loss and the multi-path fast fading gain between the typical UE and the ith interfering small cell BS, respectively.

With these definitions, we present the main result on the coverage probability, p^{cov}, in Theorem 3.3.1. Readers interested in the research article presenting these results for the first time are referred to [60].

THEOREM 3.3.1   *Considering the single-slope path loss model in equation (3.1), the coverage probability, p^{cov}, can be derived as*

$$p^{\text{cov}}(\lambda, \gamma_0) = \int_0^{+\infty} \Pr\left[\frac{P\zeta(r)h}{I_{\text{agg}} + P^{\text{N}}} > \gamma_0\right] f_R(r)dr, \tag{3.5}$$

where

- *the probability density function (PDF), $f_R(r)$, is given by*

$$f_R(r) = \exp(-\pi r^2 \lambda)2\pi r\lambda \tag{3.6}$$

and
- *the probability, $\Pr\left[\frac{P\zeta(r)h}{I_{\text{agg}}+P^{\text{N}}} > \gamma_0\right]$, is written as*

$$\Pr\left[\frac{P\zeta(r)h}{I_{\text{agg}} + P^{\text{N}}} > \gamma_0\right] = \exp\left(-\frac{\gamma_0 P^{\text{N}}}{P\zeta(r)}\right)\mathscr{L}_{I_{\text{agg}}}(s), \tag{3.7}$$

where

☐ $\mathcal{L}_{I_{\text{agg}}}(s)$ *is the Laplace transform of the random variable,* I_{agg}*, evaluated at point,* $s = \frac{\gamma_0}{P\zeta(r)}$*, which can be further written as*

$$\mathcal{L}_{I_{\text{agg}}}(s) = \exp\left(-2\pi\lambda \int_r^{+\infty} \frac{u}{1+(sP\zeta(u))^{-1}} du\right)$$

$$= \exp\left(-2\pi\lambda \int_r^{+\infty} \frac{u}{1+(\gamma_0 r^\alpha)^{-1} u^\alpha} du\right)$$

$$= \exp\left(-2\pi\lambda\rho\left(\alpha, 1, (\gamma_0 r^\alpha)^{-1}, r\right)\right), \tag{3.8}$$

where

○ $\rho(\alpha,\beta,t,d)$ *is an auxiliary term given by*

$$\rho(\alpha,\beta,t,d) = \int_d^\infty \frac{u^\beta}{1+tu^\alpha} du$$

$$= \frac{1}{\alpha} \int_{d^\alpha}^\infty \frac{y^{\left(\frac{\beta+1}{\alpha}-1\right)}}{1+ty} dy$$

$$= \left[\frac{d^{-(\alpha-\beta-1)}}{t(\alpha-\beta-1)}\right] {}_2F_1\left[1, 1-\frac{\beta+1}{\alpha}; 2-\frac{\beta+1}{\alpha}; -\frac{1}{td^\alpha}\right], \tag{3.9}$$

where

◇ $\alpha > \beta + 1$ *and*
◇ ${}_2F_1[\cdot, \cdot; \cdot; \cdot]$ *is the hypergeometric function [156].*

Proof See Appendix A.   ☐

To ease the understanding of this theorem, Figure 3.2 illustrates a flow chart code that depicts the steps needed to compute the theoretical results of Theorem 3.3.1. Here, it is important to note that the coverage probability, p^{cov}, is computed as the product of two probabilities, i.e.

- the probability of the UE's signal power being larger than the aggregated inter-cell interference power times the threshold, γ_0 and
- the probability of the UE's signal power being larger than the noise power times the threshold, γ_0,

as illustrated in the dashed block of the figure.

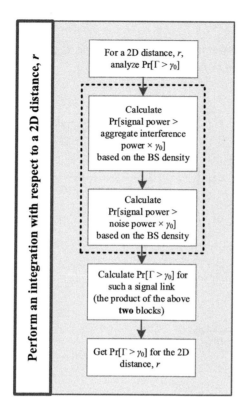

Figure 3.2 Logical steps within the standard stochastic geometry framework to obtain the results in Theorem 3.3.1.

3.3.2   Area Spectral Efficiency

In this subsection, we explain how the ASE, A^{ASE}, can be calculated from the coverage probability, p^{cov}, obtained in Section 3.3.1.

Following the method presented in Section 2.1.12, we have to plug the coverage probability, p^{cov} – obtained from Theorem 3.3.1 – into equation (2.24) to compute the PDF, $f_{\Gamma}(\gamma)$, of the SINR, Γ, of the typical UE. Then, we can derive the ASE, A^{ASE}, by solving equation (2.25). See Table 3.2 for further reference on the ASE formulation.

Here, it is important to clarify that the definition of the ASE, A^{ASE}, in equation (2.23) is different from that in [60], where the transmission rate of the typical UE is

- independent of the SINR, Γ, of such a typical UE in the scenario of study and
- directly calculated based on the SINR threshold, γ_0.

Instead, the definition of the ASE, A^{ASE}, in this book captures the dependence of the transmission rate on the SINR, Γ, of the typical UE, thus realizing the effect of an ideal link adaption operation. However, this definition is less tractable, requiring one more fold of numerical integration compared with that in [60].

3.4   Discussion

In this section, we use numerical results from static system-level simulations to evaluate the accuracy of the above theoretical performance analysis and study the performance of a small cell network in terms of the coverage probability, p^{cov}, and the ASE, A^{ASE}.

3.4.1   Case Study

To investigate the small cell network performance as a function of the small cell BS density, λ, and establish the accuracy of this theoretical performance analysis, realistic parameters are used for the calculation of the coverage probability, p^{cov}, and the ASE, A^{ASE}. Similarly, as in [60], we adopt such parameters from a well-established 3rd Generation Partnership Project (3GPP) system model, in this case, that of the 3GPP outdoor small cell deployment model presented in Section 2.1.6, which is based on [153]. In more detail, and according to Table A.1–3, Table A.1–4 and Table A.1–7 in [153], the following parameters are used for this baseline case:

- maximum antenna gain, $G_{\mathrm{M}} = 0\,\mathrm{dB}$,
- reference path loss, $A = 10^{-14.54}$ – considering a carrier frequency of $2\,\mathrm{GHz}$,
- path loss exponent, $\alpha = 3.75$,
- transmit power, $P = 24\,\mathrm{dBm}$ and
- noise power, $P^{\mathrm{N}} = \sigma^2 = -95\,\mathrm{dBm}$ – including a noise figure of $9\,\mathrm{dB}$ at the UE.

Small Cell BS Densities
Importantly, it should be noted that the selected path loss model was constructed based on measurements, and that due to the characteristics of the measurement campaign, the resulting path loss exponents and reference path losses apply to transmitter-to-receiver distances no smaller than $10\,\mathrm{m}$. To embrace this minimum distance, while still using a simple system model based on HPPP deployments, we will only consider small cell BS densities up to $\lambda = 10^4\,\mathrm{BSs/km}^2$ in our studies.

3.4.2   Coverage Probability

In this subsection, the results in terms of coverage probability, p^{cov}, are presented. In more detail, Figure 3.3 shows the coverage probability, p^{cov}, as a function of the small cell BS density, λ, while considering a SINR threshold, $\gamma_0 = 0\,\mathrm{dB}$, for the four following configurations:

- Conf. a): single-slope path loss model with a path loss exponent, $\alpha = 2.09$ (analytical results).
- Conf. b): single-slope path loss model with a path loss exponent, $\alpha = 2.09$ (simulated results).

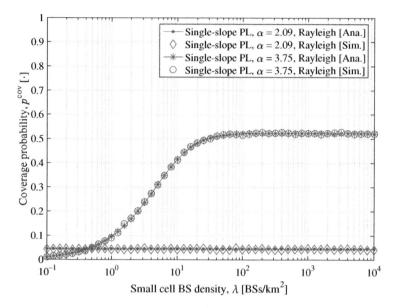

Figure 3.3 Coverage probability, p^{cov}, versus the small cell BS density, λ, for various path loss exponents, α.

- Conf. c): single-slope path loss model with a path loss exponent, $\alpha = 3.75$ (analytical results).
- Conf. d): single-slope path loss model with a path loss exponent, $\alpha = 3.75$ (simulated results).

Note that theoretical results are presented with lines and simulated ones with markers, and that for the investigated single-slope path loss model, two path loss exponents are analyzed,

- the path loss exponent, $\alpha = 3.75$, as introduced earlier, and
- the path loss exponent, $\alpha = 2.09$, for comparison purposes.

As can be observed from Figure 3.3, the theoretical results match with a high degree of accuracy the simulated ones. Due to this significant accuracy on the computation of the coverage probability, p^{cov}, and because the results of the ASE, A^{ASE}, are computed based on those of the coverage probability, p^{cov}, only analytical results are considered in our discussion hereafter in this chapter.

From Figure 3.3, we can observe that the coverage probability, p^{cov}, resulting from the single-slope path loss function with a path loss exponent, $\alpha = 3.75$,

- first increases quickly with the small cell BS density, λ, because more small cell BSs provide a better coverage in noise-limited networks.
- Then, when the small cell BS density, λ, is large enough, e.g. $\lambda > 10^2$ BSs/km², the coverage probability, p^{cov}, becomes independent of the small cell BS density, λ, as the network is pushed into the interference-limited regime. The intuition

behind this observation is that the increase in the inter-cell interference power due to a smaller distance between the typical UE and the interfering small cell BSs is exactly counterbalanced by the increase in the signal power due to the smaller distance between the typical UE and its serving small cell BS in an interference-limited network. As a result, the SINR, Γ, of the typical UE – and in turn, the coverage probability, p^{cov} – remain constant, as the small cell BS density, λ, increases.

When comparing the results with the two different path loss exponents, $\alpha = 3.75$ and $\alpha = 2.09$, we can also see that

- when considering the lower path loss exponent, $\alpha = 2.09$, the small cells are not isolated enough in terms of inter-cell electromagnetic wave attenuation, and the aggregated inter-cell interference is so high that the SINR, Γ, of the typical UE is most of the times below $0\,dB$. This results in a poor coverage probability, p^{cov}.
- When considering the higher path loss exponent, $\alpha = 3.75$, for generic NLoS conditions, the small cells are better isolated in terms of inter-cell electromagnetic wave attenuation. As a result, when the small cell BS density, λ, is relatively large, e.g. $\lambda > 10\,BSs/km^2$, the SINR, Γ, of the typical UE – and the coverage probability, p^{cov} – is much better with the path loss exponent, $\alpha = 3.75$, than that with the path loss exponent, $\alpha = 2.09$.

NetVisual Analysis

To visualize the fundamental behaviour of the coverage probability, p^{cov}, with a path loss exponent, $\alpha = 3.75$, in a more intuitive manner, Figure 3.4b illustrates the coverage probability heat maps of three different scenarios with different small cell BS densities, i.e. 50, 250 and 2500 BSs/km^2 (see Figure 3.4a). These heat maps are computed using NetVisual – a tool, which is able to capture, not only the mean, but also the standard deviation of the coverage probability, p^{cov}, as explained in Section 2.4.

Figure 3.4b visually shows that the sizes of the bright areas (high SINR coverage probabilities) and the dark areas (low SINR coverage probabilities) are approximately the same for the second and the third subfigures, indicating that the average of the SINR, Γ, of the typical UE is the same in both scenarios. This corroborates the previous conclusion, i.e. the SINR, Γ, of the typical UE – and in turn, the coverage probability, p^{cov} – remain constant, as the small cell BS density, λ, increases in the dense and the ultra-dense regimes. The size of dark areas is larger in the first subfigure due to the poor coverage probability, p^{cov}.

Summary of Findings: Coverage Probability Remark

Remark 3.1   Considering a single-slope path loss model, the SINR, Γ, of the typical UE – and in turn, the coverage probability, p^{cov} – remain constant, as the small cell BS density, λ, increases in the dense and the ultra-dense regimes.

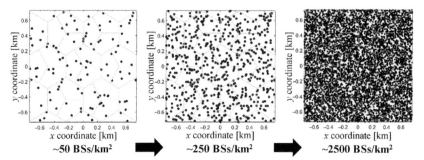

(a) Random small cell BS deployments, where small cell BSs are shown as dots. Note that the small cell BS densities of a typical 4G LTE network, a typical 5G dense network and an ultra-dense one are here considered around 50, 250, 2500 BSs/km², respectively.

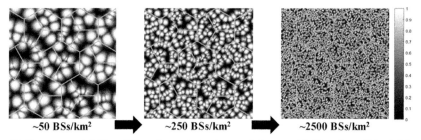

(b) Coverage probability, p^{cov}, under a single-slope path loss model. The sizes of the bright areas (high SINR coverage probabilities) and the dark areas (low SINR coverage probabilities) are approximately the same as the network densifies, indicating an invariant SINR performance, and thus a constant coverage probability, p^{cov}.

Figure 3.4 NetVisual plot of the coverage probability, p^{cov}, versus the small cell BS density, λ [Bright area: high probability, Dark area: low probability].

3.4.3   Area Spectral Efficiency

In this subsection, the results in terms of the ASE, A^{ASE}, are presented. In more detail, Figure 3.5 shows the ASE, A^{ASE}, as function of the small cell BS density, λ, while considering a SINR threshold, $\gamma_0 = 0\,\mathrm{dB}$, for the following two configurations.

- Conf. a): single-slope path loss model with a path loss exponent, $\alpha = 2.09$ (analytical results).
- Conf. b): single-slope path loss model with a path loss exponent, $\alpha = 3.75$ (analytical results).

Before commenting on the results, and as a general note for reading the results on the ASE, A^{ASE}, it should be noted that Figure 3.5 is a log–log plot using logarithmic scales on both the horizontal and the vertical axes. Importantly, a straight line – or a linear increase – in such a log–log plot indicates that a power relationship exists between the variables in the horizontal and the vertical axes, and in more detail, the slope of such a straight line describes the power relationship. For example, a unitary

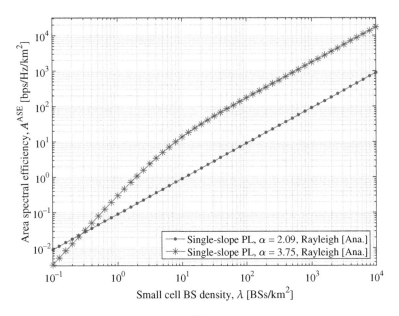

Figure 3.5 Area spectral efficiency, A^{ASE}, versus the small cell BS density, λ, for various path loss exponents, α.

slope indicates a linear relationship, $y \propto x$, a slope of two a quadratic relationship, $y \propto x^2$, etc.

With this in mind, from Figure 3.5, we can observe that the ASE, A^{ASE}, resulting from the single-slope path loss function with a path loss exponent, $\alpha = 3.75$,

- first increases quickly with the small cell BS density, λ, because more small cell BSs provide both a higher signal quality and a better spatial reuse in noise-limited networks. Note that the slope of the increase of the ASE, A^{ASE}, in this log–log plot is larger than 1 in this small cell BS density regime, indicating, according to equation (2.25), that the increase of the small cell BS density, λ, does not only help to increase the ASE, A^{ASE}, through the spatial reuse – the direct multiplier in such an equation – but also through a better coverage probability, p^{cov}, as it was shown in our earlier discussions.
- Then, when the small cell BS density, λ, is large enough, e.g. $\lambda > 10^2$ BSs/km^2, the ASE, A^{ASE}, linearly increases with the BS density, λ, once the network enters the interference-limited regime. This is shown by the slope of the increase of the ASE, A^{ASE}, in this log–log plot, which is exactly 1. This follows from the result in Section 3.4.2, which showed that the coverage probability, p^{cov}, does not vary with respect to the small BS density, λ, in such a regime. Therefore, every new cell deployed into the network introduces an equal growth of capacity, thus leading to the mentioned linear increase of the ASE, A^{ASE}. This behaviour – or capacity scaling law – with the small cell BS density, λ, is referred to as *the linear capacity scaling law* hereafter in this book.

When comparing the results with the two different path loss exponents, $\alpha = 3.75$ and $\alpha = 2.09$, we can also see that, when the small cell BS density, λ, is large enough, e.g. $\lambda > 10\,\text{BSs/km}^2$,

- the slope of the increase of the ASE, A^{ASE}, in the log–log plot is 1 for both configurations. This indicates that the ASE, A^{ASE}, linearly increases with the small cell BS density, λ, as mentioned before. Moreover, it should be noted that the gap between the two ASE straight and parallel lines in this log–log plot translates into a difference between the slopes of the corresponding straight lines in the equivalent linear plot, i.e. a faster or slower linear increase.
- The ASE, A^{ASE}, is larger with the path loss exponent, $\alpha = 3.75$, than that with the path loss exponent, $\alpha = 2.09$. This follows from the smaller aggregated inter-cell interference, and in turn, better coverage probability, p^{cov}, in the former.

Summary of Findings: Area Spectral Efficiency Remark

Remark 3.2   Considering a single-slope path loss model, the ASE, A^{ASE}, linearly increases with the increase of the small cell BS density, λ, with each small cell BS adding an equal contribution to it.

3.4.4   Impact of the Small Cell Base Station Deployment – and Modelling – on the Main Results

Modelling assumptions can have a large impact on the final results, not only quantitatively, but also qualitatively. Thus, one may wonder whether the HPPP assumption on the small cell BS deployment, which is at the core of the stochastic geometry toolbox used in this book, may have an impact on the above derived conclusions for dense and ultra-dense networks, i.e. the coverage probability, p^{cov}, does not depend on the small cell BS density, and the ASE, A^{ASE}, linearly grows with it.

As a reminder and broadly speaking, the HPPP assumption considers that small cell BSs are uniformly deployed in the scenario of interest. However, more practical models, like those used by the 3GPP for some performance evaluations [153], assume that small cell BSs may be deployed non-uniformly, e.g. in clusters. Industry models also assume that any two small cell BSs cannot be too close to each other. Such an assumption is in line with a realistic network planning rule of thumb. Two small cell BSs should not be deployed very close to each other to avoid overwhelming inter-cell interference levels.

To assess the performance impact of the small cell BS deployment model, while considering a widely used industrial one, we adopt the deterministic hexagonal layout presented in Section 2.1.1 in the following, where small cell BSs are deployed on a hexagonal lattice. Using this deployment model, we analyze the coverage probability, p^{cov}, and the ASE, A^{ASE}, performance through system-level simulations using the same configuration setup as in the previous sections. We then compare the obtained results with those previously presented based on an HPPP deployment. For

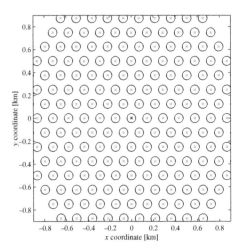

Figure 3.6 Illustration of an ideal small cell BS deployment in a hexagonal network layout. The small cell BS density is around 50 BSs/km^2.

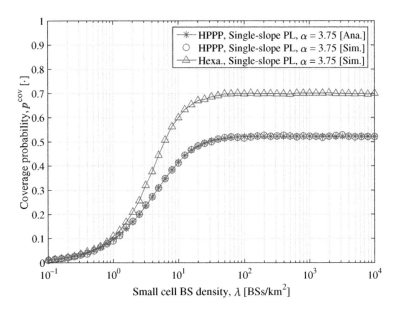

Figure 3.7 Coverage probability, p^{cov}, versus the small cell BS density, λ, considering HPPP and hexagonal deployment models.

completeness, note that more theoretical approaches to the performance analysis of a hexagonal network layout can be found at [139, 200, 201].

Figure 3.6 illustrates the hexagonal layout used in this study, and Figures 3.7 and 3.8 present the coverage probability, p^{cov}, and the ASE, A^{ASE}, for both the hexagonal- and the HPPP-based small cell BS deployments, respectively.

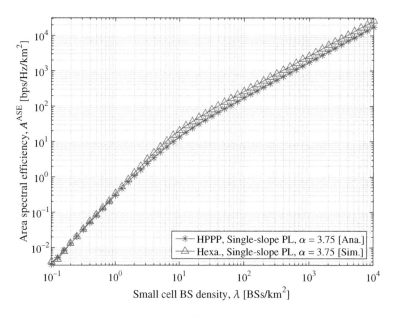

Figure 3.8 Area spectral efficiency, A^{ASE}, versus the small cell BS density, λ, considering HPPP and hexagonal deployment models.

From the results, it can be seen that the hexagonal small cell BS deployment leads to a better performance than the HPPP one. Indeed, the hexagonal small cell BS deployment leads to an upper-bound of the SINR, Γ, of the typical UE and in turn, of the coverage probability, p^{cov}. This is because the small cell BSs are evenly distributed on the scenario, maximizing the minimum distance between any two small cell BSs, and thus mitigating the worst-case inter-cell interference. This naturally results in a better ASE, A^{ASE}, too. Importantly, it should be noted that although the hexagonal small cell BS deployment changed the quantitative results, it did not change the qualitative ones. As a result, one can still observe that the coverage probability, p^{cov}, does not depend on the small cell BS density, λ, and that the ASE, A^{ASE}, linearly grows with it in the ultra-dense regime. This justifies the assumptions of an HPPP small cell deployment modelling, to ease the theoretical performance analysis.

3.4.5 Impact of the Shadow Fading and the Multi-Path Fading – and Modelling – on the Main Results

Similarly to the previous section, one may also wonder whether

- having neglected the shadow fading and
- the choice of a Rayleigh multi-path fast fading model

may change the conclusions derived in this chapter, i.e. that the coverage probability, p^{cov}, does not depend on the small cell BS density, λ, and the ASE, A^{ASE}, linearly grows with it.

In the following, we study – through system-level simulations – the performance of an HPPP-based small cell deployment, while considering

- the cross-correlated shadow fading model described in Section 2.1.7, with a variance, $\sigma_s^2 = 10\,\text{dB}$, and a cross-correlation coefficient, $\tau = 0.5$, as suggested in [153] and
- the Rician multi-path fast fading model presented in Section 2.1.8 for the line-of-sight (LoS) component, which is definitely more accurate, capturing the ratio between the power in the direct path and the power in the other, scattered, paths as a function of the distance, r, between the typical UE and its serving small cell BSs.

We then compare these new results with the baseline ones, analyzed in the main matter of this chapter. Note that more theoretical approaches to the performance analysis of cross-correlated shadow fading and Rician multi-path fast fading can be found at [96, 160], respectively.

Figures 3.9 and 3.10 present the coverage probability, p^{cov}, and the ASE, A^{ASE}, results for three cases,

- the case with no shadow fading and Rayleigh multi-path fast fading (analytical and simulation-based results shown),
- the case with cross-correlated shadow fading and Rayleigh multi-path fast fading[2] and
- the case with no shadow fading and Rician multi-path fast fading,

respectively.

From Figure 3.9, and with respect to the benchmark – the case with no shadow fading and Rayleigh multi-path fast fading – it can be seen that

- when considering the cross-correlated shadow fading gain, $s_{b,u}$, between the bth small cell BS and the uth UE, the coverage probability, p^{cov}, increases with the increase of the small cell BS density, λ, in the noise-limited scenarios, while such an increase disappears, with no gain or loss, in the inter-cell interference-limited ones. The gains in the noise-limited scenarios arise because the shadow fading gain offers the opportunity to connect to a better serving small cell BSs, taking advantage of an opportunistically larger shadow fading gain, $s_{b,u}$, in the signal power. This gain vanishes in the inter-cell interference-limited scenarios, as both the signal link and all the inter-cell interference links are subject to the same i.i.d. shadow fading gain, $s_{b,u}$, and thus such gains statistically cancel each other out in the calculation of the SINRs, Γ, of the UEs when averaging across the entire network.
- When considering the multi-path fast fading channel gain, h, following a Rician distribution, no change is observed in terms of coverage probability, p^{cov}. Since

---

[2] Note that in this case, and in contrast to the others, the UE to small cell BS association is based on the strongest received signal strength and not on the shortest distance, as the shadow fading is a large-scale fading, and varies slowly trough time. Thus, it should be considered in the association process.

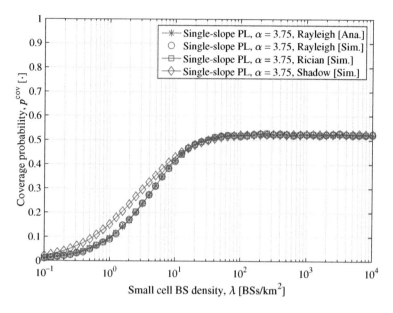

Figure 3.9 Coverage probability, p^{cov}, versus the small cell BS density, λ, considering Rayleigh, Rician and shadow fading models.

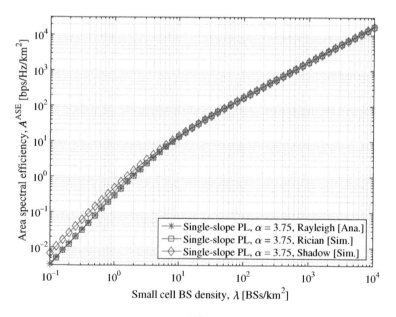

Figure 3.10 Area spectral efficiency, A^{ASE}, versus the small cell BS density, λ, considering Rayleigh, Rician and shadow fading models.

the user association strategy does not depend on the multi-path fast fading channel gain, h, and because both the signal link and all the inter-cell interference links are also subject to the same i.i.d. multi-path fast fading channel gain, h, – whether Rayleigh or Rician – no gain or loss is observed when comparing the results with the different models. Similarly as before, the i.i.d. multi-path fast fading channel gains, h, of the signal and the inter-cell interference power levels statistically cancel each other out in the calculation of the SINRs, Γ, of the UEs when averaging across the entire network.

Importantly, it should be noted that neither the cross-correlated shadow fading nor the Rician multi-path fast fading changed the qualitative results, and in turn, one can still observe that the coverage probability, p^{cov}, does not depend on the small cell BS density, λ, and that the ASE, A^{ASE}, linearly grows with it in the ultra-dense regime. This justifies the assumptions of

- having neglected the shadow fading and
- the choice of a Rayleigh multi-path fast fading model[3]

in the work carried out in [60], and we will embrace these assumptions in the rest of this book, unless otherwise stated.

3.5 Conclusions

In this chapter, we have presented the theoretical performance analysis framework for small cell networks developed by J. G. Andrews et al. in [60], which is probably the most widely used tool for such a task in the community. More precisely, we have described the system model and the assumptions taken. We have also provided details on the derivations performed in the theoretical performance analysis, and presented the resulting expressions for the coverage probability, p^{cov}, and the ASE, A^{ASE}. Moreover, we have shared numerical results for small cell BS deployments with different small cell BS densities and characteristics. To finalize, we have also discussed the important conclusions drawn from this work.

Importantly, the conclusions of this work, which are summarized in Remarks 3.1 and 3.2, indicate that the ASE, A^{ASE}, linearly increases with the increase of the small cell BS density, λ, in the ultra-dense regime. This capacity scaling law, referred to as the linear capacity scaling law hereafter in this book, leads to an exciting line of thinking, which presented the small cell BS as the ultimate mechanism for providing a superior broadband experience, thus sending an important message to network operators and service providers:

Just deploy more cells and everything will be fine!

Unfortunately, this conclusion will be shown not to hold in the ultra-dense regime in Chapters 4–11 of this book.

---

[3] Note that this statement applies for the studied case with single antenna UEs and small cell BSs, but may not apply to multi-antenna cases, where the fading correlation among the antennas of an array plays a key role.

Appendix A Proof of Theorem 3.3.1

For the sake of clarity, and following the guidelines provided in Section 2.3.2, let us first describe the main idea behind the proof of Theorem 3.3.1, and then proceed with a more detailed explanation.

Formally, Theorem 3.3.1 solves the question posed in equation (3.2) for the system model presented in this chapter, where, as depicted in Section 2.3.2, equation (3.2) can be further expanded as

$$
p^{\mathrm{cov}}(\lambda, \gamma_0) = \Pr\left[\Gamma > \gamma_0\right] = \int_{0}^{+\infty} \Pr\left[\Gamma > \gamma_0 \,\middle|\, r\right] f_R(r) \, dr. \tag{3.10}
$$

To solve equation (3.10), we basically need proper expressions for

- the PDF, $f_R(r)$, of the random variable, R, characterizing the distance between the typical UE and its serving small cell BS for the event that the typical UE is associated to the nearest small cell BS and
- the conditional probability, $\Pr[\Gamma > \gamma_0 \,|\, r]$, where
  - $\square$ $r = R(\omega)$ is a realization of the random variable, R.

Once such two expressions – both function of the random variable, R – are known, we can then derive the coverage probability, p^{cov}, by proceeding with the corresponding integrals over the distance, $r = R(\omega)$. The key here is to find such two expressions in suitable forms, as a function of such a variable, $r = R(\omega)$.

Calculation of the Probability Density Function, $f_R(r)$

Let us first show how to calculate the PDF, $f_R(r)$.
To that end, we define the following event:

- Event B: The nearest small cell BS to the typical UE is located at a distance, $r = R(\omega)$, defined by the distance random variable, R.

According to the results presented in Section 2.3.2, the PDF, $f_R(r)$, of the random variable, R, in Event B can be calculated by

$$
f_R(r) = \frac{\partial \left(1 - \bar{F}_R(r)\right)}{\partial r} = 2\pi \lambda r \exp\left(-\pi \lambda r^2\right). \tag{3.11}
$$

Let us stress here the UE association based on the nearest small cell BS.

For completeness, let us also note that the complementary cumulative distribution function (CCDF), $\bar{F}_R(r)$, of the random variable, R, in Event B can be computed as

$$
\bar{F}_R(r) = \exp\left(-\pi \lambda r^2\right). \tag{3.12}
$$

As a general comment, note that the PDF, $f_R(r)$, and the CCDF, $\bar{F}_R(r)$, are intimately related, as one can find the former by taking the derivative, $\frac{\partial(1-\bar{F}_R(r))}{\partial r}$, with respect to the distance, r.

Using these results, we can compute the PDF, $f_R(r)$, as a function of the distance, $r = R(\omega)$.

Calculation of the Probability, $\Pr[\Gamma > \gamma_0 \mid r]$

Let us now show how to calculate the probability, $\Pr[\Gamma > \gamma_0 \mid r]$.

First of all, we use the definition of the SINR, Γ, of the typical UE in equation (3.3) to find that

$$\Pr[\Gamma > \gamma_0 \mid r] = \Pr\left[\frac{P\zeta(r)h}{I_{\text{agg}} + P^{\text{N}}} > \gamma_0\right]. \tag{3.13}$$

Then, working on expression, $\Pr\left[\frac{P\zeta(r)h}{I_{\text{agg}}+P^{\text{N}}} > \gamma_0\right]$, we can further obtain that

$$\Pr\left[\frac{P\zeta(r)h}{I_{\text{agg}} + P^{\text{N}}} > \gamma_0\right] = \mathbb{E}_{[I_{\text{agg}}]}\left\{\Pr\left[h > \frac{\gamma_0\left(I_{\text{agg}} + P^{\text{N}}\right)}{P\zeta(r)}\right]\right\}$$

$$= \mathbb{E}_{[I_{\text{agg}}]}\left\{\bar{F}_H\left(\frac{\gamma_0\left(I_{\text{agg}} + P^{\text{N}}\right)}{P\zeta(r)}\right)\right\}, \tag{3.14}$$

where

- $\bar{F}_H(h)$ is the CCDF of the multi-path fast fading random variable, h.

Since we assumed that the random variable, h, is an exponentially distributed random variable – Rayleigh fading – we can define the CCDF, $\bar{F}_H(h)$, as

$$\bar{F}_H(h) = \exp(-h). \tag{3.15}$$

Using this result, we can further elaborate on equation (3.14), and find that

$$\Pr\left[\frac{P\zeta(r)h}{I_{\text{agg}} + P^{\text{N}}} > \gamma_0\right] = \mathbb{E}_{[I_{\text{agg}}]}\left\{\bar{F}_H\left(\frac{\gamma_0\left(I_{\text{agg}} + P^{\text{N}}\right)}{P\zeta(r)}\right)\right\}$$

$$= \exp\left(-\frac{\gamma_0 P^{\text{N}}}{P\zeta(r)}\right)\mathbb{E}_{[I_{\text{agg}}]}\left\{\exp\left(-\frac{\gamma_0}{P\zeta(r)}I_{\text{agg}}\right)\right\}$$

$$= \exp\left(-\frac{\gamma_0 P^{\text{N}}}{P\zeta(r)}\right)\mathscr{L}_{I_{\text{agg}}}(s), \tag{3.16}$$

where

- $\mathscr{L}_{I_{\text{agg}}}(s)$ is the Laplace transform of the PDF of the aggregated inter-cell interference random variable, I_{agg}, evaluated at point, $s = \frac{\gamma_0}{P\zeta(r)}$, and the equality,

$$\mathscr{L}_X(s) = \mathbb{E}_{[X]}\left\{\exp(-sX)\right\}, \tag{3.17}$$

follows from the definitions and derivations leading to equation (2.46) in Chapter 2.

Having made such clarification, and based on the considered user association strategy – the shortest distance – the Laplace transform, $\mathscr{L}_{I_{\text{agg}}}(s)$, can be shown to be,

$$\mathscr{L}_{I_{\text{agg}}}(s) = \mathbb{E}_{[I_{\text{agg}}]}\left\{ \exp(-sI_{\text{agg}}) \middle| 0 < r \le +\infty \right\}$$

$$= \mathbb{E}_{[\Phi,\{\beta_i\},\{g_i\}]}\left\{ \exp\left(-s\sum_{i\in\Phi/b_o} P\beta_i g_i \right) \middle| 0 < r \le +\infty \right\}$$

$$= \mathbb{E}_{[\Phi,\{\beta_i\},\{g_i\}]}\left\{ \prod_{i\in\Phi/b_o} \exp(-sP\beta_i g_i) \middle| 0 < r \le +\infty \right\}$$

$$\overset{(a)}{=} \mathbb{E}_{[\Phi,\{\beta_i\}]}\left\{ \prod_{i\in\Phi/b_o} \mathbb{E}_{[g]}\left\{ \exp(-sP\beta_i g) \right\} \middle| 0 < r \le +\infty \right\}$$

$$\overset{(b)}{=} \exp\left(-2\pi\lambda \int_r^{+\infty} \left(1 - \mathbb{E}_{[g]}\left\{ \exp(-sP\zeta(u)g) \right\} \right) u\, du \right), \tag{3.18}$$

where

- Φ is the set of small cell BSs,
- b_o is the small cell BS serving the typical UE,
- b_i is the ith interfering small cell BS,
- β_i and g_i are the path loss and the multi-path fast fading gain between the typical UE and the ith interfering small cell BS, respectively,
- step (a) is true because the interfering multi-path fast fading gains, g_i, are i.i.d. random variables, in this case, Rayleigh random variables and
- step (b) follows from Campbell's formula, presented in equation (2.47) of Section 2.3.1, where the change, $\Lambda(u)\, du = 2\pi\lambda u\, du$, has been used considering our focus on the two-dimensional plane.

Since the distance, r, should be evaluated in the finite range, $0 < r \le +\infty$, the Laplace transform, $\mathscr{L}_{I_{\text{agg}}}(s)$, can be further developed as,

$$\mathscr{L}_{I_{\text{agg}}}(s) = \exp\left(-2\pi\lambda \int_r^{+\infty} \left(1 - \mathbb{E}_{[g]}\left\{ \exp(-sP\zeta(u)g) \right\} \right) u\, du \right)$$

$$\overset{(a)}{=} \exp\left(-2\pi\lambda \int_r^{+\infty} \frac{u}{1 + (sP\zeta(u))^{-1}}\, du \right)$$

$$= \exp\left(-2\pi\lambda \int_r^{+\infty} \frac{u}{1 + (\gamma_0 r^\alpha)^{-1} u^\alpha}\, du \right), \tag{3.19}$$

where

- in step (a), the PDF, $f_G(g) = 1 - \exp(-g)$, of the Rayleigh random variable, g, has been used to calculate the expectation, $\mathbb{E}_{[g]}\{\exp(-sP\zeta(u)g)\}$.

Finally, the last step to derive the Laplace transform, $\mathcal{L}_{I_{\mathrm{agg}}}(s)$, is to calculate the integral, $\int_r^{+\infty} \frac{u}{1+(\gamma_0 r^\alpha)^{-1} u^\alpha} du$, in equation (3.19). This can be done by using the definition of the hypergeometric function. For completeness, note that the hypergeometric function is defined as [156]

$$2F_1[a,b;c;z] = \sum_{n=0}^{+\infty} \frac{(a)_n (b)_n}{(c)_n} \frac{z^n}{n!}, \tag{3.20}$$

where

- $(q)_n$ is defined as

$$(q)_n = \begin{cases} 1 & \text{if } n = 0 \\ q(q+1)\cdots(q+n-1) & \text{if } n > 0 \end{cases}, \tag{3.21}$$

and that the resulting expressions for the Laplace transform, $\mathcal{L}_{I_{\mathrm{agg}}}(s)$, based on the hypergeometric function are formally presented in Theorem 3.3.1. Please note that the definition and interpretation of the hypergeometric function are complicated, and thus readers interested in further details are referred to [156].

Using these results, we can compute the probability, $\Pr[\Gamma > \gamma_0 \,|\, r]$.

This concludes the proof.

Part II

Fundamentals of Ultra-Dense Small Cell Networks

4 The Impact of Line-of-Sight Transmissions on Ultra-Dense Wireless Networks

4.1 Introduction

Deploying and operating a large network is expensive, and thus requires careful network dimensioning and planning to ensure a high radio resource utilization, and in turn, a large network performance. Due to its complexity, however, the manual design of a network to improve such radio resource utilization is prone to failure. There are many parameters and hidden trade-offs involved in the process. This called out for the development of automated tools and optimization algorithms that are able to assist network operators and service providers in this difficult task.

Formally, network planning refers to the process of designing a network deployment with its respective network architecture and elements, subject to various design requirements. Importantly, network planning must ensure that the network deployment has a sufficient amount of radio resources, and enable its effective utilization, to achieve a certain level of network performance at a given cost. In more detail, network planning is associated with two main activities: network dimensioning and detailed planning, i.e. two network life phases, both of which are of crucial importance. This is because once the network is planned and deployed, such implementation imposes hard constraints on the future network performance. In cellular networks, examples of such hard constraints are the frequency band of operation, the cell site location, etc. These are characteristic of the network, which are hard, i.e. costly, to change once the network is deployed, and are likely to remain unchanged for several years.

Once the network is deployed, another network life phase of vast importance is that of network optimization. Network optimization is aimed at finding the network configuration, which can achieve the best possible network performance during network operation, including the best possible radio resource utilization. Another important task of network optimization is to identify and solve eventual network performance issues in particular situations, which can be temporally originated due to, e.g. unusual radio resource exhaustion due to equipment failure, high inter-cell interference and/or unpredicted heavy load.

The applicability of optimization algorithms for ensuring a good network performance is quite intuitive for both network planning and optimization. However, their performance strictly depends on a reasonable trade-off between the accuracy of the model with respect to the reality and its complexity. Good references on network planning and optimization can be found in [202–204].

While reviewing the literature on the subject, it can be easily seen that two important characteristics that strongly influence network performance and should be taken into account while designing a network are

- the trade-off between coverage and capacity and
- the impact of the radio channel.

The coverage of a network indicates the fraction of traffic-weighted area that can be offered a service with a minimum guaranteed quality of service, where both the offered service and the minimum guaranteed quality of service are open for specific definition. For example, some may define their network coverage as the ability of decoding a given pilot signal with a given received signal strength for a given fraction of time, while others may choose more sophisticated definitions, such as the capability of receiving a voice service with a given mean opinion score (MOS).

The capacity of a network, instead, is the maximum amount of traffic that the network can serve with a given set of radio resources, e.g. with a given bandwidth. This definition is network-wide, and automatically captures the degree of load balancing between the actual traffic and the available radio resources, as well as interference-related aspects.

It is important to note that these two metrics – the coverage and the capacity of the network – depend on many factors, such as the orography and the buildings, the technology, the radio frequency, the transmit power and the receiver sensitivity, to cite a few. Not only that, the coverage and the capacity of the network are also tightly interrelated. For example, the coverage of the network – the ability of the user equipment (UE) to connect to the base station (BS) in short for the sake of explanation – is a function of the received signal strength at the former. This may be boosted by a higher transmit power, a taller antenna mast, a better antenna, better propagation characteristics, etc. However, in an interconnected network with many neighbouring cells, such as those that are the subject of this book, increasing the received signal strength towards the intended UE can significantly increase the inter-cell interference to the UEs of neighbouring BSs in the same area or direction. This, in turn, has a negative effect on the signal-to-interference-plus-noise ratio (SINR) of such neighbouring UEs, and thus on the capacity of the network. This trade-off can be simply explained by the Shannon–Hartley theorem [9], presented in equation (1.1).

Importantly, better propagation characteristics, such as those of a line-of-sight (LoS) transmission with respect to those of a non-line-of-sight (NLoS) transmission, can enhance the received signal strength towards the intended UE. This enhancement can translate into a significant improvement of the SINR of such a UE, if its inter-cell interference can be kept constant – or at least, its increase rate can be maintained to a reasonable level. This is a primary target of traditional network planning tools, where macrocells in urban areas are intelligently deployed above rooftops, and their antennas carefully down-tilted towards the streets, such that they

- can provide an LoS coverage towards most of the intended UEs in the cell while
- keeping most of the UEs of neighbouring BSs in NLoS.

Given the difference of 20 dB – or more – between an LoS and an NLoS transmission, this planning driver provides a major performance boost.

Similar performance improvements through network planning can be expected in small cell networks, if the same rule of thumb can be applied, i.e. keep intended UEs in LoS and non-intended ones in NLoS. However, as one can imagine, this is increasingly difficult when entering the realm of ultra-dense networks, where the average inter-site distance between any two outdoor small cell BSs can be short. For example, two small cell BSs can be deployed in the same street, 100 m apart from each other. As a result, both such small cells can have a decreased path loss towards their corresponding UEs in a given location, that of the aforementioned street, providing LoS communications, and enhancing the received signal strength towards such UEs in close proximity. Unfortunately, it is also likely that both small cell BSs will "see" each other's UEs in LoS, thus generating large inter-cell interference towards the UEs of the neighbouring small cell BS in their same such street. This violates the previously mentioned network design rule, and can significantly degrade the SINRs of the UEs at the cell-edge, making one doubt the benefits of deploying the second small cell BS in such a location, in LoS with the UEs of the neighbouring small cell BS.

To illustrate this, Figure 4.1 provides an example with four deployment cases. The two deployments at the top correspond to two traditional macrocell network scenarios, and the two deployments at the bottom correspond to two small cell ones, where the scenario at the bottom left represents a sparse small cell network and that at the bottom right shows an ultra-dense network.

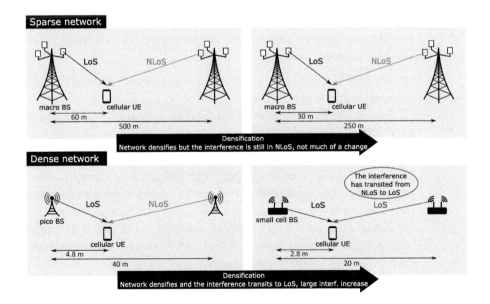

Figure 4.1 NLoS-to-LoS transition.

Let us pay attention to the two macrocell network scenarios at the top. In the figure on the left-hand side, we can see a deployment following the previous rule of thumb, with an inter-site distance of 500 m, where the UE is located at 60 m from its serving macrocell BS, and benefits from an LoS connection, while the closest neighbouring macrocell BS is at 440 m and in NLoS. In the figure at the right-hand side, the network density has been increased, reducing the inter-site distance to 250 m. However, since such distance is still large enough, the same generic conditions are kept. The two-dimensional distances are halved, and the UE still benefits from an LoS connection to its serving macrocell BS, while the closest neighbouring macrocell BS is in NLoS. Thus, the same performance should be roughly expected, as both the inter-cell interference and the signal power are scaled in the same proportion. We should note that this resembles the system model analyzed and the conclusions drawn in Chapter 3.

Let us pay attention now to the two small cell networks at the bottom. In the figure on the left-hand side, we can see a deployment still following the same rule of thumb of the macrocell ones, but this time with an inter-site distance of 40 m, where the UE is located at 4.8 m from its serving small cell BS, and benefits from an LoS connection, while the closest neighbouring small cell BS is at 35.2 m and in NLoS. In the figure at the right-hand side, the network enters the ultra-dense network realm. The distances are also halved now, but the situation changes, and the rule of thumb breaks. The closest small cell BS has transitioned from NLoS to LoS, and the inter-cell interference power grows at a faster rate than the signal power in this case. The 20 dB – or more – of protection between an LoS and an NLoS transmission disappears. This clearly leads to a degradation of the SINR of the UE, and thus of the capacity of the network. We should also note that this case was not captured in the network investigated in Chapter 3, as its single-slope path loss model did not embrace this NLoS-to-LoS inter-cell interference transition.

Although the exact numbers provided in this example may not apply to all scenarios, the transition of a large number of interfering links from NLoS to LoS will occur in dense networks, both

- *outdoors*, in dense urban areas, with a large variety of buildings and man-made structures, as well as
- *indoors*, in offices, enterprises and factories with wide-open spaces and large corridors.

With all these in mind, and particularly considering that the model used in Chapter 3 did not consider a path loss model with LoS and NLoS transmissions, it is fair to wonder about the applicability of its results to ultra-dense networks. In more detail, one may fear that the important and optimistic conclusions of Chapter 3, summarized in Remarks 3.1 and 3.2, may after all not apply in the ultra-dense regime, and thus deploying an increasing number of small cell BSs may not be the ultimate mechanism in providing a superior broadband experience.

This gives rise to a fundamental question:

Will the linear capacity scaling law *presented in [60] hold in the ultra-dense regime?*

In this chapter, we answer this fundamental question by means of in-depth theoretical analyses.

The rest of this chapter is organized as follows:

- Section 4.2 introduces the system model and the assumptions taken in the theoretical performance analysis framework presented in this chapter, which considers, among others, both LoS and NLoS transmissions.
- Section 4.3 presents the theoretical expressions for the coverage probability and the area spectral efficiency (ASE) under the new system model.
- Section 4.4 provides results for a number of small cell BS deployments with different densities and characteristics. It also presents the conclusions drawn by this theoretical performance analysis, shedding new light on the importance of the transition of a large number of interfering links from NLoS to LoS in dense networks. Importantly, and for completeness, this section also studies via system-level simulations the impact of (i) multiple different LoS probability functions as wells as (ii) a Rician multi-path fast fading – instead of a Rayleigh one – on the derived results and conclusions.
- Section 4.5, finally, summarizes the key takeaways of this chapter.

4.2   Updated System Model

To assess the impact of the channel characteristics described in Section 4.1 on an ultra-dense network, in this section, we upgrade the system model presented in Section 3.2, which has accounted for

- a single-slope path loss model,

with a new one, considering

- both LoS and NLoS transmissions as well as
- a new user association strategy, based on the strongest received signal strength.

As in Chapter 3, Table 4.1 provides a concise and updated summary of the system model used in this chapter.

In the following, the two new upgrades are presented in detail.

4.2.1   Multi-Slope Path Loss Model

To consider both LoS and NLoS transmissions, we embrace the more complex but practical path loss model presented in Section 2.1.6, which accounts for both LoS and NLoS transmissions, and provide in the following a more general piecewise multi-slope definition for the sake of completeness and tractability.

Formally, in the rest of this book, the path loss, $\zeta(r)$, between the typical UE and an arbitrary small cell BS at a distance, r, is modelled for theoretical performance analysis as

Table 4.1. System model

Model	Description	Reference
Transmission link		
Link direction	Downlink only	Transmissions from the small cell BSs to the UEs
Deployment		
Small cell BS deployment	HPPP with a finite density, $\lambda < +\infty$	Section 3.2.2
UE deployment	HPPP with full load, resulting in at least one UE per cell	Section 3.2.3
UE to small cell BS association		
Strongest small cell BS	UEs connect to the small cell BS providing the strongest received signal strength	Section 3.2.8 and references therein, equation (2.14)
Path loss		
3GPP UMi [153]	Multi-slope path loss with probabilistic LoS and NLoS transmissions	Section 4.2.1, equation (4.1)
	☐ LoS component	equation (4.2)
	☐ NLoS component	equation (4.3)
	☐ exponential probability of LoS*	equation (4.19)
Multi-path fast fading		
Rayleigh**	Highly scattered scenario	Section 3.2.7 and references therein equation (2.15)
Shadow fading		
Not modelled	For tractability reasons, shadowing is not modelled since it has no qualitative impact on the results, see Section 3.4.5	Section 3.2.6 references therein
Antenna		
Small cell BS antenna	Isotropic single-antenna element with 0 dBi gain	Section 3.2.4
UE antenna	Isotropic single-antenna element with 0 dBi gain	Section 3.2.4
Small cell BS antenna height	Not considered	—
UE antenna height	Not considered	—
Idle mode capability at small cell BSs		
Always on	Small cell BSs always transmit control signals	—
Scheduler at small cell BSs		
Round robin	UEs take turns in accessing the radio channel	Section 7.1

*   Section 4.4.4 presents simulated results with a linear probability of LoS function to demonstrate its impact on the obtained results.

** Section 4.4.5 presents simulated results with Rician multi-path fast fading to demonstrate its impact on the obtained results.

$$
\zeta(r) = \begin{cases}
\zeta_1(r) = \begin{cases} \zeta_1^{L}(r), & \text{with prob. } \mathrm{Pr}_1^{L}(r) \\ \zeta_1^{NL}(r), & \text{with prob. } \left(1 - \mathrm{Pr}_1^{L}(r)\right) \end{cases}, & \text{when } 0 \le r \le d_1 \\[2ex]
\zeta_2(r) = \begin{cases} \zeta_2^{L}(r), & \text{with prob. } \mathrm{Pr}_2^{L}(r) \\ \zeta_2^{NL}(r), & \text{with prob. } \left(1 - \mathrm{Pr}_2^{L}(r)\right) \end{cases}, & \text{when } d_1 < r \le d_2 \\[2ex]
\vdots & \vdots \\[1ex]
\zeta_N(r) = \begin{cases} \zeta_N^{L}(r), & \text{with prob. } \mathrm{Pr}_N^{L}(r) \\ \zeta_N^{NL}(r), & \text{with prob. } \left(1 - \mathrm{Pr}_N^{L}(r)\right) \end{cases}, & \text{when } r > d_{N-1}
\end{cases}
$$

$$(4.1)$$

where

- ζ is the path loss in linear units,
- r is the two-dimensional distance between the typical UE and the arbitrary small cell BS in kilometres and
- the path loss function, $\zeta(r)$, is divided into N pieces, where each piece, $\zeta_n(r)$, $n \in \{1, 2, \ldots, N\}$, has different propagation characteristics, defined by
  - □ the nth LoS component, $\zeta_n^{L}(r)$,
  - □ the nth NLoS component, $\zeta_n^{NL}(r)$, and
  - □ the nth LoS probability function, $\mathrm{Pr}_n^{L}(r)$.

Before diving into the details of the nth LoS component, $\zeta_n^{L}(r)$, the nth NLoS component, $\zeta_n^{NL}(r)$, and the nth LoS probability function, $\mathrm{Pr}_n^{L}(r)$, let us note some of the properties of this path loss model.

This path loss model is general, able to accommodate a number of widely used models in the literature. For example, it can encompass any of the 3rd Generation Partnership Project (3GPP) multi-slope path loss models defined in [153], group to which the model presented Section 2.1.6 belongs. Moreover, this model can also realize most of the path loss models adopted in seminal works in academia, such as those in [205–207], by setting the number of pieces and the number of slopes accordingly.

This model is also accurate and tractable. A realistic LoS component, a realistic NLoS component and/or a realistic LoS probability function may take complicated mathematical forms, as it will be shown, as an example, for the LoS probability function in Section 4.4.4. To achieve both accurate results and analytical tractability, this model can approximate such a complicated function into a piecewise one, with each piece represented by an elementary function. In this way, the accuracy can be maintained while breaking through its complexity.

Line-of-Sight and Non-Line-of-Sight Path Loss Functions

Let us now formally present the nth LoS component, $\zeta_n^{L}(r)$ and the nth NLoS component, $\zeta_n^{NL}(r)$.

The nth LoS component, $\zeta_n^{L}(r)$, characterizes how the signal power decays as a function of the distance, r, when the transmission path is unobstructed, and can be modelled as

$$\zeta_n^L(r) = A_n^L r^{-\alpha_n^L},\tag{4.2}$$

where

- A_n^L is the path loss at a reference distance, $r = 1$ km, specified at a given carrier frequency for the LoS case and
- α_n^L is its path loss exponent.

Similarly, the nth NLoS component, $\zeta_n^{NL}(r)$, characterizes how the signal power decays as a function of the distance, r, when the transmission path is obstructed, and can be modelled as

$$\zeta_n^{NL}(r) = A_n^{NL} r^{-\alpha_n^{NL}},\tag{4.3}$$

where

- A_n^{NL} is the path loss at a reference distance, $r = 1$ km, specified at a given carrier frequency for the NLoS case and
- α_n^{NL} is its path loss exponent.

For the sake of presentation, the nth LoS and the nth NLoS components, $\zeta_n^L(r)$ and $\zeta_n^{NL}(r)$, $n \in \{1, 2, \ldots, N\}$, are stacked into piecewise functions as

$$\zeta^{Path}(r) = \begin{cases} \zeta_1^{Path}(r), & \text{when } 0 \le r \le d_1 \\ \zeta_2^{Path}(r), & \text{when } d_1 < r \le d_2 \\ \vdots & \vdots \\ \zeta_N^{Path}(r), & \text{when } r > d_{N-1} \end{cases},\tag{4.4}$$

where

- the string variable, *Path*, takes the value of "L" (for the LoS case) or "NL" (for the NLoS case).

Specific values for all these functions will be provided in Section 4.4.1, when constructing the specific case studies.

Probability of Line-of-Sight Function

With respect to the nth LoS probability function, $\mathrm{Pr}_n^L(r)$, in equation (4.1), it should be noted that it characterizes whether the transmission path is unobstructed or not as a function of the distance, r, and is typically modelled as a monotonically decreasing function with respect to such distance.

For notational convenience, the nth LoS probability function, $\mathrm{Pr}_n^L(r)$, $n \in \{1, 2, \ldots, N\}$, is also stacked into a piecewise LoS probability function as

$$\mathrm{Pr}^L(r) = \begin{cases} \mathrm{Pr}_1^L(r), & \text{when } 0 \le r \le d_1 \\ \mathrm{Pr}_2^L(r), & \text{when } d_1 < r \le d_2 \\ \vdots & \vdots \\ \mathrm{Pr}_N^L(r), & \text{when } r > d_{N-1} \end{cases}.\tag{4.5}$$

Since we will study the performance of an ultra-dense network with multiple LoS probability functions, $\Pr_n^L(r)$, in this chapter, their specific expression for each piece will be provided in the corresponding sections later.

4.2.2 User Association Strategy

Each UE is associated with the small cell BS that exhibits the smallest path loss, i.e. with the smallest $\zeta(r)$. Note that in Chapter 3, each UE was associated with the small cell BS at the closest proximity. However, such an assumption does not reflect a realistic behaviour in light of the new path loss model considering both LoS and NLoS transmissions. A UE associating with the small cell BS providing the strongest received signal strength may associate with a small cell BS that is further away but in LoS, rather than with a small cell BS that is closer but in NLoS. This behaviour is captured now, through this user association strategy.

4.3 Theoretical Performance Analysis and Main Results

The definitions of the coverage probability, p^{cov}, and the ASE, A^{ASE}, were presented in equations (2.22) and (2.25), respectively, and are summarized for convenience in Table 4.2. In this section, we analyze these two key performance indicators for the new path loss model and the new user association strategy presented in Section 4.2. Using the expressions derived in this section, we will show the important impact of both LoS and NLoS transmissions on an ultra-dense network, Moreover, we elaborate on how they change the traditional understanding and conclusions presented in Chapter 3.

4.3.1 Coverage Probability

In this subsection, we present the main result on the coverage probability, p^{cov}, through Theorem 4.3.1, considering the above introduced system model, where

- the multi-slope path loss model with both LoS and NLoS transmissions and
- the strongest received signal strength UE association strategy

Table 4.2. Key performance indicators

Metric	Formulation	Reference
Coverage probability, p^{cov}	$p^{cov}(\lambda, \gamma_0) = \Pr[\Gamma > \gamma_0] = \bar{F}_\Gamma(\gamma_0)$	Section 2.1.12, equation (2.22)
ASE, A^{ASE}	$A^{ASE}(\lambda, \gamma_0) = \frac{\lambda}{\ln 2} \int_{\gamma_0}^{+\infty} \frac{p^{cov}(\lambda, \gamma)}{1+\gamma} d\gamma$ $\qquad + \lambda \log_2(1 + \gamma_0) p^{cov}(\lambda, \gamma_0)$	Section 2.1.12, equation (2.25)

should be highlighted. Readers interested in the research article originally presenting these results are referred to [208].

THEOREM 4.3.1   *Considering the path loss model in equation (4.1), the coverage probability, p^{cov}, can be derived as*

$$p^{\text{cov}}(\lambda, \gamma_0) = \sum_{n=1}^{N} \left(T_n^{\text{L}} + T_n^{\text{NL}} \right), \tag{4.6}$$

where

-

$$T_n^{\text{L}} = \int_{d_{n-1}}^{d_n} \Pr\left[\frac{P\zeta_n^{\text{L}}(r)h}{I_{\text{agg}} + P^{\text{N}}} > \gamma_0 \right] f_{R,n}^{\text{L}}(r)\, dr, \tag{4.7}$$

-

$$T_n^{\text{NL}} = \int_{d_{n-1}}^{d_n} \Pr\left[\frac{P\zeta_n^{\text{NL}}(r)h}{I_{\text{agg}} + P^{\text{N}}} > \gamma_0 \right] f_{R,n}^{\text{NL}}(r)\, dr \tag{4.8}$$

and
- *d_0 and d_N are defined as 0 and $+\infty$, respectively.*

Moreover, the probability density function (PDF), $f_{R,n}^{\text{L}}(r)$, and the PDF, $f_{R,n}^{\text{NL}}(r)$, are given by

$$f_{R,n}^{\text{L}}(r) = \exp\left(-\int_0^{r_1} \left(1 - \Pr^{\text{L}}(u)\right) 2\pi u\lambda\, du \right) \exp\left(-\int_0^{r} \Pr^{\text{L}}(u) 2\pi u\lambda\, du \right)$$

$$\times \; \Pr_n^{\text{L}}(r) 2\pi r\lambda, \quad (d_{n-1} < r \le d_n) \tag{4.9}$$

and

$$f_{R,n}^{\text{NL}}(r) = \exp\left(-\int_0^{r_2} \Pr^{\text{L}}(u) 2\pi u\lambda\, du \right) \exp\left(-\int_0^{r} \left(1 - \Pr^{\text{L}}(u)\right) 2\pi u\lambda\, du \right)$$

$$\times \left(1 - \Pr_n^{\text{L}}(r)\right) 2\pi r\lambda, \quad (d_{n-1} < r \le d_n), \tag{4.10}$$

respectively, where

- *r_1 and r_2 are determined by*

$$r_1 = \arg_{r_1} \left\{ \zeta^{\text{NL}}(r_1) = \zeta_n^{\text{L}}(r) \right\} \tag{4.11}$$

and

$$r_2 = \arg_{r_2} \left\{ \zeta^{\text{L}}(r_2) = \zeta_n^{\text{NL}}(r) \right\}, \tag{4.12}$$

respectively.

In addition, the probability, $\Pr\left[\frac{P\zeta_n^L(r)h}{I_{\text{agg}}+P^N} > \gamma_0\right]$, *and the probability,* $\Pr\left[\frac{P\zeta_n^{\text{NL}}(r)h}{I_{\text{agg}}+P^N} > \gamma_0\right]$, *are computed as*

$$\Pr\left[\frac{P\zeta_n^L(r)h}{I_{\text{agg}} + P^N} > \gamma_0\right] = \exp\left(-\frac{\gamma_0 P^N}{P\zeta_n^L(r)}\right)\mathscr{L}_{I_{\text{agg}}}^L\left(\frac{\gamma_0}{P\zeta_n^L(r)}\right), \qquad (4.13)$$

where

- $\mathscr{L}_{I_{\text{agg}}}^L(s)$ *is the Laplace transform of the aggregated inter-cell interference random variable,* I_{agg}, *for an LoS signal transmission evaluated at the variable value,* $s = \frac{\gamma_0}{P\zeta_n^L(r)}$, *which can be expressed as*

$$\mathscr{L}_{I_{\text{agg}}}^L(s) = \exp\left(-2\pi\lambda \int_r^{+\infty} \frac{\Pr^L(u)u}{1 + \left(sP\zeta^L(u)\right)^{-1}} du\right)$$

$$\times \exp\left(-2\pi\lambda \int_{r_1}^{+\infty} \frac{\left[1 - \Pr^L(u)\right]u}{1 + \left(sP\zeta^{\text{NL}}(u)\right)^{-1}} du\right) \qquad (4.14)$$

and

$$\Pr\left[\frac{P\zeta_n^{\text{NL}}(r)h}{I_{\text{agg}} + P^N} > \gamma_0\right] = \exp\left(-\frac{\gamma_0 P^N}{P\zeta_n^{\text{NL}}(r)}\right)\mathscr{L}_{I_{\text{agg}}}^{\text{NL}}\left(\frac{\gamma_0}{P\zeta_n^{\text{NL}}(r)}\right), \qquad (4.15)$$

where

- $\mathscr{L}_{I_{\text{agg}}}^{\text{NL}}(s)$ *is the Laplace transform of the aggregated inter-cell interference random variable,* I_{agg}, *for an NLoS signal transmission evaluated at the variable value,* $s = \frac{\gamma_0}{P\zeta_n^{\text{NL}}(r)}$, *which can be written as*

$$\mathscr{L}_{I_{\text{agg}}}^{\text{NL}}(s) = \exp\left(-2\pi\lambda \int_{r_2}^{+\infty} \frac{\Pr^L(u)u}{1 + \left(sP\zeta^L(u)\right)^{-1}} du\right)$$

$$\times \exp\left(-2\pi\lambda \int_r^{+\infty} \frac{\left[1 - \Pr^L(u)\right]u}{1 + \left(sP\zeta^{\text{NL}}(u)\right)^{-1}} du\right). \qquad (4.16)$$

Proof   See Appendix A.   □

To facilitate the understanding of this theorem, Figure 4.2 illustrates a flow chart that depicts the necessary enhancements to the traditional stochastic geometry framework presented in Chapter 3 to compute the new results of Theorem 4.3.1. On top of the logic illustrated by Figure 3.2, the probability of the signal power of the typical UE being larger than the aggregated inter-cell interference power times the SINR threshold, γ_0, is now further broken into LoS and NLoS for both the signal and the inter-cell interference parts, as illustrated by the dashed blocks of Figure 4.2.

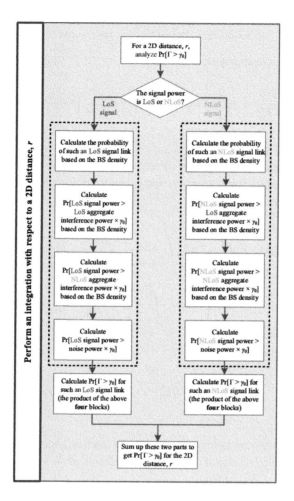

Figure 4.2 Logical steps within the standard stochastic geometry framework to obtain the results in Theorem 4.3.1, considering NLoS and LoS transmissions.

4.3.2   Area Spectral Efficiency

Similarly as in Chapter 3, to compute the theoretical results on the ASE, A^{ASE}, we have to plug the coverage probability, p^{cov} – obtained from Theorem 4.3.1 – into equation (2.24) to compute the PDF, $f_{\Gamma}(\gamma)$, of the SINR, Γ, of the typical UE. Then, we can obtain the ASE, A^{ASE}, by solving equation (2.25). See Table 4.2 for further reference on the ASE formulation.

4.3.3   Computational Complexity

To calculate the coverage probability, p^{cov}, presented in Theorem 4.3.1, three folds of integrals are generally required for the calculation of $\left\{f_{R,n}^{Path}(r)\right\}$, $\left\{\mathscr{L}_{I_{\mathrm{agg}}}\left(\frac{\gamma_0}{P\zeta_n^{Path}(r)}\right)\right\}$ and $\left\{T_n^{Path}\right\}$, respectively, where the string variable, *Path*, takes the value of "L" (for the LoS case) or "NL" (for the NLoS case). Note that an additional fold of integral is needed for the calculation of the ASE, A^{ASE}, making it a four-fold integral computation.

4.4   Discussion

In this section, we use numerical results from static system-level simulations to evaluate the accuracy of the above theoretical performance analysis and study the impact of both LoS and NLoS transmissions on an ultra-dense network in terms of the coverage probability, p^{cov}, and the ASE, A^{ASE}.

4.4.1   Case Study

As can be observed from Theorem 4.3.1,

- the nth LoS component, $\zeta_n^{\mathrm{L}}(r)$,
- the nth NLoS component, $\zeta_n^{\mathrm{NL}}(r)$, and
- the nth LoS probability function, $\mathrm{Pr}_n^{\mathrm{L}}(r)$,

all play an active role in determining the coverage probability, p^{cov}, and in turn, the ASE, A^{ASE}. These active roles are investigated in more detail in the remainder of this chapter.

As a case study of Theorem 4.3.1, and to give a more practical sense to the calculation of the coverage probability, p^{cov}, and the ASE, A^{ASE}, in this chapter, we use the realistic assumptions of the 3GPP outdoor small cell deployment presented in Section 2.1.6, which is based on [153], considering both

- the path loss function, $\zeta(r)$,

$$\zeta(r) = \begin{cases} A^{\mathrm{L}} r^{-\alpha^{\mathrm{L}}}, & \text{with probability } \mathrm{Pr}^{\mathrm{L}}(r) \\ A^{\mathrm{NL}} r^{-\alpha^{\mathrm{NL}}}, & \text{with probability } \left(1 - \mathrm{Pr}^{\mathrm{L}}(r)\right) \end{cases} \quad (4.17)$$

and
- the exponential LoS probability function, $\mathrm{Pr}^{\mathrm{L}}(r)$,

$$\mathrm{Pr}^{\mathrm{L}}(r) = 0.5 - \min\left\{0.5, 5\exp\left(-\frac{R_1}{r}\right)\right\} + \min\left\{0.5, 5\exp\left(-\frac{r}{R_2}\right)\right\}, \quad (4.18)$$

where
- ☐ R_1 and R_2 are shape parameters to ensure the continuity of this LoS probability function, $\mathrm{Pr}^{\mathrm{L}}(r)$.

Importantly, this LoS probability function, $\mathrm{Pr}^{\mathrm{L}}(r)$, can be reformulated to fit the presented general path loss model presented in equation (4.5) as

$$\mathrm{Pr}^{\mathrm{L}}(r) = \begin{cases} 1 - 5\exp(-R_1/r), & 0 < r \le d_1 \\ 5\exp(-r/R_2), & r > d_1 \end{cases}, \quad (4.19)$$

where

$\square$ $d_1 = \frac{R_1}{\ln 10}$.

Considering the general path loss model presented in equation (4.1), the particular path loss model resulting from the combination of equations (4.17) and (4.19) can be deemed as a special case of equation (4.1) with the following assignments:

- $N = 2$,
- $\zeta_1^L(r) = \zeta_2^L(r) = A^L r^{-\alpha^L}$,
- $\zeta_1^{NL}(r) = \zeta_2^{NL}(r) = A^{NL} r^{-\alpha^{NL}}$,
- $\mathrm{Pr}_1^L(r) = 1 - 5\exp(-R_1/r)$ and
- $\mathrm{Pr}_2^L(r) = 5\exp(-r/R_2)$.

For clarity, this 3GPP-based use case is referred to as the 3GPP case study in this book, which, in more detail, and according to Tables A.1–3, Table A.1–4 and Table A.1–7 of [153], uses the following parameters:

- maximum antenna gain, $G_M = 0\,\mathrm{dB}$,
- path loss exponents, $\alpha^L = 2.09$ and $\alpha^{NL} = 3.75$ – considering a carrier frequency of 2 GHz,
- reference path losses, $A^L = 10^{-10.38}$ and $A^{NL} = 10^{-14.54}$,
- transmit power, $P = 24\,\mathrm{dBm}$, and
- noise power, $P^N = -95\,\mathrm{dBm}$ – including a noise figure of 9 dB at the UE.

With this system model in mind, it should be noted that to obtain the results of the coverage probability, p^{cov}, and the ASE, A^{ASE} for this 3GPP case study, one needs to plug equations (4.17) and (4.19) into Theorem 4.3.1.

Small Cell BS Densities
With regard to the small cell BS densities of study, and similar to Chapter 3, we consider small cell BS densities up to $\lambda = 10^4\,\mathrm{BSs/km^2}$ in our studies. This allows us to embrace the minimum transmitter-to-receiver distance of the selected path loss model of 10 m, while still using a simple system model based on homogeneous Poisson point process (HPPP) deployments.

Benchmark
In this performance evaluation, we use the results of the analysis with a single-slope path loss model, presented earlier in Chapter 3, as the benchmark, with a reference path loss, $A^{NL} = 10^{-14.54}$, and a path loss exponent, $\alpha = 3.75$.

4.4.2   Coverage Probability Performance

First of all, Figure 4.3 shows the coverage probability, p^{cov}, for an SINR threshold, $\gamma_0 = 0\,\mathrm{dB}$, and the following four configurations:

- Conf. (a): single-slope path loss model (analytical results).
- Conf. (b): single-slope path loss model (simulated results).

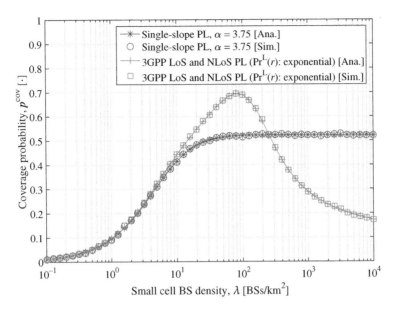

Figure 4.3 Coverage probability, p^{cov}, versus the small cell BS density, λ.

- Conf. (c): multi-slope path loss model with both LoS and NLoS transmissions (analytical results).
- Conf. (d): multi-slope path loss model with both LoS and NLoS transmissions (simulated results).

Note that Conf. a) and b) were already presented and discussed in Chapter 3, and that they are incorporated for comparison purposes in Figure 4.3.

As can be observed from Figure 4.3, the new analytical results, Conf. c), match well with the new simulation results, Conf. d). This corroborates the accuracy of Theorem 4.3.1 – and the derivations therein. Due to such significant accuracy, and since the results of the ASE, A^{ASE}, are computed based on the results of the coverage probability, p^{cov}, only analytical results are considered in our discussion hereafter in this chapter.

From Figure 4.3, we can observe that the coverage probability, p^{cov}, resulting from the single-slope path loss model – the benchmark, Conf. a) –

- first increases with the small cell BS density, λ, because more small cell BSs provide a better coverage in noise-limited networks.
- Then, when the small cell BS density, λ, is large enough, e.g. $\lambda > 10^2$ BSs/km², the coverage probability, p^{cov}, becomes independent of the small cell BS density, λ, as the network is pushed into the interference-limited regime. The intuition behind this observation is that, with the simplistic assumption of a single-slope path loss function, the increase in the inter-cell interference power due to a smaller distance between the typical UE and the interfering small cell BSs is exactly counterbalanced by the increase in the signal power due to the smaller distance

between the typical UE and its serving small cell BS in an interference-limited network. As a result, the SINR, Γ, of the typical UE – and in turn, the coverage probability, p^{cov} – remain constant, as the small cell BS density, λ, increases.

In contrast, from Figure 4.3, we can also observe that the coverage probability, p^{cov}, resulting from the multi-slope path loss model, incorporating both LoS and LoS transmissions, Conf. c), exhibits a different behaviour. In this 3GPP case study,

- when the network is sparse, e.g. $\lambda \leq 10^2$ BSs/km^2, the coverage probability, p^{cov}, grows as the small cell BS density, λ, increases, similarly as with the single-slope path loss function.
- However, when the small cell BS density, λ, is large enough, e.g. $\lambda > 10^2$ BSs/km^2, the coverage probability, p^{cov}, decreases as the small cell BS density, λ, increases, due to the transition of a large number of interfering links from NLoS to LoS in dense networks. In other words, when the small cell BS density, λ, increases, the average distance between the transmitters and the receivers decreases, and LoS transmissions occur with an increasingly higher probability than NLoS transmissions, significantly enhancing the inter-cell interference. Note that a maximum coverage probability, p^{cov}, exists, and that it is obtained for a certain small cell BS density, λ_0, in this 3GPP case study, which can be readily obtained by setting the partial derivative of the coverage probability, p^{cov}, with respect to the small cell BS density, λ, to 0, i.e. $\lambda_0 = \arg_\lambda \left\{ \frac{\partial p^{\mathrm{cov}}}{\partial \lambda} = 0 \right\}$.
  The solution to this equation can be numerically calculated using a standard bisection searching method [209]. In Figure 4.3, the small cell BS density, $\lambda_0 = 79.43$ BSs/km^2.
- When the small cell BS density, λ, is considerably large, e.g. $\lambda \geq 10^3$ BSs/km^2, the coverage probability, p^{cov}, decreases at a slower pace, asymptotically tending towards a constant value. This is because both the signal and the inter-cell interference power levels are both increasingly more LoS dominated, thus ultimately trending towards the same path loss exponent, that of the LoS component, and growing at the same pace.

When comparing the results with the two different path loss models, single-slope and multi-slope, it is also important to note from Figure 4.3 that

- the results of the analysis with the multi-slope path loss model, incorporating both LoS and NLoS transmissions, exhibit a better coverage probability, p^{cov}, with respect to the single-slope path loss analysis, when the small cell BS density is in the range, $\lambda \in [20, 200]$ BSs/km^2. This is because, in such a small cell BS density range, for the multi-slope path loss case, the inter-cell interference power is NLoS dominated, while the signal power is LoS dominated, whereas for the single-slope path loss case, both the inter-cell interference and the signal power are always modelled as NLoS. Thus, the signal power benefits from a smaller path loss exponent in the former case, which enhances the SINR, Γ, of the typical UE and in turn, the coverage probability, p^{cov}.

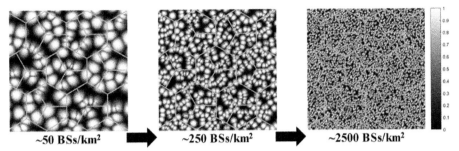

(a) Coverage probability, p^{cov}, under a single-slope path loss model, NLoS only. The sizes of the bright areas (high SINR coverage probabilities) and the dark areas (low SINR coverage probabilities) are approximately the same as the network densifies, indicating an invariant SINR performance.

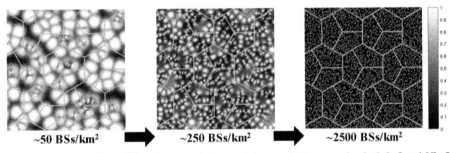

(b) Coverage probability, p^{cov}, under a multi-slope path loss model incorporating both LoS and NLoS transmissions. Compared with Figure 4.4a, the SINR heat map becomes darker as the small cell BS density, λ, increases, showing a significant performance degradation because of the transition of a large number of interfering links from NLoS to LoS in dense networks.

Figure 4.4 NetVisual plot of the coverage probability, p^{cov}, versus the small cell BS density, λ [Bright area: high probability, Dark area: low probability].

NetVisual Analysis

To visualize the fundamental behaviour of the coverage probability, p^{cov}, in a more intuitive manner, Figure 4.4 shows the coverage probability heat maps of three different scenarios with different small cell BS densities, i.e. 50, 250 and 2500 BSs/km^2, while considering the single-slope and the multi-slope path loss functions – this last one with both LoS and NLoS transmissions. These heat maps are computed using NetVisual. As explained in Section 2.4, this tool is able to capture, not only the mean, but also the standard deviation of the coverage probability, p^{cov}. It also provides striking graphs to assess performance.

From Figure 4.4, we can see that, compared with the single-slope path loss model,

- the coverage probability heat map becomes brighter for the multi-slope path loss model, when the small cell BS density, λ, is around 50 BSs/km^2. This is because LoS transmissions enhance the signal power, while most of the interfering links remain dominated by NLoS transmissions. In more detail, the average and standard deviation go from 0.54 and 0.31 to 0.71 and 0.22, respectively, i.e. a 31.48% improvement in mean.

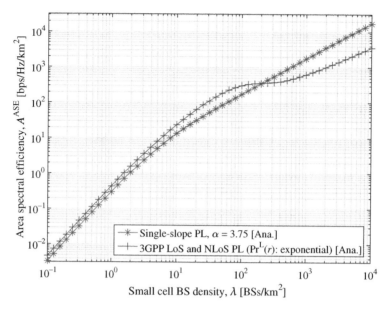

Figure 4.5 Area spectral efficiency, A^{ASE}, versus the small cell BS density, λ.

- The coverage probability heat map becomes darker for the multi-slope path loss model, when the small cell BS density, λ, is equal to or larger than 250 BSs/km^2, thus showing a performance degradation. This is due to the transition of a large number of interfering links from NLoS to LoS in dense networks. In more detail, for the case with 250 BSs/km^2, the average and standard deviation go from 0.54 and 0.31 to 0.51 and 0.23, respectively, i.e. a 5.56% degradation in mean.
- The degradation of the coverage probability heat map due to this NLoS-to-LoS inter-cell interference transition is larger with the largest small cell BS density, λ, tested in this figure. For the case with 2500 BSs/km^2, the average and standard deviation go from 0.53 and 0.32 to 0.22 and 0.26, respectively, i.e. a 58.49% degradation in mean.

Summary of Findings: Coverage Probability Remark

Remark 4.1   Considering a multi-slope path loss model with LoS and NLoS transmissions, the SINR, Γ, of the typical UE suffers in dense networks due to the transition of a large number of interfering links from NLoS to LoS. As a result, the coverage probability, p^{cov}, does not remain constant as in Chapter 3, but decreases with the increase of the small cell BS density, λ, in the dense and the ultra-dense regimes.

4.4.3   Area Spectral Efficiency Performance

Let us now explore the behaviour of the ASE, A^{ASE}.

Figure 4.5 shows the ASE, A^{ASE}, for an SINR threshold, $\gamma_0 = 0\,\mathrm{dB}$, for the following two configurations:

- Conf. a): single-slope path loss model (analytical results).
- Conf. b): multi-slope path loss model with both LoS and NLoS transmissions (analytical results).

From Figure 4.5, we can observe that the ASE, A^{ASE}, resulting from the single-slope path loss function – the benchmark, Conf. a) –

- increases linearly with the small cell density, λ, once the network enters the interference-limited regime, as explained in Section 3.4.3. This follows from the coverage probability results, which showed that the coverage probability, p^{cov}, does not vary with respect to the small cell BS density, λ, in such a small cell BS density regime.

In contrast, from Figure 4.5, we can also observe that the ASE, A^{ASE}, resulting from the multi-slope path loss function, incorporating both LoS and NLoS transmissions, Conf. b), reveals a more complicated trend. In this 3GPP case study,

- when the network is sparse, e.g. $\lambda \leq \lambda_0$ BSs/km^2, the ASE, A^{ASE}, quickly increases with the small cell BS density, λ, because the network is generally noise-limited, and thus adding an increasing number of small cell BSs significantly benefits both the signal quality and the spatial reuse. Importantly, note that the slope of the increase of the ASE, A^{ASE}, in this log–log plot is larger than 1 in this small cell BS density regime, indicating, according to equation (2.25), that the increase of the small cell BS density, λ, does not only help to increase the ASE, A^{ASE}, through the spatial reuse – the direct multiplier in such an equation – but also through a better coverage probability, p^{cov}, as was shown in our earlier discussions.
- However, when the small cell BS density, λ, is large enough, e.g. $\lambda > 10^2$ BSs/km^2, the ASE, A^{ASE}, shows a slowdown in the rate of growth or even a decrease as the small cell BS density, λ, increases. This is driven by the decrease of the coverage probability, p^{cov}, at these small cell BS densities shown in Figure 4.3, caused by the transition of a large number of interfering links from NLoS to LoS in dense networks.
- When the small cell BS density, λ, is tremendously large, e.g. $\lambda \geq 10^3$ BSs/km^2, the ASE, A^{ASE}, grows again, and exhibits a nearly linear increase with respect to the increase of small cell BS density, λ. This is shown by the behaviour of the slope of the increase of the ASE, A^{ASE}, in the log–log plot, which trends towards 1.[1] This follows from the decrease of the degradation rate of the coverage probability, p^{cov}, in this ultra-dense regime, with the coverage probability, p^{cov}, asymptotically tending towards a constant value. This is because both the signal and the inter-cell interference power levels are both increasingly more LoS dominated, thus ultimately depending on the same path loss exponent, that of the LoS component, and as a result, growing at the same pace.

---

[1] A unitary slope in a log–log plot indicates a linear relationship in the equivalent linear plot.

The ASE Crawl

Overall, the fact that the ASE, A^{ASE}, crawls as the small cell BS density, λ, increases in a practical range of small cell BS densities shows that the small cell BS density matters, and that selecting the right densification level is a problem that should not be taken lightly. Moreover, these results indicate the importance of an accurate path loss modelling, incorporating both LoS and NLoS transmissions, as their consideration makes a difference on the final results, not only quantitatively, but also qualitatively.

With this in mind, let us make the definition.

DEFINITION 4.4.1   The ASE Crawl.

A much shorter distance between a UE and its serving small cell BS in an ultra-dense network implies high probabilities of LoS transmissions. Broadly speaking, these LoS transmissions are helpful to improve the signal power, but they also aggravate the inter-cell interference power, coming from the more and closer neighbouring small cell BSs in a densifying network. The ASE Crawl is defined as the undesired slow growth – or even decrease – of the ASE, A^{ASE}, due to the transition of a large number of interfering links from NLoS to LoS in dense networks.

It is worth noting that, even if the conclusion on the ASE Crawl has been obtained from a particular path loss model and a particular set of parameters, a number of studies using other path loss models and other sets of parameters corroborate the generality of these conclusions. We are going to show examples of this in the following two sections. Importantly, the impact of such an NLoS-to-LoS inter-cell interference transition and the conclusion behind this theoretical result is also confirmed by the real-world experiments in [80], in which a densification factor of $100\times$ (from 9 to 1 107 BSs/km^2) led to a network capacity increase of $40\times$ (from 16 to 1 107 bps/Hz/km^2), clearly, not a linear one.

Summary of Findings: Area Spectral Efficiency Remark

Remark 4.2   Considering a multi-slope path loss model with LoS and NLoS transmissions, the ASE, A^{ASE}, does not increase linearly with the increase of the small cell BS density, λ, as in Chapter 3, but shows a slowdown in the rate of growth or even a decrease as the small cell BS density, λ, increases. Due to the increasing inter-cell interference, the contribution of each small cell BS to the ASE, A^{ASE}, cannot be kept constant. Once most interfering small cell BSs are in LoS, the ASE, A^{ASE}, of the ultra-dense network starts picking up its linear growth, but with a smaller slope.

4.4.4   Impact of the Probability of Line-of-Sight – and Modelling – on the Main Results

For completeness, it should be noted that other LoS probability functions different than the exponential one presented in equation (4.18) exist, and thus one may wonder whether a different modelling of the NLoS-to-LoS inter-cell interference transition may change the above conclusions.

To answer this question, we analyze the coverage probability, p^{cov}, and the ASE, A^{ASE}, with a different LoS probability function, $\mathrm{Pr}^{\mathrm{L}}(r)$, in this section. In more detail, we use the LoS probability function, $\mathrm{Pr}^{\mathrm{L}}(r)$, defined in section 5.5.3 of [157], also for outdoor small cell deployments, which takes the mathematical form given by

$$\mathrm{Pr}^{\mathrm{L}}(r) = \begin{cases} 1 - \frac{r}{d_1}, & 0 < r \le d_1 \\ 0, & r > d_1 \end{cases}, \tag{4.20}$$

where

- d_1 is a parameter that determines the steepness of this LoS probability function, $\mathrm{Pr}^{\mathrm{L}}(r)$. In [157], $d_1 = 0.3\,\mathrm{km}$ is used for urban cases.

It is important to note that this LoS probability function, $\mathrm{Pr}^{\mathrm{L}}(r)$, does not have an exponential form, but a linear one, thus allowing for further derivations and more compact results.

Considering the general path loss model presented in equation (4.1), the particular path loss model resulting from the combination of equations (4.17) and (4.20) can be deemed as another special case of equation (4.1) with the following assignments:

- $N = 2$,
- $\zeta_1^{\mathrm{L}}(r) = \zeta_2^{\mathrm{L}}(r) = A^{\mathrm{L}} r^{-\alpha^{\mathrm{L}}}$,
- $\zeta_1^{\mathrm{NL}}(r) = \zeta_2^{\mathrm{NL}}(r) = A^{\mathrm{NL}} r^{-\alpha^{\mathrm{NL}}}$,
- $\mathrm{Pr}_1^{\mathrm{L}}(r) = 1 - \frac{r}{d_1}$ and
- $\mathrm{Pr}_2^{\mathrm{L}}(r) = 0$.

For clarity, this 3GPP-based use case is referred to as the alternative 3GPP case study in this chapter.

In the following, we derive the coverage probability, p^{cov}, and the ASE, A^{ASE}, for this alternative 3GPP case study, taking advantage of the new LoS probability function, $\mathrm{Pr}^{\mathrm{L}}(r)$, presented in equation (4.20), and its better tractability.

Coverage Probability

Building from Theorem 4.3.1, the coverage probability, p^{cov}, for this alternative 3GPP case study can be calculated as

$$p^{\mathrm{cov}}(\lambda, \gamma_0) = \sum_{n=1}^{2} \left(T_n^{\mathrm{L}} + T_n^{\mathrm{NL}} \right) = T_1^{\mathrm{L}} + T_1^{\mathrm{NL}} + T_2^{\mathrm{L}} + T_2^{\mathrm{NL}}. \tag{4.21}$$

In the following, we describe how to compute the different components, T_1^{L}, T_1^{NL}, T_2^{L} and T_2^{NL}, using numerically tractable integral-form expressions.

The Computation of Component, T_1^{L}
From Theorem 4.3.1, the component, T_1^{L}, can be derived as

$$T_1^L = \int_0^{d_1} \exp\left(-\frac{\gamma_0 P^N}{P\zeta_1^L(r)}\right) \mathscr{L}_{I_{agg}}\left(\frac{\gamma_0}{P\zeta_1^L(r)}\right) f_{R,1}^L(r) dr$$

$$\stackrel{(a)}{=} \int_0^{d_1} \exp\left(-\frac{\gamma_0 r^{\alpha^L} P^N}{P A^L}\right) \mathscr{L}_{I_{agg}}\left(\frac{\gamma_0 r^{\alpha^L}}{P A^L}\right) f_{R,1}^L(r) dr, \tag{4.22}$$

where

- $\zeta_1^L(r) = A^L r^{-\alpha^L}$ from equation (4.17) is substituted into step (a) of equation (4.22) and

- $\mathscr{L}_{I_{agg}}(s)$ is the Laplace transform of the aggregated inter-cell interference random variable, I_{agg}, evaluated at the variable value, $s = \frac{\gamma_0 r^{\alpha^L}}{P A^L}$.

In equation (4.22), according to Theorem 4.3.1 and equation (4.20), the PDF, $f_{R,1}^L(r)$, can be obtained as

$$f_{R,1}^L(r) = \exp\left(-\int_0^{r_1} \lambda \frac{u}{d_1} 2\pi u\, du\right) \exp\left(-\int_0^r \lambda\left(1 - \frac{u}{d_1}\right) 2\pi u\, du\right)\left(1 - \frac{r}{d_1}\right) 2\pi r\lambda$$

$$= \exp\left(-\pi\lambda r^2 + 2\pi\lambda\left(\frac{r^3}{3d_1} - \frac{r_1^3}{3d_1}\right)\right)\left(1 - \frac{r}{d_1}\right) 2\pi r\lambda, \quad (0 < r \le d_1), \tag{4.23}$$

where

- $r_1 = \left(\frac{A^{NL}}{A^L}\right)^{\frac{1}{\alpha^{NL}}} r^{\frac{\alpha^L}{\alpha^{NL}}}$, according to equation (4.11).

Besides, to compute the Laplace transform, $\mathscr{L}_{I_{agg}}\left(\frac{\gamma_0 r^{\alpha^L}}{P A^L}\right)$, in equation (4.22) for the range, $0 < r \le d_1$, we provide Lemma (4.4.2).

LEMMA 4.4.2   *The Laplace transform, $\mathscr{L}_{I_{agg}}\left(\frac{\gamma_0 r^{\alpha^L}}{P A^L}\right)$ in the range, $0 < r \le d_1$, can be calculated by*

$$\mathscr{L}_{I_{agg}}\left(\frac{\gamma_0 r^{\alpha^L}}{P A^L}\right) =$$

$$\exp\left(-2\pi\lambda\left(\rho_1\left(\alpha^L, 1, \left(\gamma_0 r^{\alpha^L}\right)^{-1}, d_1\right) - \rho_1\left(\alpha^L, 1, \left(\gamma_0 r^{\alpha^L}\right)^{-1}, r\right)\right)\right)$$

$$\times \exp\left(\frac{2\pi\lambda}{d_0}\left(\rho_1\left(\alpha^L, 2, \left(\gamma_0 r^{\alpha^L}\right)^{-1}, d_1\right) - \rho_1\left(\alpha^L, 2, \left(\gamma_0 r^{\alpha^L}\right)^{-1}, r\right)\right)\right)$$

$$\times \exp\left(-\frac{2\pi\lambda}{d_0}\left(\rho_1\left(\alpha^{NL}, 2, \left(\frac{\gamma_0 A^{NL}}{A^L} r^{\alpha^L}\right)^{-1}, d_1\right) - \rho_1\left(\alpha^{NL}, 2, \left(\frac{\gamma_0 A^{NL}}{A^L} r^{\alpha^L}\right)^{-1}, r_1\right)\right)\right)$$

$$\times \exp\left(-2\pi\lambda\rho_2\left(\alpha^{NL}, 1, \left(\frac{\gamma_0 A^{NL}}{A^L} r^{\alpha^L}\right)^{-1}, d_1\right)\right), \quad (0 < r \le d_1) \tag{4.24}$$

where

$$\rho_1(\alpha, \beta, t, d) = \left[\frac{d^{(\beta+1)}}{\beta+1}\right] {}_2F_1 \left[1, \frac{\beta+1}{\alpha}; 1 + \frac{\beta+1}{\alpha}; -td^\alpha\right] \qquad (4.25)$$

and

$$\rho_2(\alpha, \beta, t, d) = \left[\frac{d^{-(\alpha-\beta-1)}}{t(\alpha-\beta-1)}\right] {}_2F_1 \left[1, 1 - \frac{\beta+1}{\alpha}; 2 - \frac{\beta+1}{\alpha}; -\frac{1}{td^\alpha}\right], \quad (\alpha > \beta + 1), \qquad (4.26)$$

where

- $_2F_1[\cdot, \cdot; \cdot; \cdot]$ *is the hypergeometric function [156].*

Proof   See Appendix B.   □

To sum up, the component, T_1^{L}, can be evaluated as

$$T_1^{\mathrm{L}} = \int_0^{d_1} \exp\left(-\frac{\gamma_0 r^{\alpha^{\mathrm{L}}} P^{\mathrm{N}}}{P A^{\mathrm{L}}}\right) \mathscr{L}_{I_{\mathrm{agg}}} \left(\frac{\gamma_0 r^{\alpha^{\mathrm{L}}}}{P A^{\mathrm{L}}}\right) f_{R,1}^{\mathrm{L}}(r) dr, \qquad (4.27)$$

where

- the PDF, $f_{R,1}^{\mathrm{L}}(r)$, is given by equation (4.23) and
- the Laplace transform, $\mathscr{L}_{I_{\mathrm{agg}}} \left(\frac{\gamma_0 r^{\alpha^{\mathrm{L}}}}{P A^{\mathrm{L}}}\right)$, is given by Lemma (4.4.2).

The Computation of Component, T_1^{NL}
From Theorem 4.3.1, the component, T_1^{NL}, can be derived as

$$T_1^{\mathrm{NL}} = \int_0^{d_1} \exp\left(-\frac{\gamma_0 P^{\mathrm{N}}}{P \zeta_1^{\mathrm{NL}}(r)}\right) \mathscr{L}_{I_{\mathrm{agg}}} \left(\frac{\gamma_0}{P \zeta_1^{\mathrm{NL}}(r)}\right) f_{R,1}^{\mathrm{NL}}(r) dr$$

$$\overset{(a)}{=} \int_0^{d_1} \exp\left(-\frac{\gamma_0 r^{\alpha^{\mathrm{NL}}} P^{\mathrm{N}}}{P A^{\mathrm{NL}}}\right) \mathscr{L}_{I_{\mathrm{agg}}} \left(\frac{\gamma_0 r^{\alpha^{\mathrm{NL}}}}{P A^{\mathrm{NL}}}\right) f_{R,1}^{\mathrm{NL}}(r) dr, \qquad (4.28)$$

where

- $\zeta_1^{\mathrm{NL}}(r) = A^{\mathrm{NL}} r^{-\alpha^{\mathrm{NL}}}$ from equation (4.17) is substituted into step (a) of equation (4.28) and
- $\mathscr{L}_{I_{\mathrm{agg}}}(s)$ is the Laplace transform of the random variable, I_{agg}, evaluated at the variable value, $s = \frac{\gamma_0 r^{\alpha^{\mathrm{NL}}}}{P A^{\mathrm{NL}}}$.

In equation (4.28), according to Theorem 4.3.1 and equation (4.20), the PDF, $f_{R,1}^{\mathrm{NL}}(r)$, can be obtained as

$$f_{R,1}^{NL}(r) = \exp\left(-\int_0^{r_2} \lambda \Pr^L(u) 2\pi u \, du\right)$$

$$\times \exp\left(-\int_0^{r} \lambda\left(1 - \Pr^L(u)\right) 2\pi u \, du\right)\left(\frac{r}{d_1}\right) 2\pi r\lambda, \quad (0 < r \le d_1), \quad (4.29)$$

where

- $r_2 = \left(\frac{A^L}{A^{NL}}\right)^{\frac{1}{\alpha^L}} r^{\frac{\alpha^{NL}}{\alpha^L}}$, according to equation (4.12).

Since the numerical relationship between the distance, r_2, and the distance, d_1, affects the calculation of the first multiplier in equation (4.29), i.e. $\exp\left(-\int_0^{r_2} \lambda \Pr^L(u)\right.$ $\left. 2\pi u \, du\right)$, we separately discuss the range, $0 < r_2 \le d_1$, and the range, $r_2 > d_1$, in the following.

If $0 < r_2 \le d_1$, i.e. $0 < r \le y_1 = d_1^{\frac{\alpha^L}{\alpha^{NL}}}\left(\frac{A^{NL}}{A^L}\right)^{\frac{1}{\alpha^{NL}}}$, the PDF, $f_{R,1}^{NL}(r)$, can be derived as

$$f_{R,1}^{NL}(r) = \exp\left(-\int_0^{r_2} \lambda\left(1 - \frac{u}{d_1}\right) 2\pi u \, du\right) \exp\left(-\int_0^{r} \lambda \frac{u}{d_1} 2\pi u \, du\right)\left(\frac{r}{d_1}\right) 2\pi r\lambda$$

$$= \exp\left(-\pi\lambda r_2^2 + 2\pi\lambda\left(\frac{r_2^3}{3d_1} - \frac{r^3}{3d_1}\right)\right)\left(\frac{r}{d_1}\right) 2\pi r\lambda, \quad (0 < r \le y_1). \quad (4.30)$$

Otherwise, if $r_2 > d_1$, i.e. $y_1 < r \le d_1$, such a PDF can be obtained as

$$f_{R,1}^{NL}(r) = \exp\left(-\int_0^{d_1} \lambda\left(1 - \frac{u}{d_1}\right) 2\pi u \, du\right) \exp\left(-\int_0^{r} \lambda \frac{u}{d_1} 2\pi u \, du\right)\left(\frac{r}{d_1}\right) 2\pi r\lambda$$

$$= \exp\left(-\frac{\pi\lambda d_1^2}{3} - \frac{2\pi\lambda r^3}{3d_1}\right)\left(\frac{r}{d_1}\right) 2\pi r\lambda, \quad (y_1 < r \le d_1). \quad (4.31)$$

Besides, to compute the Laplace transform, $\mathscr{L}_{I_{agg}}\left(\frac{\gamma_0 r^{\alpha^{NL}}}{P_A^{NL}}\right)$, in equation (4.28) for the range, $0 < r \le d_1$, we provide Lemma (4.4.3). Note that since the calculation of the PDF, $f_{R,1}^{NL}(r)$, is divided into two cases, shown in equations (4.30) and (4.31), respectively, the calculation of the Laplace transform, $\mathscr{L}_{I_{agg}}\left(\frac{\gamma_0 r^{\alpha^{NL}}}{P_A^{NL}}\right)$, in the range, $0 < r \le d_1$, should also be divided into those two cases, as the inter-cell interference is integrated from the distance, r, to infinity.

LEMMA 4.4.3 *The Laplace transform, $\mathscr{L}_{I_{agg}}\left(\frac{\gamma_0 r^{\alpha^{NL}}}{P_A^{NL}}\right)$, in the range, $0 < r \le d_1$, can be divided into two cases, i.e. that of range, $0 < r \le y_1$ and that of range, $y_1 < r \le d_1$. The results are as follows:*

$$\mathscr{L}_{I_{\text{agg}}}\left(\frac{\gamma_0 r^{\alpha^{\text{NL}}}}{P_A{}^{\text{NL}}}\right) =$$

$$\exp\left(-2\pi\lambda\left(\rho_1\left(\alpha^{\text{L}},1,\left(\frac{\gamma_0 A^{\text{L}}}{A^{\text{NL}}}r^{\alpha^{\text{NL}}}\right)^{-1},d_1\right)-\rho_1\left(\alpha^{\text{L}},1,\left(\frac{\gamma_0 A^{\text{L}}}{A^{\text{NL}}}r^{\alpha^{\text{NL}}}\right)^{-1},r_2\right)\right)\right)$$

$$\times\exp\left(\frac{2\pi\lambda}{d_0}\left(\rho_1\left(\alpha^{\text{L}},2,\left(\frac{\gamma_0 A^{\text{L}}}{A^{\text{NL}}}r^{\alpha^{\text{NL}}}\right)^{-1},d_1\right)-\rho_1\left(\alpha^{\text{L}},2,\left(\frac{\gamma_0 A^{\text{L}}}{A^{\text{NL}}}r^{\alpha^{\text{NL}}}\right)^{-1},r_2\right)\right)\right)$$

$$\times\exp\left(-\frac{2\pi\lambda}{d_0}\left(\rho_1\left(\alpha^{\text{NL}},2,\left(\gamma_0 r^{\alpha^{\text{NL}}}\right)^{-1},d_1\right)-\rho_1\left(\alpha^{\text{NL}},2,\left(\gamma_0 r^{\alpha^{\text{NL}}}\right)^{-1},r\right)\right)\right)$$

$$\times\exp\left(-2\pi\lambda\rho_2\left(\alpha^{\text{NL}},1,\left(\gamma_0 r^{\alpha^{\text{NL}}}\right)^{-1},d_1\right)\right), \qquad (0<r\le y_1) \tag{4.32}$$

and

$$\mathscr{L}_{I_{\text{agg}}}\left(\frac{\gamma_0 r^{\alpha^{\text{NL}}}}{P_A{}^{\text{NL}}}\right)$$

$$= \exp\left(-\frac{2\pi\lambda}{d_0}\left(\rho_1\left(\alpha^{\text{NL}},2,\left(\gamma_0 r^{\alpha^{\text{NL}}}\right)^{-1},d_1\right)-\rho_1\left(\alpha^{\text{NL}},2,\left(\gamma_0 r^{\alpha^{\text{NL}}}\right)^{-1},r\right)\right)\right)$$

$$\times\exp\left(-2\pi\lambda\rho_2\left(\alpha^{\text{NL}},1,\left(\gamma_0 r^{\alpha^{\text{NL}}}\right)^{-1},d_1\right)\right), \qquad (y_1<r\le d_1),x \tag{4.33}$$

where

- $\rho_1\left(\alpha,\beta,t,d\right)$ *and* $\rho_2\left(\alpha,\beta,t,d\right)$ *are defined in equations (4.25) and (4.26), respectively.*

Proof See Appendix C. □

To sum up, the component, T_1^{NL}, can be evaluated as

$$T_1^{\text{NL}} = \int_0^{y_1}\exp\left(-\frac{\gamma_0 r^{\alpha^{\text{NL}}}P^{\text{N}}}{P_A{}^{\text{NL}}}\right)\left[\mathscr{L}_{I_{\text{agg}}}\left(\frac{\gamma_0 r^{\alpha^{\text{NL}}}}{P_A{}^{\text{NL}}}\right)f_{R,1}^{\text{NL}}(r)\Big|0<r\le y_1\right]dr$$

$$+ \int_{y_1}^{d_1}\exp\left(-\frac{\gamma_0 r^{\alpha^{\text{NL}}}P^{\text{N}}}{P_A{}^{\text{NL}}}\right)\left[\mathscr{L}_{I_{\text{agg}}}\left(\frac{\gamma_0 r^{\alpha^{\text{NL}}}}{P_A{}^{\text{NL}}}\right)f_{R,1}^{\text{NL}}(r)\Big|y_1<r\le d_1\right]dr, \tag{4.34}$$

where

- the PDF, $f_{R,1}^{\text{NL}}(r)$, is derived using equations (4.30) and (4.31) and
- the Laplace transform, $\mathscr{L}_{I_{\text{agg}}}\left(\frac{\gamma_0 r^{\alpha^{\text{NL}}}}{P_A{}^{\text{NL}}}\right)$, is given by Lemma (4.4.3).

The Computation of Component, T_2^{L}
From Theorem 4.3.1, the component, T_2^{L}, can be derived as

$$T_2^{\text{L}} = \int_{d_1}^{\infty}\exp\left(-\frac{\gamma_0 P^{\text{N}}}{P\zeta_2^{\text{L}}(r)}\right)\mathscr{L}_{I_{\text{agg}}}\left(\frac{\gamma_0}{P\zeta_2^{\text{L}}(r)}\right)f_{R,2}^{\text{L}}(r)dr. \tag{4.35}$$

According to Theorem 4.3.1 and equation (4.20), the PDF, $f_{R,1}^{NL}(r)$, can be obtained by

$$f_{R,2}^{L}(r) = \exp\left(-\int_0^{r_1} \lambda\left(1 - \Pr^{L}(u)\right) 2\pi u \, du\right) \exp\left(-\int_0^{r} \lambda\Pr^{L}(u) 2\pi u \, du\right) \times 0 \times 2\pi r\lambda$$

$$= 0, \quad (r > d_1).$$ (4.36)

Plugging equation (4.36) into equation (4.35), we can find that

$$T_2^{L} = 0.$$ (4.37)

The Computation of Component, T_2^{NL}

From Theorem 4.3.1, the component, T_2^{NL}, can be derived as

$$T_2^{NL} = \int_{d_1}^{\infty} \exp\left(-\frac{\gamma_0 P^{N}}{P\zeta_2^{NL}(r)}\right) \mathscr{L}_{I_{agg}}\left(\frac{\gamma_0}{P\zeta_2^{NL}(r)}\right) f_{R,2}^{NL}(r) dr$$

$$\overset{(a)}{=} \int_{d_1}^{\infty} \exp\left(-\frac{\gamma_0 r^{\alpha^{NL}} P^{N}}{PA^{NL}}\right) \mathscr{L}_{I_{agg}}\left(\frac{\gamma_0 r^{\alpha^{NL}}}{PA^{NL}}\right) f_{R,2}^{NL}(r) dr,$$ (4.38)

where

- $\zeta_2^{NL}(r) = A^{NL} r^{-\alpha^{NL}}$ from equation (4.17) is substituted into step (a) of equation (4.38) and
- $\mathscr{L}_{I_{agg}}(s)$ is the Laplace transform of the aggregated inter-cell interference random variable, I_{agg}, evaluated at the variable value, $s = \frac{\gamma_0 r^{\alpha^{NL}}}{PA^{NL}}$.

In equation (4.38), according to Theorem 4.3.1 and equation (4.20), the PDF, $f_{R,2}^{NL}(r)$, can be obtained as

$$f_{R,2}^{NL}(r) = \exp\left(-\int_0^{d_1} \lambda\left(1 - \frac{u}{d_1}\right) 2\pi u \, du\right) \exp\left(-\int_0^{d_1} \lambda\frac{u}{d_1} 2\pi u \, du - \int_{d_1}^{r} \lambda 2\pi u \, du\right) 2\pi r\lambda$$

$$= \exp\left(-\pi\lambda r^2\right) 2\pi r\lambda, \quad (r > d_1).$$ (4.39)

Besides, to compute the Laplace transform, $\mathscr{L}_{I_{agg}}\left(\frac{\gamma_0 r^{\alpha^{NL}}}{PA^{NL}}\right)$, in equation (4.38) for the range, $r > d_1$, we provide Lemma (4.4.4).

LEMMA 4.4.4   *The Laplace transform,* $\mathscr{L}_{I_{\mathrm{agg}}} \left(\frac{\gamma_0 r^{\alpha^{\mathrm{NL}}}}{PA^{\mathrm{NL}}} \right)$, *in the range,* $r > d_1$, *can be calculated by*

$$\mathscr{L}_{I_{\mathrm{agg}}} \left(\frac{\gamma_0 r^{\alpha^{\mathrm{NL}}}}{PA^{\mathrm{NL}}} \right) = \exp \left(-2\pi \lambda \rho_2 \left(\alpha^{\mathrm{NL}}, 1, \left(\gamma_0 r^{\alpha^{\mathrm{NL}}} \right)^{-1}, r \right) \right), \quad (r > d_1),$$

(4.40)

where

- $\rho_2 \left(\alpha, \beta, t, d \right)$ *is defined in equation (4.26).*

Proof   See Appendix D.   ☐

To sum up, the component, T_2^{NL}, can be evaluated as

$$T_2^{\mathrm{NL}} = \int_{d_1}^{\infty} \exp \left(-\frac{\gamma_0 r^{\alpha^{\mathrm{NL}}} P^{\mathrm{N}}}{PA^{\mathrm{NL}}} \right) \mathscr{L}_{I_{\mathrm{agg}}} \left(\frac{\gamma_0 r^{\alpha^{\mathrm{NL}}}}{PA^{\mathrm{NL}}} \right) f_{R,2}^{\mathrm{NL}}(r) dr, \quad (4.41)$$

where

- the PDF, $f_{R,2}^{\mathrm{NL}}(r)$, is given by equation (4.39) and
- the Laplace transform, $\mathscr{L}_{I_{\mathrm{agg}}} \left(\frac{\gamma_0 r^{\alpha^{\mathrm{NL}}}}{PA^{\mathrm{NL}}} \right)$, is given by Lemma (4.4.4).

Area Spectral Efficiency

Similarly as before, plugging the coverage probability, p^{cov} – obtained from equation (4.21) – into equation (2.24) to compute the PDF, $f_{\Gamma}(\gamma)$, of the SINR, Γ, of the typical UE, we can obtain the ASE, A^{ASE}, solving equation (2.25).

Regarding the computational process to calculate the coverage probability, p^{cov}, in this specific case, note that only one fold of integral is required for the calculation of $\{T_n^{\mathrm{L}}\}$ and $\{T_n^{\mathrm{NL}}\}$ in equation (4.21), with respect to the three folds of integrals required for the general case in Theorem 4.3.1. Note that an additional fold of integrals is also needed for the calculation of the ASE, A^{ASE}, making it a two-fold integral computation. Thus, the results for this alternative 3GPP case study are more tractable than those of the 3GPP case presented in Section 4.4 because of the simpler LoS probability function, $\mathrm{Pr}^{\mathrm{L}}(r)$, in equation (4.20).

Discussion

As can be seen from Figures 4.6 and 4.7, all the observations with respect to the coverage probability, p^{cov}, and the ASE, A^{ASE}, obtained with the exponential LoS probability function, $\mathrm{Pr}^{\mathrm{L}}(r)$, in Section 4.4 are qualitatively valid for the linear LoS probability function, $\mathrm{Pr}^{\mathrm{L}}(r)$, in this section. The same trends are present.

The coverage probability, p^{cov}, first increases and then decreases, and only a quantitative deviation exists. Specifically, in Figure 4.6, the small cell BS density, λ_0, for which the coverage probability, p^{cov}, is maximized around $20\,\mathrm{BSs/km}^2$, instead of around $80\,\mathrm{BSs/km}^2$. Since there is more LoS probability in this alternative LoS

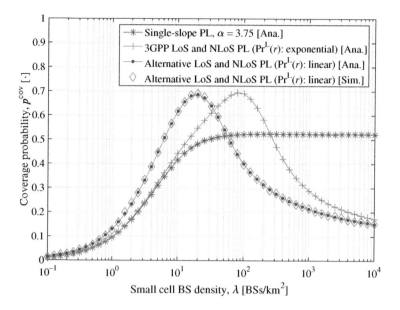

Figure 4.6 Coverage probability, p^{cov}, versus the small cell BS density, λ, for an alternative case study with a linear LoS probability function.

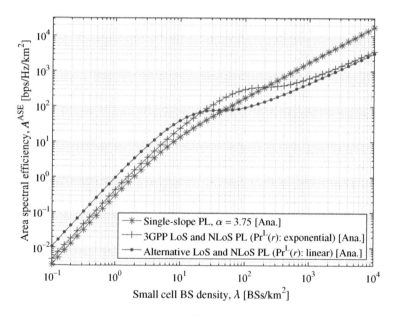

Figure 4.7 Area spectral efficiency, A^{ASE}, versus the small cell BS density, λ, for an alternative case study with a linear LoS probability function.

probability function, i.e. the transition of a signal from NLoS to LoS occurs at larger distances, the transition of a large number of interfering links from NLoS to LoS occurs at sparser small cell BS densities. This is why the small cell BS density, λ_0, for which the coverage probability, p^{cov}, is maximized is smaller now than before. Note that the maximum of the coverage probability, p^{cov}, is the same in both cases, around 0.7.

The same logic applies to the performance of the ASE, A^{ASE}, with the ASE Crawl appearing earlier.

Overall, changing the LoS probability function, $\mathrm{Pr}^{\mathrm{L}}(r)$, may cause quantitative differences on the coverage probability, p^{cov}, and the ASE, A^{ASE}, but the observed network performance trend remains the same. This corroborates the generality of the results shown in this chapter with respect to the impact on performance of the transition of a large number of interfering links from NLoS to LoS in dense networks.

In the remainder of this book, we stick to the original exponential LoS probability function, $\mathrm{Pr}^{\mathrm{L}}(r)$, presented in equation (4.19).

4.4.5   Impact of the Multi-Path Fading – and Modelling – on the Main Results

For completeness, and similarly to Section 4.4.4, it should also be noted that other multi-path fast fading channel models different to the Rayleigh one previously used exist, and are actually better suited for strong LoS transmissions. In fact, the assumption of a Rayleigh multi-path fast fading is a simplification for LoS transmissions, quite handy for analytical tractability, but inaccurate. Rayleigh fading is most applicable when there is no LoS propagation between the transmitter and the receiver, and the signal is expected to arrive at the receiver from a multitude of directions of space (see Section 2.1.8)

This opens the door to questions with regard to the accuracy of the results of this chapter, where one may wonder whether

- a more accurate multi-path fading channel model may change the obtained conclusions so far and whether
- such a more accurate multi-path fading channel model can mitigate or exacerbate the SINR degradation caused by the NLoS-to-LoS inter-cell interference transition.

Due to the close proximity between the UEs and the small cell BSs in an ultra-dense network, to answer such a question in this section, we consider the more accurate Rician multi-path fast fading model presented in Section 2.1.8 for the LoS component, which is able to capture the ratio between the power in the direct path and the power in the other scattered paths as a function of the distance, r, between a UE and a small cell BS.

For the sake of mathematical complexity, we only present system-level simulation results with this alternative Rician-based multi-path fast fading model hereafter.

Figure 4.8 illustrates the results on the coverage probability, p^{cov}, while Figure 4.9 shows those on the ASE, A^{ASE}. Readers interested in the mathematical modelling and derivations of the coverage probability, p^{cov}, and the ASE, A^{ASE}, for this case with Rician multi-path fast fading are referred to [96]. Moreover, and as a reminder, note

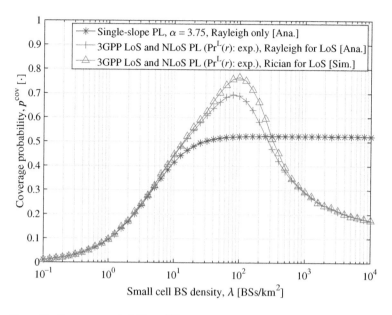

Figure 4.8 Coverage probability, p^{cov}, versus the small cell BS density, λ, for an alternative case study with multi-path Rician fading.

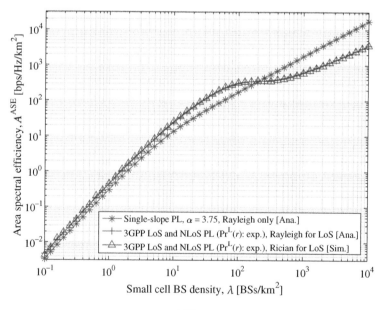

Figure 4.9 Area spectral efficiency, A^{ASE}, versus the small cell BS density, λ, for an alternative case study with multi-path Rician fading.

that the assumptions and the parameters used to obtain the results in these two figures are the same as those used to get the previous results shown in this chapter, except for the distance-dependent Rician multi-path fast fading considered here.

As can be seen from Figures 4.6 and 4.7, all the observations with respect to the coverage probability, p^{cov}, and the ASE, A^{ASE}, obtained under the Rayleigh-based multi-path fast fading in Section 4.4 are qualitatively valid under the Rician-based multi-path fast fading model. The same trends are present.

The coverage probability, p^{cov}, first increases and then decreases with the small cell BS density, λ, and only a quantitative deviation exists. Specifically,

- in Figure 4.8, the maximum coverage probability, p^{cov}, is slightly larger under the Rician-based multi-path fast fading model, 0.77, than under the Rayleigh-based one, 0.7.
- Moreover, the small cell BS density, λ_0, for which such maximum coverage probability, p^{cov}, is obtained, is also slightly larger under the Rician-based multi-path fast fading model, $100\,\mathrm{BSs/km^2}$, than under the Rayleigh-based one, $80\,\mathrm{BSs/km^2}$.

    This is because, as shown in Figure 2.4, the dynamic range of the multi-path fast fading gain, h, is smaller under the Rician multi-path fast fading model than that under the Rayleigh one – the former has larger minimum and smaller maximum values. Importantly, however, the mean of the Rician multi-path fast fading gain, h, increases with the decrease of the coverage radius of the small cell BS, and in turn, with the increase of the small cell BS density, λ. As a result, this increase of the mean signal and inter-cell interference power levels as the small cell BS density, λ, increases leads to

    □ an enhancement of the SINR, Γ, of the UEs when the signal power transits from NLoS to LoS in the small cell BS range, $\lambda \in [10, 100]$ BSs/km^2, and
    □ a degradation of the SINR, Γ, of the UEs when the inter-cell interference power transits from NLoS to LoS in the small cell BS range, $\lambda \in [100, 1000]$ BSs/ km^2.

    This phenomenon – the increase of the mean signal power due to Rician multi-path fast fading gain, h, in the LoS component – enhanced the peak of the coverage probability, p^{cov}, and delayed its decrease in terms of the small cell BS density, λ_0, when comparing the Rician-based multi-path fast fading model with the Rayleigh-based one.

- Once the deployment is ultra-dense, and both the signal and the inter-cell interference power levels are LoS dominated, thus using the Rician-based multi-path fast fading model in both the signal and the inter-cell interference power calculations (or the Rayleigh-based one in the case of the benchmark), the advantage of the Rician over the Rayleigh multi-path fast fading disappears, as the independent and identical distributed (i.i.d.) multi-path fast fading channel gains, h, statistically cancel each other out in the calculation of the SINR, Γ, of the UEs when averaging across the entire network.

As for the ASE, A^{ASE}, note that the behaviours under the Rician- and the Rayleigh-based multi-path fast fading models are almost identical. This is because the slight change in the coverage probability, p^{cov}, discussed earlier, results in an even slighter change in the ASE, A^{ASE}.

Overall, and despite the quantitative differences, these results on the coverage probability, p^{cov}, and the ASE, A^{ASE}, confirm our statement, i.e. the main conclusions in this chapter are general, and do not qualitatively change due to the assumption on the multi-path fast fading model.[2]

In the remainder of this book, since the modelling of the multi-path fading did not have a profound impact in terms of qualitative results, we stick to Rayleigh multi-path fast fading modelling for mathematical tractability, unless otherwise stated.

4.5   Conclusions

In this chapter, we have brought to attention the importance of the channel model in network performance analysis. More precisely, we have described the necessary upgrades that one has to perform over the system model presented in Chapter 3 of this book, in terms of LoS and NLoS transmissions, multi-path fast fading and user association strategy, to realize a more practical and close-to-reality study. We have also provided details on the new derivations conducted in the theoretical performance analysis, and presented the resulting expressions for the coverage probability, p^{cov}, and the ASE, A^{ASE}. Moreover, we have shared numerical results for small cell BS deployments with different small cell BS densities and characteristics. To finalize, we have also discussed the important conclusions drawn from this work, which are significantly different from those in Chapter 3.

Importantly, such conclusions, which are summarized in Remarks 4.1 and 4.2, have shown a new performance behaviour in the ultra-dense regime, referred to as the ASE Crawl hereafter in this book. This new behaviour shows how a network operator or a service provider should carefully consider the densification level of its network and/or implement inter-cell interference coordination techniques to avoid inter-cell interference issues. Otherwise, they may invest exponentially more money to densify their network to get a diminishing network performance gain. This sends an important message:

Both the channel characteristics and the small cell BS density matter!

In this chapter, we have also shown that this behaviour – the ASE Crawl – is quantitatively – but not qualitatively – affected by (i) the small cell deployment layout and/or (ii) the multi-path fast fading model. This indicated the generality of the findings.

Appendix A Proof of Theorem 4.3.1

For the sake of clarity, let us first describe the main idea behind the proof of Theorem 4.3.1, and then proceed with a more detailed explanation.

---

[2] Note that this statement applies for the studied case with single antenna UEs and small cell BSs, but may not apply to multi-antenna cases, where the fading correlation among the antennas of an array plays a key role.

In line with the guidelines provided in Section 2.3.2, to evaluate the coverage probability, p^{cov}, we need proper expressions for

- the PDF, $f_R(r)$, of the random variable, R, characterizing the distance between the typical UE and its serving small cell BS for the event that the typical UE is associated to the strongest small cell BS through either an LoS or an NLoS path and
- the conditional probability, $\Pr[\Gamma > \gamma_0 \mid r]$, where
  - $\square$ $r = R(\omega)$ is a realization of the distance random variable, R, for both the LoS and the NLoS transmission.

Once such expressions are known, we can finally derive, p^{cov}, by performing the corresponding appropriate integrals, some of which will be shown in the following.

Before proceeding with the more detailed calculations, however, it is important to note that, from equations (2.22) and (3.3), we can derive the coverage probability, p^{cov}, as

$$
p^{\text{cov}}(\lambda, \gamma_0) \overset{(a)}{=} \int_{r>0} \Pr\left[\text{SINR} > \gamma_0 \mid r\right] f_R(r) dr
$$

$$
= \int_{r>0} \Pr\left[\frac{P\zeta(r)h}{I_{\text{agg}} + P^{\text{N}}} > \gamma_0\right] f_R(r) dr
$$

$$
= \int_{0}^{d_1} \Pr\left[\frac{P\zeta_1^{\text{L}}(r)h}{I_{\text{agg}} + P^{\text{N}}} > \gamma_0\right] f_{R,1}^{\text{L}}(r) dr + \int_{0}^{d_1} \Pr\left[\frac{P\zeta_1^{\text{NL}}(r)h}{I_{\text{agg}} + P^{\text{N}}} > \gamma_0\right] f_{R,1}^{\text{NL}}(r) dr
$$

$$
+ \cdots
$$

$$
+ \int_{d_{N-1}}^{\infty} \Pr\left[\frac{P\zeta_N^{\text{L}}(r)h}{I_{\text{agg}} + P^{\text{N}}} > \gamma_0\right] f_{R,N}^{\text{L}}(r) dr + \int_{d_{N-1}}^{\infty} \Pr\left[\frac{P\zeta_N^{\text{NL}}(r)h}{I_{\text{agg}} + P^{\text{N}}} > \gamma_0\right] f_{R,N}^{\text{NL}}(r) dr
$$

$$
\overset{\Delta}{=} \sum_{n=1}^{N} \left(T_n^{\text{L}} + T_n^{\text{NL}}\right),
\tag{4.42}
$$

where

- R_n^{L} and R_n^{NL} are the piecewise distributions of the distances over which the typical UE is associated to a small cell BS through an LoS or an NLoS path, respectively:
  - $\square$ Note that these two events, i.e. the typical UE being associated with a small cell BS through an LoS or an NLoS path, are disjoint, and thus the coverage probability, p^{cov}, is the sum of the corresponding probabilities of these two events.
- $f_{R,n}^{\text{L}}(r)$ and $f_{R,n}^{\text{NL}}(r)$ are the piecewise PDFs of the distance random variables, R_n^{L} and R_n^{NL}, respectively:
  - $\square$ For clarity, both piecewise PDFs, $f_{R,n}^{\text{L}}(r)$ and $f_{R,n}^{\text{NL}}(r)$, are stacked into the PDF, $f_R(r)$, in step (a) of equation (4.42), where the stacked PDF, $f_R(r)$, takes a similar form as in equation (4.1), and is defined in equation (4.45).
  - $\square$ Moreover, since the two events, i.e. the typical UE being associated with a small cell BS through an LoS or an NLoS path, are disjoint, as mentioned

before, we can rely on the following equality,

$$\sum_{n=1}^{N}\int_{d_{n-1}}^{d_n} f_{R,n}(r)dr = \sum_{n=1}^{N}\int_{d_{n-1}}^{d_n} f_{R,n}^{L}(r)dr + \sum_{n=1}^{N}\int_{d_{n-1}}^{d_n} f_{R,n}^{NL}(r)dr = 1.$$

- T_n^{L} and T_n^{NL} are two piecewise functions, defined as

$$T_n^{L} = \int_{d_{n-1}}^{d_n} \Pr\left[\frac{P\zeta_n^{L}(r)h}{I_{agg} + P^{N}} > \gamma_0\right] f_{R,n}^{L}(r)dr \qquad (4.43)$$

and

$$T_n^{NL} = \int_{d_{n-1}}^{d_n} \Pr\left[\frac{P\zeta_n^{NL}(r)h}{I_{agg} + P^{N}} > \gamma_0\right] f_{R,n}^{NL}(r)dr, \qquad (4.44)$$

respectively, and
- d_0 and d_N are equal to 0 and $+\infty$, respectively.

$$f_R(r) = \begin{cases} f_{R,1}(r) = \begin{cases} f_{R,1}^{L}(r), & \text{UE associated to an LoS BS} \\ f_{R,1}^{NL}(r), & \text{UE associated to an NLoS BS} \end{cases}, & 0 \le r \le d_1 \\ \\ f_{R,2}(r) = \begin{cases} f_{R,2}^{L}(r), & \text{UE associated to an LoS BS} \\ f_{R,2}^{NL}(r), & \text{UE associated to an NLoS BS} \end{cases}, & d_1 < r \le d_2 \\ \vdots & \vdots \\ f_{R,N}(r) = \begin{cases} f_{R,N}^{L}(r), & \text{UE associated to an LoS BS} \\ f_{R,N}^{NL}(r), & \text{UE associated to an NLoS BS} \end{cases}, & r > d_{N-1} \end{cases} \qquad (4.45)$$

Following this method, let us now dive into the more detailed derivations.

LoS-Related Calculations

Let us first look into the LoS transmissions, and show how to first calculate the PDF, $f_{R,n}^{L}(r)$, and then the probability, $\Pr\left[\frac{P\zeta_n^{L}(r)h}{I_{agg}+P^{N}} > \gamma_0\right]$, in equation (4.42).

To that end, we define first the two following events:

- Event B^{L}: The nearest small cell BS to the typical UE with an LoS path is located at a distance, $x = X^{L}(\omega)$, defined by the distance random variable, X^{L}. According to the results presented in Section 2.3.2, the PDF, $f_R(r)$, of the random variable, R, in Event B^{L} can be calculated by

$$f_X^{L}(x) = \exp\left(-\int_0^x \Pr^{L}(u)2\pi u\lambda \, du\right) \Pr^{L}(x)2\pi x\lambda. \qquad (4.46)$$

This is because, according to [52], the complementary cumulative distribution function (CCDF), $\bar{F}_X^{L}(x)$, of the random variable, X^{L}, in Event B^{L} is given by

$$\bar{F}_X^{L}(x) = \exp\left(-\int_0^x \Pr^{L}(u)2\pi u\lambda \, du\right) \qquad (4.47)$$

and taking the derivative of the cumulative distribution function (CDF), $\left(1 - \bar{F}_X^{\mathrm{L}}(x)\right)$, of the random variable, X^{L}, with respect to the distance, x, we can get the PDF, $f_X^{\mathrm{L}}(x)$, of the random variable, X^{L}, shown in equation (4.46). It is important to note that the derived equation (4.46) is more complex than equation (2.42) in Section 2.3.2. This is because the LoS small cell BS deployment is inhomogeneous due to the fact that the closer small cell BSs to the typical UE are more likely to establish LoS links than the farther away ones. Compared with equation (2.42), we can see two important changes in equation (4.46).

☐ The probability of a disc of a radius, r, containing exactly 0 points, which took the form, $\exp\left(-\pi\lambda r^2\right)$, as presented in equation (2.38) of Section 2.3.2, has changed now, and is replaced by the expression,

$$\exp\left(-\int_0^x \mathrm{Pr}^{\mathrm{L}}(u)2\pi u\lambda\, du\right),$$

in equation (4.46). This is because the equivalent intensity of the LoS small cell BSs in an inhomogeneous Poisson point process (PPP) is distance-dependent, i.e. $\mathrm{Pr}^{\mathrm{L}}(u)\lambda$. This adds a new integral to equation (4.46) with respect to the distance, u.

☐ The intensity, λ, of the homogeneous PPP in equation (2.42) has been replaced by the equivalent intensity, $\mathrm{Pr}^{\mathrm{L}}(x)\lambda$, of the inhomogeneous PPP in equation (4.46).

- Event C^{NL} conditioned on the value, $x = X^{\mathrm{L}}(\omega)$, of the random variable, X^{L}: The typical UE is associated to the nearest small cell BS located at a distance, $x = X^{\mathrm{L}}(\omega)$, through an LoS path. If the typical UE is associated to the nearest LoS small cell BS located at a distance, $x = X^{\mathrm{L}}(\omega)$, this LoS small cell BS must be at the smallest path loss, $\zeta(r)$. As a result, there must be no NLoS small cell BS inside the disc

☐ centred on the typical UE,
☐ with a radius, x_1,

where such a radius, x_1, satisfies the following condition, $x_1 = \arg_{x_1}\left\{\zeta^{\mathrm{NL}}(x_1) = \zeta^{\mathrm{L}}(x)\right\}$. Otherwise, this NLoS small cell BS would outperform the LoS small cell BS at the distance, $x = X^{\mathrm{L}}(\omega)$.

According to [52], the conditional probability, $\mathrm{Pr}\left[C^{\mathrm{NL}}\,|\,X^{\mathrm{L}} = x\right]$, of the Event, C^{NL}, conditioned on the realization, $x = X^{\mathrm{L}}(\omega)$, of the random variable, X^{L}, is given by

$$\mathrm{Pr}\left[C^{\mathrm{NL}}\,|\,X^{\mathrm{L}} = x\right] = \exp\left(-\int_0^{x_1}(1 - \mathrm{Pr}^{\mathrm{L}}(u))2\pi u\lambda\, du\right). \qquad (4.48)$$

As a summary, note that Event B^{L} ensures that the path loss, $\zeta^{\mathrm{L}}(x)$, associated with *an arbitrary LoS small cell BS* is always larger than that associated with *the considered LoS small cell BS* at the distance, $x = X^{\mathrm{L}}(\omega)$. Besides, conditioned on such a distance, $x = X^{\mathrm{L}}(\omega)$, Event C^{NL} guarantees that the path loss, $\zeta^{\mathrm{NL}}(x)$, associated with *an arbitrary NLoS small cell BS* is also always larger than that associated with *the considered LoS small cell BS at the distance, $x = X^L(\omega)$. With this, we can guarantee that the typical UE is associated with the strongest LoS small cell BS.*

Let us thus now consider the resulting new event, in which the typical UE is associated with an LoS small cell BS, and such a small cell BS is located at the distance, $r = R^L(\omega)$. The CCDF, $\bar{F}_R^L(r)$, of such a random variable, R^L, can be derived as

$$
\begin{aligned}
\bar{F}_R^L(r) &= \Pr\left[R^L > r\right] \\
&\stackrel{(a)}{=} \mathbb{E}_{[X^L]}\left\{\Pr\left[R^L > r \mid X^L\right]\right\} \\
&= \int_0^{+\infty} \Pr\left[R^L > r \mid X^L = x\right] f_X^L(x)\, dx \\
&\stackrel{(b)}{=} \int_0^r 0 \times f_X^L(x)\, dx + \int_r^{+\infty} \Pr\left[C^{NL} \mid X^L = x\right] f_X^L(x)\, dx \\
&= \int_r^{+\infty} \Pr\left[C^{NL} \mid X^L = x\right] f_X^L(x)\, dx,
\end{aligned}
\tag{4.49}
$$

where

- $\mathbb{E}_{[X]}\{\cdot\}$ in step (a) of equation (4.49) is the expectation operation over the random variable, X, and
- step (b) of equation (4.49) is valid because
  - □ $\Pr\left[R^L > r \mid X^L = x\right] = 0$ for $0 < x \leq r$ and
  - □ the conditional event, $\left[R^L > r \mid X^L = x\right]$, is equivalent to the conditional event, $\left[C^{NL} \mid X^L = x\right]$, in the range, $x > r$.

Now, given the CCDF, $\bar{F}_R^L(r)$, one can find its PDF, $f_R^L(r)$, by taking the derivative, $\frac{\partial\left(1 - \bar{F}_R^L(r)\right)}{\partial r}$, with respect to the distance, r, i.e.

$$
f_R^L(r) = \Pr\left[C^{NL} \mid X^L = r\right] f_X^L(r).
\tag{4.50}
$$

Considering the distance range, $(d_{n-1} < r \leq d_n)$, we can find the PDF of such a segment, $f_{R,n}^L(r)$, from such a PDF, $f_R^L(r)$, as

$$
f_{R,n}^L(r) = \exp\left(-\int_0^{r_1}\left(1 - \Pr^L(u)\right) 2\pi u \lambda\, du\right)
$$

$$
\times \exp\left(-\int_0^r \Pr^L(u) 2\pi u \lambda\, du\right) \Pr_n^L(r) 2\pi r \lambda, \quad (d_{n-1} < r \leq d_n), \quad (4.51)
$$

where

- $r_1 = \underset{r_1}{\arg}\left\{\zeta^{NL}(r_1) = \zeta_n^L(r)\right\}.$

Having obtained the PDF of a segment, $f_{R,n}^{L}(r)$, we can evaluate the probability, $\Pr\left[\frac{P\zeta_n^{L}(r)h}{I_{agg}+P^{N}} > \gamma_0\right]$, in equation (4.42) as

$$\Pr\left[\frac{P\zeta_n^{L}(r)h}{I_{agg}+P^{N}} > \gamma_0\right] = \mathbb{E}_{[I_{agg}]}\left\{\Pr\left[h > \frac{\gamma_0\left(I_{agg}+P^{N}\right)}{P\zeta_n^{L}(r)}\right]\right\}$$
$$= \mathbb{E}_{[I_{agg}]}\left\{\bar{F}_H\left(\frac{\gamma_0\left(I_{agg}+P^{N}\right)}{P\zeta_n^{L}(r)}\right)\right\}, \qquad (4.52)$$

where

- $\bar{F}_H(h)$ is the CCDF of the multi-path fast fading channel gain, h, which is assumed to be drawn from a Rayleigh fading distribution (see Section 2.1.8).

Since the CCDF, $\bar{F}_H(h)$, of the multi-path fast fading channel gain, h, follows an exponential distribution with unitary mean given by the expression,

$$\bar{F}_H(h) = \exp(-h),$$

equation (4.52) can be further derived as

$$\Pr\left[\frac{P\zeta_n^{L}(r)h}{I_{agg}+P^{N}} > \gamma_0\right] = \mathbb{E}_{[I_{agg}]}\left\{\exp\left(-\frac{\gamma_0\left(I_{agg}+P^{N}\right)}{P\zeta_n^{L}(r)}\right)\right\}$$
$$\overset{(a)}{=} \exp\left(-\frac{\gamma_0 P^{N}}{P\zeta_n^{L}(r)}\right)\mathbb{E}_{[I_{agg}]}\left\{\exp\left(-\frac{\gamma_0}{P\zeta_n^{L}(r)}I_{agg}\right)\right\}$$
$$= \exp\left(-\frac{\gamma_0 P^{N}}{P\zeta_n^{L}(r)}\right)\mathscr{L}_{I_{agg}}^{L}\left(\frac{\gamma_0}{P\zeta_n^{L}(r)}\right), \qquad (4.53)$$

where

- $\mathscr{L}_{I_{agg}}^{L}(s)$ is the Laplace transform of the aggregated inter-cell interference random variable, I_{agg}, conditioned on the LoS signal transmission, evaluated at the variable value, $s = \frac{\gamma_0}{P\zeta_n^{L}(r)}$.

For the sake of clarity, it should be noted that the use of the Laplace transform is to ease the mathematical representation. By definition, $\mathscr{L}_{I_{agg}}^{L}(s)$ is the Laplace transform of the PDF of the aggregated inter-cell interference random variable, I_{agg}, evaluated at the variable value, s, and the equality,

$$\mathscr{L}_X(s) = \mathbb{E}_{[X]}\left\{\exp\left(-sX\right)\right\},$$

follows from the definitions and derivations leading to equation (2.46) in Chapter 2.

Based on the condition of LoS signal transmission, the Laplace transform, $\mathcal{L}^{\mathrm{L}}_{I_{\mathrm{agg}}}(s)$, can be derived as

$$
\begin{aligned}
\mathcal{L}^{\mathrm{L}}_{I_{\mathrm{agg}}}(s) &= \mathbb{E}_{[I_{\mathrm{agg}}]}\left\{\exp\left(-sI_{\mathrm{agg}}\right)\right\} \\
&= \mathbb{E}_{[\Phi,\{\beta_i\},\{g_i\}]}\left\{\exp\left(-s\sum_{i\in\Phi/b_o}P\beta_i g_i\right)\right\} \\
&\overset{(a)}{=} \exp\left(-2\pi\lambda\int\left(1 - \mathbb{E}_{[g]}\left\{\exp\left(-sP\beta(u)g\right)\right\}\right)u\,du\right),
\end{aligned}
\tag{4.54}
$$

where

- Φ is the set of small cell BSs,
- b_o is the small cell BS serving the typical UE,
- b_i is the ith interfering small cell BS,
- β_i and g_i are the path loss and the multi-path fast fading gain between the typical UE and the ith interfering small cell BS and
- step (a) of equation (4.54) has been explained in detail in equation (3.18), and further derived in equation (3.19) of Chapter 3.

Compared with equation (3.19), which considers the inter-cell interference from a single-slop path loss model, the expression, $\mathbb{E}_{[g]}\left\{\exp\left(-sP\beta(u)g\right)\right\}$, in equation (4.54) should consider the inter-cell interference from both the LoS and the NLoS paths. As a result, the Laplace transform, $\mathcal{L}^{\mathrm{L}}_{I_{\mathrm{agg}}}(s)$, can be further developed as

$$
\begin{aligned}
\mathcal{L}^{\mathrm{L}}_{I_{\mathrm{agg}}}(s) &= \exp\left(-2\pi\lambda\int\left(1 - \mathbb{E}_{[g]}\left\{\exp\left(-sP\beta(u)g\right)\right\}\right)u\,du\right) \\
&\overset{(a)}{=} \exp\left(-2\pi\lambda\int\left[\mathrm{Pr}^{\mathrm{L}}(u)\left(1 - \mathbb{E}_{[g]}\left\{\exp\left(-sP\zeta^{\mathrm{L}}(u)g\right)\right\}\right)\right.\right. \\
&\qquad\left.\left. + \left(1 - \mathrm{Pr}^{\mathrm{L}}(u)\right)\left(1 - \mathbb{E}_{[g]}\left\{\exp\left(-sP\zeta^{\mathrm{NL}}(u)g\right)\right\}\right)\right]u\,du\right) \\
&\overset{(b)}{=} \exp\left(-2\pi\lambda\int_{r}^{+\infty}\mathrm{Pr}^{\mathrm{L}}(u)\left(1 - \mathbb{E}_{[g]}\left\{\exp\left(-sP\zeta^{\mathrm{L}}(u)g\right)\right\}\right)u\,du\right) \\
&\qquad\times\exp\left(-2\pi\lambda\int_{r_1}^{+\infty}\left(1 - \mathrm{Pr}^{\mathrm{L}}(u)\right)\left(1 - \mathbb{E}_{[g]}\left\{\exp\left(-sP\zeta^{\mathrm{NL}}(u)g\right)\right\}\right)u\,du\right) \\
&\overset{(c)}{=} \exp\left(-2\pi\lambda\int_{r}^{+\infty}\frac{\mathrm{Pr}^{\mathrm{L}}(u)u}{1 + \left(sP\zeta^{\mathrm{L}}(u)\right)^{-1}}\,du\right) \\
&\qquad\times\exp\left(-2\pi\lambda\int_{r_1}^{+\infty}\frac{\left[1 - \mathrm{Pr}^{\mathrm{L}}(u)\right]u}{1 + \left(sP\zeta^{\mathrm{NL}}(u)\right)^{-1}}\,du\right),
\end{aligned}
\tag{4.55}
$$

where

- in step (a), the integration is probabilistically divided into two parts considering the LoS and the NLoS inter-cell interference,
- in step (b), the LoS and the NLoS inter-cell interference come from a distance larger than distances, r and r_1, respectively, and
- in step (c), the PDF, $f_G(g) = 1 - \exp(-g)$, of the Rayleigh random variable, g, has been used to calculate the expectations, $\mathbb{E}_{[g]}\left\{\exp\left(-sP\zeta^L(u)g\right)\right\}$ and $\mathbb{E}_{[g]}\left\{\exp\left(-sP\zeta^{NL}(u)g\right)\right\}$.

To give some intuition for equation (4.53), it should be noted that

- the exponential factor, $\exp\left(-\frac{\gamma_0 P^N}{P\zeta_n^L(r)}\right)$, measures the probability that the signal power exceeds the noise power by at least a factor, γ_0, while
- the Laplace transform, $\mathscr{L}_{I_{agg}}\left(\frac{\gamma_0}{P\zeta_n^L(r)}\right)$, measures the probability that the signal power exceeds the aggregated inter-cell interference power by at least a factor, γ_0.

As a result, and because the multi-path fast fading channel gain, h, follows an exponential distribution, the product of the above probabilities, shown in step (a) of equation (4.53), yields the probability that the signal power exceeds the sum power of the noise and the aggregated inter-cell interference by at least a factor, γ_0.

NLoS-Related Calculations

Let us now look into the NLoS transmissions, and show how to first calculate the PDF, $f_{R,n}^{NL}(r)$, and then the probability, $\Pr\left[\frac{P\zeta_n^{NL}(r)h}{I_{agg}+P^N} > \gamma_0\right]$, in equation (4.42).
To that end, we define the two following events:

- Event B^{NL}: The nearest small cell BS to the typical UE with an NLoS path is located at a distance, $x = X^{NL}(\omega)$, defined by the distance random variable, X^{NL}. Similarly to equation (4.46), one can find the PDF, $f_X^{NL}(x)$, as

$$f_X^{NL}(x) = \exp\left(-\int_0^x \left(1 - \Pr^L(u)\right) 2\pi u\lambda \, du\right)\left(1 - \Pr^L(x)\right) 2\pi x\lambda. \qquad (4.56)$$

It is important to note that the derived equation (4.56) is more complex than equation (2.42) of Section 2.3.2. This is because the NLoS small cell BS deployment is an inhomogeneous one due to the fact that the further away small cell BSs to the typical UE are more likely to establish NLoS links than the closer ones. Compared with equation (2.42), we can see two important changes in equation (4.56):

☐ The probability of a disc of a radius, r, containing exactly 0 points, which took the form, $\exp\left(-\pi\lambda r^2\right)$, as presented in equation (2.38) of Section 2.3.2, has changed now, and is replaced by the expression,

$$\exp\left(-\int_0^x \left(1 - \text{Pr}^{\text{L}}(u)\right) 2\pi u\lambda \, du\right),$$

in equation (4.56). This is because the equivalent intensity of the NLoS small cell BSs in an inhomogeneous PPP is distance-dependent, i.e. $\left(1 - \text{Pr}^{\text{L}}(u)\right)\lambda$. This adds a new integral to equation (4.46) with respect to the distance, u.
  □ The intensity, λ, of the homogeneous PPP in equation (2.42), has been replaced by the equivalent intensity, $\left(1 - \text{Pr}^{\text{L}}(u)\right)\lambda$, of the inhomogeneous PPP in equation (4.56).
• Event C^{L} conditioned on the value, $x = X^{\text{NL}}(\omega)$, of the random variable, X^{NL}: The typical UE is associated to the nearest small cell BS located at a distance, $x = X^{\text{NL}}(\omega)$, through an NLoS path. If the typical UE is associated to the nearest NLoS small cell BS located at a distance, $x = X^{\text{NL}}(\omega)$, this NLoS small cell BS must be at the smallest path loss, $\zeta(r)$. As a result, there must be no LoS small cell BS inside the disc
  □ centred on the typical UE,
  □ with a radius, x_2,
where such a radius, x_2, satisfies the following condition,
$x_2 = \underset{x_2}{\arg}\left\{\zeta^{\text{L}}(x_2) = \zeta^{\text{NL}}(x)\right\}$. Otherwise, this LoS small cell BS would outperform the NLoS small cell BS at the distance, $x = X^{\text{NL}}(\omega)$.
  Similarly to equation (4.48), the conditional probability, $\text{Pr}\left[C^{\text{L}} \mid X^{\text{NL}} = x\right]$, of Event C^{L} conditioned on the realization, $x = X^{\text{NL}}(\omega)$, of the random variable, X^{NL}, is given by

$$\text{Pr}\left[C^{\text{L}} \mid X^{\text{NL}} = x\right] = \exp\left(-\int_0^{x_2} \text{Pr}^{\text{L}}(u)2\pi u\lambda \, du\right). \tag{4.57}$$

Let us thus now consider the resulting new event, in which the typical UE is associated with an NLoS small cell BS, and such a small cell BS is located at the distance, $r = R^{\text{NL}}(\omega)$. The CCDF, $\bar{F}_R^{\text{NL}}(r)$, of such a random variable, R^{NL}, can be derived as

$$\bar{F}_R^{\text{NL}}(r) = \text{Pr}\left[R^{\text{NL}} > r\right]$$
$$= \int_r^{+\infty} \text{Pr}\left[C^{\text{L}} \mid X^{\text{NL}} = x\right] f_X^{\text{NL}}(x) \, dx. \tag{4.58}$$

Now, given the CCDF, $\bar{F}_R^{\text{NL}}(r)$, one can find its PDF, $f_R^{\text{NL}}(r)$, by taking the derivative, $\frac{\partial\left(1 - \bar{F}_R^{\text{NL}}(r)\right)}{\partial r}$, with respect to the distance, r, i.e.

$$f_R^{\text{NL}}(r) = \text{Pr}\left[C^{\text{L}} \mid X^{\text{NL}} = r\right] f_X^{\text{NL}}(r). \tag{4.59}$$

Considering the distance range, $(d_{n-1} < r \leq d_n)$, we can find the PDF of such a segment, $f_{R,n}^{\text{NL}}(r)$, from such a PDF, $f_R^{\text{NL}}(r)$, as

$$f_{R,n}^{NL}(r) = \exp\left(-\int_0^{r_2} \Pr^L(u) 2\pi u\lambda \, du\right)$$

$$\times \exp\left(-\int_0^{r} \left(1 - \Pr^L(u)\right) 2\pi u\lambda \, du\right) \left(1 - \Pr_n^L(r)\right) 2\pi r\lambda, \quad (d_{n-1} < r \le d_n),$$

$$(4.60)$$

where

- $r_2 = \underset{r_2}{\arg} \left\{\zeta^L(r_2) = \zeta_n^{NL}(r)\right\}.$

Having obtained the PDF of a segment, $f_{R,n}^{NL}(r)$, we can evaluate the probability, $\Pr\left[\frac{P\zeta_n^{NL}(r)h}{I_{agg} + P^N} > \gamma_0\right]$, in equation (4.42) as

$$\Pr\left[\frac{P\zeta_n^{NL}(r)h}{I_{agg} + P^N} > \gamma_0\right] = \mathbb{E}_{[I_{agg}]}\left\{\Pr\left[h > \frac{\gamma_0\left(I_{agg} + P^N\right)}{P\zeta_n^{NL}(r)}\right]\right\}$$

$$= \mathbb{E}_{[I_{agg}]}\left\{\bar{F}_H\left(\frac{\gamma_0\left(I_{agg} + P^N\right)}{P\zeta_n^{NL}(r)}\right)\right\}. \qquad (4.61)$$

Since the CCDF, $\bar{F}_H(h)$, of the multi-path fast fading channel gain, h, follows an exponential distribution with unitary mean given by the expression,

$$\bar{F}_H(h) = \exp(-h),$$

equation (4.61) can be further derived as

$$\Pr\left[\frac{P\zeta_n^{NL}(r)h}{I_{agg} + P^N} > \gamma_0\right] = \mathbb{E}_{[I_{agg}]}\left\{\exp\left(-\frac{\gamma_0\left(I_{agg} + P^N\right)}{P\zeta_n^{NL}(r)}\right)\right\}$$

$$= \exp\left(-\frac{\gamma_0 P^N}{P\zeta_n^{NL}(r)}\right)\mathbb{E}_{[I_{agg}]}\left\{\exp\left(-\frac{\gamma_0}{P\zeta_n^{NL}(r)}I_{agg}\right)\right\}$$

$$= \exp\left(-\frac{\gamma_0 P^N}{P\zeta_n^{NL}(r)}\right)\mathscr{L}_{I_{agg}}^{NL}\left(\frac{\gamma_0}{P\zeta_n^{NL}(r)}\right). \qquad (4.62)$$

Based on the condition of NLoS signal transmission, the Laplace transform, $\mathscr{L}_{I_{agg}}^{NL}(s)$, can be derived as

$$\mathscr{L}_{I_{agg}}^{NL}(s) = \mathbb{E}_{[I_{agg}]}\left\{\exp\left(-sI_{agg}\right)\right\}$$

$$= \mathbb{E}_{[\Phi, \{\beta_i\}, \{g_i\}]}\left\{\exp\left(-s\sum_{i\in\Phi/b_o} P\beta_i g_i\right)\right\}$$

$$\overset{(a)}{=} \exp\left(-2\pi\lambda\int\left(1 - \mathbb{E}_{[g]}\left\{\exp\left(-sP\beta(u)g\right)\right\}\right)u \, du\right), \qquad (4.63)$$

where

- step (a) of equation (4.63) has been explained in detail in equation (3.18), and further derived in equation (3.19) of Chapter 3.

Compared with equation (3.19), which considers the inter-cell interference from a single-slop path loss model, the expression, $\mathbb{E}_{[g]}\left\{\exp\left(-sP\beta(u)g\right)\right\}$, in equation (4.63) should consider the inter-cell interference from both the LoS and the NLoS paths. As a result, similar to equation (4.64), the Laplace transform, $\mathscr{L}_{I_{agg}}^{NL}(s)$, can be further developed as

$$
\mathscr{L}_{I_{agg}}^{NL}(s) = \exp\left(-2\pi\lambda\int\left(1 - \mathbb{E}_{[g]}\left\{\exp\left(-sP\beta(u)g\right)\right\}\right)u\,du\right)
$$

$$
\stackrel{(a)}{=} \exp\left(-2\pi\lambda\int_{r_2}^{+\infty}\Pr{}^L(u)\left(1 - \mathbb{E}_{[g]}\left\{\exp\left(-sP\zeta^L(u)g\right)\right\}\right)u\,du\right)
$$

$$
\times\exp\left(-2\pi\lambda\int_{r}^{+\infty}\left(1 - \Pr{}^L(u)\right)\left(1 - \mathbb{E}_{[g]}\left\{\exp\left(-sP\zeta^{NL}(u)g\right)\right\}\right)u\,du\right)
$$

$$
\stackrel{(b)}{=} \exp\left(-2\pi\lambda\int_{r_2}^{+\infty}\frac{\Pr{}^L(u)u}{1 + \left(sP\zeta^L(u)\right)^{-1}}\,du\right)
$$

$$
\times\exp\left(-2\pi\lambda\int_{r}^{+\infty}\frac{\left[1 - \Pr{}^L(u)\right]u}{1 + \left(sP\zeta^{NL}(u)\right)^{-1}}\,du\right),
\tag{4.64}
$$

where

- in step (a), the LoS and the NLoS inter-cell interference come from a distance larger than the distances, r_2 and r, respectively, and
- in step (b), the PDF, $f_G(g) = 1 - \exp(-g)$, of the Rayleigh random variable, g, has been used to calculate the expectations, $\mathbb{E}_{[g]}\left\{\exp\left(-sP\zeta^L(u)g\right)\right\}$ and $\mathbb{E}_{[g]}\left\{\exp\left(-sP\zeta^{NL}(u)g\right)\right\}$.

Our proof of Theorem 4.3.1 is completed by plugging equations (4.51), (4.53), (4.60) and (4.62) into equation (4.42).

Appendix B Proof of Lemma (4.4.2)

Based on the considered UE association strategy, using as a metric the strongest cell, the Laplace transform, $\mathscr{L}_{I_{agg}}(s)$, can be derived in the range, $0 < r \le d_1$, as

$$
\mathscr{L}_{I_{agg}}(s) = \mathbb{E}_{[I_{agg}]}\left\{\exp\left(-sI_{agg}\right)\middle| 0 < r \le d_1\right\}
$$

$$
= \mathbb{E}_{[\Phi,\{\beta_i\},\{g_i\}]}\left\{\exp\left(-s\sum_{i\in\Phi/b_o}P\beta_ig_i\right)\middle| 0 < r \le d_1\right\}
$$

$$
\stackrel{(a)}{=} \exp\left(-2\pi\lambda\int_{r}^{\infty}\left(1 - \mathbb{E}_{[g]}\left\{\exp\left(-sP\beta(u)g\right)\right\}\right)u\,du\middle| 0 < r \le d_1\right),
\tag{4.65}
$$

where

- step (a) of equation (4.65) has been explained in detail in equation (3.18) of Chapter 3.

Importantly, given that we are considering the range, $0 < r \le d_1$, the expression, $\mathbb{E}_{[g]}\{\exp(-sP\beta(u)g)\}$, in equation (4.65) should consider the inter-cell interference from both the LoS and the NLoS paths. As a result, the Laplace transform, $\mathscr{L}_{I_{agg}}(s)$, can be further developed as

$$
\mathscr{L}_{I_{agg}}(s) = \exp\left(-2\pi\lambda \int_r^{d_1} \left(1 - \frac{u}{d_1}\right)\left[1 - \mathbb{E}_{[g]}\left\{\exp\left(-sPA^L u^{-\alpha^L} g\right)\right\}\right] u\, du\right)
$$

$$
\times \exp\left(-2\pi\lambda \int_{r_1}^{d_1} \frac{u}{d_1}\left[1 - \mathbb{E}_{[g]}\left\{\exp\left(-sPA^{NL} u^{-\alpha^{NL}} g\right)\right\}\right] u\, du\right)
$$

$$
\times \exp\left(-2\pi\lambda \int_{d_1}^{\infty} \left[1 - \mathbb{E}_{[g]}\left\{\exp\left(-sPA^{NL} u^{-\alpha^{NL}} g\right)\right\}\right] u\, du\right)
$$

$$
= \exp\left(-2\pi\lambda \int_r^{d_1} \left(1 - \frac{u}{d_1}\right)\frac{u}{1 + (sPA^L)^{-1} u^{\alpha^L}}\, du\right)
$$

$$
\times \exp\left(-2\pi\lambda \int_{r_1}^{d_1} \frac{u}{d_1}\frac{u}{1 + (sPA^{NL})^{-1} u^{\alpha^{NL}}}\, du\right)
$$

$$
\times \exp\left(-2\pi\lambda \int_{d_1}^{\infty} \frac{u}{1 + (sPA^{NL})^{-1} u^{\alpha^{NL}}}\, du\right). \tag{4.66}
$$

Plugging the variable value, $s = \frac{\gamma_0 r^{\alpha^L}}{PA^L}$, into equation (4.66), and considering the definition of the variable, $\rho_1(\alpha,\beta,t,d)$, in equation (4.25) and that of the variable, $\rho_2(\alpha,\beta,t,d)$, in equation (4.26), we can compute the Laplace transform, $\mathscr{L}_{I_{agg}}\left(\frac{\gamma_0 r^{\alpha^L}}{PA^L}\right)$, which for completeness is shown in Lemma (4.4.2).

This concludes the proof.

Appendix C Proof of Lemma (4.4.3)

Similarly to Appendix B, the Laplace transform, $\mathscr{L}_{I_{agg}}\left(\frac{\gamma_0 r^{\alpha^{NL}}}{PA^{NL}}\right)$, can be derived in the range, $0 < r \le y_1$, as

$$
\mathscr{L}_{I_{agg}}\left(\frac{\gamma_0 r^{\alpha^{NL}}}{PA^{NL}}\right) = \exp\left(-2\pi\lambda \int_{r_2}^{d_1} \left(1 - \frac{u}{d_1}\right)\frac{u}{1 + \left(\frac{\gamma_0 r^{\alpha^{NL}}}{PA^{NL}} PA^L\right)^{-1} u^{\alpha^L}}\, du\right.
$$

$$\times \exp\left(-2\pi\lambda \int_r^{d_1} \frac{u}{d_1} \frac{u}{1 + \left(\frac{\gamma_0 r^{\alpha^{NL}}}{PA^{NL}} PA^{NL}\right)^{-1} u^{\alpha^{NL}}} du\right)$$

$$\times \exp\left(-2\pi\lambda \int_{d_1}^{\infty} \frac{u}{1 + \left(\frac{\gamma_0 r^{\alpha^{NL}}}{PA^{NL}} PA^{NL}\right)^{-1} u^{\alpha^{NL}}} du\right), \quad (0 < r \le y_1),$$

$$(4.67)$$

and in the range, $y_1 < r \le d_1$, as

$$\mathcal{L}_{I_{agg}}\left(\frac{\gamma_0 r^{\alpha^{NL}}}{PA^{NL}}\right) = \exp\left(-2\pi\lambda \int_r^{d_1} \frac{u}{d_1} \frac{u}{1 + \left(\frac{\gamma_0 r^{\alpha^{NL}}}{PA^{NL}} PA^{NL}\right)^{-1} u^{\alpha^{NL}}} du\right)$$

$$\times \exp\left(-2\pi\lambda \int_{d_1}^{\infty} \frac{u}{1 + \left(\frac{\gamma_0 r^{\alpha^{NL}}}{PA^{NL}} PA^{NL}\right)^{-1} u^{\alpha^{NL}}} du\right), \quad (y_1 < r \le d_1).$$

$$(4.68)$$

The proof is completed by plugging equations (4.25) and (4.26) into equations (4.67) and (4.68), respectively.

Appendix D Proof of Lemma (4.4.4)

Similarly to Appendices B and C, the Laplace transform, $\mathcal{L}_{I_{agg}}\left(\frac{\gamma_0 r^{\alpha^{NL}}}{PA^{NL}}\right)$, can be derived in the range, $r > d_1$, as

$$\mathcal{L}_{I_{agg}}\left(\frac{\gamma_0 r^{\alpha^{NL}}}{PA^{NL}}\right) = \exp\left(-2\pi\lambda \int_r^{\infty} \frac{u}{1 + \left(\gamma_0 r^{\alpha^{NL}}\right)^{-1} u^{\alpha^{NL}}} du\right), \quad (r > d_1). \quad (4.69)$$

The proof is completed by plugging equation (4.26) into equation (4.69).

5 The Impact of Antenna Heights on Ultra-Dense Wireless Networks

5.1 Introduction

Small cells offer many advantages over macrocells, making them the right choice for network densification in the next generation of radio technology. Small cells not only offer a lower total cost of ownership than macrocells, but they also provide a great deal of deployment flexibility. Because small cell base stations (BSs) have a much lower form factor and transmit power than macrocell towers, they can be placed closer to where the end-users gather to significantly improve both the coverage and the capacity of the network. In closer proximity to the user equipment (UE), small cell BSs are able to deliver a higher quality air interface, leading to more and faster, reliable data connections, with the higher throughput and the lower delays needed to serve high-demanding applications, such as video. Small cells can also work in conjunction with macrocells to provide the UE with a seamless mobility experience, as they move between coverage areas.

As shown in Chapter 4, even though the inter-cell interference power will significantly increase in denser deployments of small cells due to the transition of a large number of interfering links from non-line-of-sight (NLoS) to line-of-sight (LoS), the vision of many small cell BSs installed on utility poles, streetlights and the sides of buildings, throughout an ultra-dense deployment, is still valid. Such ultra-dense deployments have the potential to significantly increase both the coverage probability and the area spectral efficiency (ASE) of the network. However, transforming such a vision into reality is far from simple, particularly outdoors.

Note that although deploying small cell BSs on utility poles, streetlights and the sides of buildings might appear to be a quick and easy fix to solve the challenges of delivering wireless broadband connections, the deployment of outdoor small cell BSs can be cumbersome. While it is true that covering an area with small cells can be faster than erecting a tower to deploy a macrocell site, a non-negligible amount of planning and preparation is needed for each small cell site before the network operator or the service provider can show up and plug in a radio access node. For example, planning is required to avoid the negative impact of an unplanned network as shown in Chapter 4, mostly due to channel and propagation characteristics, and the effect of the transition of a large number of interfering links from NLoS to LoS in dense networks. To practically deploy outdoor small cell BSs, one should expect the actual installation to account for about 20% of the work, while planning and site preparation account for the other 80% [210].

In small cell BS planning, the location of the small cell BS determines the channel and propagation characteristics and thus the amount of inter-cell interference in a dense network, but also the backhaul and power capabilities from which the small cell BS can avail. Access to backhaul and power is critical, and some potential small cell BS sites may have to be dropped at the planning stage, if these are not there, or cannot be cost-effectively supplied.

With this in mind, it is important to note that urban small cell BSs may be placed at various locations. Importantly, their small form factor allows for deploying them closer to the UE, but also in discreet locations to minimize the visual impact on the surrounding environment. This is capital to municipalities. Typical examples of small cell BS placement locations are [211]:

- Utility structures, such as utility poles, lampposts or traffic lights.
- Existing outdoor structures, such as building exterior walls or rooftops.
- Ceilings in indoor locations or sheltered outdoor locations, such as train stations, venues or stadiums.
- Street furniture, such as bus shelters, advertisement panels or newsstands.
- Specially designed facilities that conform to the surrounding environment, such as artificial trees in community parks and similar public areas.
- Aerial community antenna television (CATV) cable-strand mounting or using existing cable service lines.

As one can imagine, these different mountings may have different characteristics, and thus benefits and drawbacks. As radio access network planners state: no perfect deployment spot exits. However, one important common characteristic of all the above examples is that of their height. All of them generally put the small cell BS a few metres above the traditional end-user located at the ground level, from 5 m to 10 m, or even more. This follows due to the nature of such deploying structures, and we must acknowledge that this height is beneficial in terms of physical security. Being out of human reach, vandalism, as well as the tampering or even the theft of the small cell BSs, can be prevented. However, the height of the deploying structures also has a downside, imposing an important radio constraint. Contrary to the assumption of the previous two chapters, Chapters 3 and 4, the height of the small cell placement locations prevents the UE from getting closer to its serving small cell BS with the increase of the small cell BS density. This, in turn, prevents a linear increase of the signal power with the densification of the network, which does not allow compensation for the increased inter-cell interference power coming from the large number of neighbouring small cell BSs.

The height difference between the antennas of the UE and those of the serving and the interfering BSs is a cell planning issue that has not been traditionally accounted for in macrocell and sparse small cell deployments. Macrocells benefit from rooftop propagation, and thus the height difference is not a problem, but an advantage. In sparse small cell deployments, the connections to the serving small cell BS are mostly

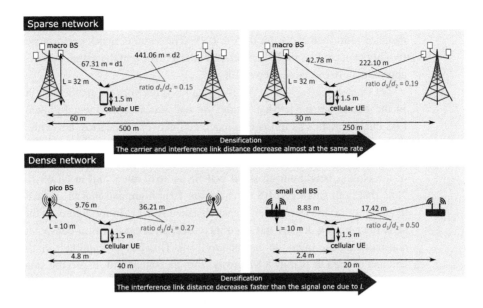

Figure 5.1 Impact of the BS antenna height in sparse and dense network scenarios.

LoS dominated, while the links towards the interfering small cell BSs are in NLoS in most cases. As a result, the signal of a UE still benefits from a significant path loss advantage with respect to interfering link, thus resulting in a good signal quality. Unfortunately, this is not the case in ultra-dense networks, where following the example in Chapter 4, two or more small cell BSs can be deployed in the same street, 100 m apart from one another, and thus in LoS. In this case, both the signal and the interfering links are LoS dominated, and the antenna height difference matters.

To illustrate the impact of the height difference between the antennas of the UE and those of the serving and interfering small cell BSs, Figure 5.1 provides an example with four deployment cases. The two deployments at the top correspond to two traditional macrocell network scenarios, and the two deployments at the bottom correspond to two small cell ones, where the scenario at the bottom left represents a sparse small cell network and that at the bottom right shows an ultra-dense network.

Let us take a look at the two macrocell network scenarios at the top in Figure 5.1. In the figure on the left-hand side, we can see a deployment following a traditional macrocell layout, with an inter-site distance of 500 m, where the antenna of the UE is 67 m away from the antenna of its serving macrocell BS, and benefits from an LoS connection, while the antenna of the closest neighbouring macrocell BS is 441 m away and in NLoS from the UE. Note that both macrocell BS antennas are at the height of 32 m, and that the ratio of the signal distance to the inter-cell interference distance is $\frac{67}{441} \approx 0.15$. In the figure on the right-hand side, the network density has been increased, reducing the inter-site distance to 250 m. The two-dimensional distances are halved, and the antenna of the UE is now 43 m away from the antenna of its serving macrocell BS, while the antenna of the closest neighbouring macrocell BS is

222 m away from the UE. This makes the signal distance to the inter-cell interference distance ratio increase from 0.15 to 0.19. The same network performance should be roughly expected, as both the inter-cell interference and the signal power are scaled in approximately the same proportion.

Let us now take a look at the two small cell networks at the bottom in Figure 5.1. In the figure on the left-hand side, we can see a deployment following the same LoS characteristics as in the two macrocell network scenarios, but with an inter-site distance of 40 m, where the antenna of the UE is 9.8 m away from the antenna of its serving small cell BS, and benefits from an LoS connection, while the antenna of the closest neighbouring small cell BS is 36 m away and in NLoS from the UE. Note that both small cell BS antennas are 10 m high, and that the ratio of the signal distance to the inter-cell interference distance is 0.27. In the figure on the right-hand side, the network density has been increased, entering the ultra-dense network realm and reducing the inter-site distance to 20 m. This has an important implication. The antenna of the UE is now 8.8 m away from the antenna of its serving small cell BS – almost the same as before – but the antenna of the closest neighbouring small cell BS is only 17.4 m away – much closer. This makes the signal distance to the inter-cell interference distance ratio increase rapidly from 0.27 to 0.50, thus showing a faster increase of the inter-cell interference power with respect to the signal power. This will naturally lead to a degradation of the signal-to-interference-plus-noise ratio (SINR) at the UE, and thus of the capacity of the network, which may be potentially large, if we account for the many interfering small cell BSs in an ultra-dense network.

Although, the numbers provided in this example may change depending on the mounting selected for the small cell BSs, e.g. utility poles, streetlights and the sides of buildings, and other characteristics, the height difference between the antennas of the UE and those of the serving and the interfering small cell BSs will exist in a dense network. This is unavoidable, since the backhaul and the power locations in such city fixtures are driving the deployment of those.

With all these in mind, and particularly considering that the models used in Chapters 3 and 4 have not considered the UE and the small cell BS antenna heights and their impact on network performance, it is fair to wonder about the applicability of the presented results so far in a realistic ultra-dense network, accounting for such practical mountings. In more detail, one may fear that the conclusions of Chapter 4, summarized in Remarks 4.1 and 4.2, may not apply in the ultra-dense regime, and that the overwhelming inter-cell interference resulting from the antenna height differences discussed earlier may have a larger negative impact than that of the NLoS-to-LoS inter-cell interference transition, and renders these ultra-dense deployments counterproductive.

This gives rise to a fundamental question:

Will the height difference between the antennas of the UE and those of the serving and the interfering BSs bring ultra-dense deployments into a complete inter-cell interference overload?

In this chapter, we answer this fundamental question by means of in-depth theoretical analyses.

The rest of this chapter is organized as follows:

- Section 5.2 introduces the system model and the assumptions taken in the theoretical performance analysis framework presented in this chapter, considering the height of the antenna of the typical UE and those of the small cell BSs, while embracing the channel model presented in Chapter 4, with both LoS and NLoS transmissions.
- Section 5.3 presents the theoretical expressions for the coverage probability and the ASE under the new system model.
- Section 5.4 provides results for a number of small cell BS deployments with different densities and characteristics. It also presents the conclusions drawn by this theoretical performance analysis, shedding new light on the importance of the height difference between the antenna of the typical UE and those of the small cell BSs in a dense network. Importantly, and for completeness, this section also studies via system-level simulations the impact of (i) multiple antenna patterns and down-tilts as wells as (ii) a Rician multi-path fast fading – instead of a Rayleigh one – on the derived results and conclusions.
- Section 5.5, finally, summarizes the key takeaways of this chapter.

5.2   Updated System Model

To assess the impact of the cell site and the antenna characteristics described in Section 5.1 on an ultra-dense network, in this section, we upgrade the system model described in Section 4.2, which has accounted for

- both LoS and NLoS transmissions and
- a practical user association strategy, based on the strongest received signal strength,

by adding the following features,

- the height of the antenna of the typical UE and those of the small cell BSs and
- the antenna patterns of the small cell BSs.

To facilitate the reader understanding the complete system model used in this chapter, Table 5.1 provides a concise and updated summary of it.

In the following, we present the incorporation of antenna height into the system model in more detail, while the incorporation of the antenna pattern is relegated to Section 5.4.4.

5.2.1   Antenna Height

To model the antenna height, we replace the two-dimensional distances in the path loss model presented in Section 4.2 with three-dimensional ones. In more detail, we replace the two-dimensional distance, r, between the typical UE and an arbitrary small cell BS with its three-dimensional equivalent, d, which can be expressed as

Table 5.1. System model

Model	Description	Reference
Transmission link		
Link direction	Downlink only	Transmissions from the small cell BSs to the UEs
Deployment		
Small cell BS deployment	HPPP with a finite density, $\lambda < +\infty$	Section 3.2.2
UE deployment	HPPP with full load, resulting in at least one UE per cell	Section 3.2.3
UE to small cell BS association		
Strongest small cell BS	UEs connect to the small cell BS providing the strongest received signal strength	Section 3.2.8 and references therein, equation (2.14)
Path loss		
3GPP UMi [153]	Multi-slope path loss with probabilistic LoS and NLoS transmissions	Section 4.2.1, equation (4.1)
	□ LoS component	equation (4.2)
	□ NLoS component	equation (4.3)
	□ Exponential probability of LoS	equation (4.19)
Multi-path fast fading		
Rayleigh*	Highly scattered scenario	Section 3.2.7 and references therein equation (2.15)
Shadow fading		
Not modelled	For tractability reasons, shadowing is not modelled since it has no qualitative impact on the results, see Section 3.4.5	Section 3.2.6 references therein
Antenna		
Small cell BS antenna**	Isotropic single-antenna element with 0 dBi gain	Section 3.2.4
UE antenna	Isotropic single-antenna element with 0 dBi gain	Section 3.2.4
Small cell BS antenna height	Variable antenna height, $1.5 + L$ m	Section 5.2.1
UE antenna height	Not considered	–
Idle mode capability at small cell BSs		
Always on	Small cell BSs always transmit control signals	–
Scheduler at small cell BSs		
Round robin	UEs take turns in accessing the radio channel	Section 7.1

*   Section 5.4.5 presents simulated results with Rician multi-path fast fading to demonstrate
      its impact on the obtained results.
** Section 5.4.4 presents simulated results with a directive antenna pattern to demonstrate
      its impact on the obtained results.

$$d = \sqrt{r^2 + L^2},\tag{5.1}$$

where

- L is the absolute height difference between the typical UE antenna and that of an arbitrary small cell BS in kilometres.

Note that the height of the antenna of the typical UE is usually larger than 0, which should be taken into account when calculating the absolute antenna height difference, $L \geq 0$.

With this change, the path loss, $\zeta(r)$, between the typical UE and an arbitrary small cell BS, which was defined in equation (4.1), and stacked into equation (4.4), is presented now as

$$\zeta(d) = \begin{cases} \zeta_1(d) = \begin{cases} \zeta_1^{\mathrm{L}}(d), & \text{with prob. } \mathrm{Pr}_1^{\mathrm{L}}(d) \\ \zeta_1^{\mathrm{NL}}(d), & \text{with prob. } \left(1 - \mathrm{Pr}_1^{\mathrm{L}}(d)\right) \end{cases}, & \text{when } 0 \leq d \leq d_1 \\ \zeta_2(d) = \begin{cases} \zeta_2^{\mathrm{L}}(d), & \text{with prob. } \mathrm{Pr}_2^{\mathrm{L}}(d) \\ \zeta_2^{\mathrm{NL}}(d), & \text{with prob. } \left(1 - \mathrm{Pr}_2^{\mathrm{L}}(d)\right) \end{cases}, & \text{when } d_1 < d \leq d_2 \\ \vdots & \vdots \\ \zeta_N(d) = \begin{cases} \zeta_N^{\mathrm{L}}(d), & \text{with prob. } \mathrm{Pr}_N^{\mathrm{L}}(d) \\ \zeta_N^{\mathrm{NL}}(d), & \text{with prob. } \left(1 - \mathrm{Pr}_N^{\mathrm{L}}(d)\right) \end{cases}, & \text{when } d > d_{N-1} \end{cases}\tag{5.2}$$

and stacked into

$$\zeta^{Path}(d) = \begin{cases} \zeta_1^{Path}(d), & \text{when } 0 \leq d \leq d_1 \\ \zeta_2^{Path}(d), & \text{when } d_1 < d \leq d_2 \\ \vdots & \vdots \\ \zeta_N^{Path}(d), & \text{when } d > d_{N-1} \end{cases}\tag{5.3}$$

while the piecewise LoS probability function, $\mathrm{Pr}_n^{\mathrm{L}}(r)$, presented in equation (4.5) can be rewritten as

$$\mathrm{Pr}^{\mathrm{L}}(d) = \begin{cases} \mathrm{Pr}_1^{\mathrm{L}}(d), & \text{when } 0 \leq d \leq d_1 \\ \mathrm{Pr}_2^{\mathrm{L}}(d), & \text{when } d_1 < d \leq d_2 \\ \vdots & \vdots \\ \mathrm{Pr}_N^{\mathrm{L}}(d), & \text{when } d > d_{N-1} \end{cases}.\tag{5.4}$$

5.3 Theoretical Performance Analysis and Main Results

In this section, given the definition of the coverage probability, p^{cov}, and the ASE, A^{ASE}, summarized for convenience in Table 5.2, we derive expressions for these two key performance indicators, accounting for the new three-dimensional path loss

Table 5.2. Key performance indicators

Metric	Formulation	Reference
Coverage probability, p^{cov}	$p^{\text{cov}}(\lambda,\gamma_0)=\Pr\left[\Gamma>\gamma_0\right]=\bar{F}_\Gamma(\gamma_0)$	Section 2.1.12, equation (2.22)
ASE, A^{ASE}	$A^{\text{ASE}}(\lambda,\gamma_0)=\frac{\lambda}{\ln 2}\int_{\gamma_0}^{+\infty}\frac{p^{\text{cov}}(\lambda,\gamma)}{1+\gamma}d\gamma$ $+\lambda\log_2\left(1+\gamma_0\right)p^{\text{cov}}(\lambda,\gamma_0)$	Section 2.1.12, equation (2.25)

model presented in the previous section. These expressions will be used to unveil the important impact of the small cell BS antenna height on an ultra-dense network.

5.3.1   Coverage Probability

In the following, we present the new main result on the coverage probability, p^{cov}, through Theorem 5.3.1, considering the above introduced system model, where

- the height of the antenna of the typical UE and those of the small cell BSs

should be highlighted. Readers interested in the research article originally presenting these results are referred to [110].

THEOREM 5.3.1   *Considering the new path loss model in equation (5.2), and the strongest small cell BS association presented in Section 4.2, the coverage probability, p^{cov}, can be derived as*

$$p^{\text{cov}}(\lambda,\gamma_0)=\sum_{n=1}^{N}\left(T_n^{\text{L}}+T_n^{\text{NL}}\right),\tag{5.5}$$

where

-

$$T_n^{\text{L}}=\int_{\sqrt{d_{n-1}^2-L^2}}^{\sqrt{d_n^2-L^2}}\Pr\left[\frac{P\zeta_n^{\text{L}}\left(\sqrt{r^2+L^2}\right)h}{I_{\text{agg}}+P^{\text{N}}}>\gamma_0\right]f_{R,n}^{\text{L}}(r)dr,\tag{5.6}$$

-

$$T_n^{\text{NL}}=\int_{\sqrt{d_{n-1}^2-L^2}}^{\sqrt{d_n^2-L^2}}\Pr\left[\frac{P\zeta_n^{\text{NL}}\left(\sqrt{r^2+L^2}\right)h}{I_{\text{agg}}+P^{\text{N}}}>\gamma_0\right]f_{R,n}^{\text{NL}}(r)dr\tag{5.7}$$

and
- d_0 and d_N are defined as $L\geq 0$ and $+\infty$, respectively.

Moreover, the probability density function (PDF), $f_{R,n}^{L}(r)$, and the PDF, $f_{R,n}^{NL}(r)$, within the range, $\sqrt{d_{n-1}^{2} - L^{2}} < r \leq \sqrt{d_{n}^{2} - L^{2}}$, are given by

$$
f_{R,n}^{L}(r) = \exp\left(-\int_{0}^{r_{1}} \left(1 - \Pr^{L}\left(\sqrt{u^{2} + L^{2}}\right)\right) 2\pi u\lambda \, du\right)
$$

$$
\times \exp\left(-\int_{0}^{r} \Pr^{L}\left(\sqrt{u^{2} + L^{2}}\right) 2\pi u\lambda \, du\right)
$$

$$
\times \Pr_{n}^{L}\left(\sqrt{r^{2} + L^{2}}\right) 2\pi r\lambda, \tag{5.8}
$$

and

$$
f_{R,n}^{NL}(r) = \exp\left(-\int_{0}^{r_{2}} \Pr^{L}\left(\sqrt{u^{2} + L^{2}}\right) 2\pi u\lambda \, du\right)
$$

$$
\times \exp\left(-\int_{0}^{r} \left(1 - \Pr^{L}\left(\sqrt{u^{2} + L^{2}}\right)\right) 2\pi u\lambda \, du\right)
$$

$$
\times \left(1 - \Pr_{n}^{L}\left(\sqrt{r^{2} + L^{2}}\right)\right) 2\pi r\lambda, \tag{5.9}
$$

where

- *r_1 and r_2 are determined by*

$$
r_{1} = \underset{r_{1}}{\arg}\left\{\zeta^{NL}\left(\sqrt{r_{1}^{2} + L^{2}}\right) = \zeta_{n}^{L}\left(\sqrt{r^{2} + L^{2}}\right)\right\} \tag{5.10}
$$

and

$$
r_{2} = \underset{r_{2}}{\arg}\left\{\zeta^{L}\left(\sqrt{r_{2}^{2} + L^{2}}\right) = \zeta_{n}^{NL}\left(\sqrt{r^{2} + L^{2}}\right)\right\}, \tag{5.11}
$$

respectively.

The probability, $\Pr\left[\dfrac{P\zeta_{n}^{L}\left(\sqrt{r^{2}+L^{2}}\right)h}{I_{agg}+P^{N}} > \gamma_{0}\right]$, is computed by

$$
\Pr\left[\frac{P\zeta_{n}^{L}\left(\sqrt{r^{2} + L^{2}}\right)h}{I_{agg} + P^{N}} > \gamma_{0}\right] = \exp\left(-\frac{\gamma_{0}P^{N}}{P\zeta_{n}^{L}\left(\sqrt{r^{2} + L^{2}}\right)}\right) \mathcal{L}_{I_{agg}}^{L}(s), \tag{5.12}
$$

where

- *$\mathcal{L}_{I_{agg}}^{L}(s)$ is the Laplace transform of the aggregated inter-cell interference random variable, I_{agg}, for the LoS signal transmission evaluated at point, $s = \dfrac{\gamma_{0}}{P\zeta_{n}^{L}(r)}$,* *which can be expressed as*

$$\mathscr{L}_{I_{agg}}^{L}(s) = \exp\left(-2\pi\lambda \int_{r}^{+\infty} \frac{\Pr^{L}\left(\sqrt{u^2+L^2}\right)u}{1+\left(sP\zeta^{L}\left(\sqrt{u^2+L^2}\right)\right)^{-1}}du\right)$$

$$\times \exp\left(-2\pi\lambda \int_{r_1}^{+\infty} \frac{\left[1-\Pr^{L}\left(\sqrt{u^2+L^2}\right)\right]u}{1+\left(sP\zeta^{NL}\left(\sqrt{u^2+L^2}\right)\right)^{-1}}du\right). \tag{5.13}$$

Moreover, the probability, $\Pr\left[\frac{P\zeta_n^{NL}\left(\sqrt{r^2+L^2}\right)h}{I_{agg}+P^N} > \gamma_0\right]$, *is computed by*

$$\Pr\left[\frac{P\zeta_n^{NL}\left(\sqrt{r^2+L^2}\right)h}{I_{agg}+P^N} > \gamma_0\right] = \exp\left(-\frac{\gamma_0 P^N}{P\zeta_n^{NL}\left(\sqrt{r^2+L^2}\right)}\right)\mathscr{L}_{I_{agg}}^{NL}(s), \tag{5.14}$$

where

- $\mathscr{L}_{I_{agg}}^{NL}(s)$ *is the Laplace transform of the aggregated inter-cell interference random variable,* I_{agg}, *for the NLoS signal transmission evaluated at point,* $s = \frac{\gamma_0}{P\zeta_n^{NL}(r)}$, *which can be written as*

$$\mathscr{L}_{I_{agg}}^{NL}(s) = \exp\left(-2\pi\lambda \int_{r_2}^{+\infty} \frac{\Pr^{L}\left(\sqrt{u^2+L^2}\right)u}{1+\left(sP\zeta^{L}\left(\sqrt{u^2+L^2}\right)\right)^{-1}}du\right)$$

$$\times \exp\left(-2\pi\lambda \int_{r}^{+\infty} \frac{\left[1-\Pr^{L}\left(\sqrt{u^2+L^2}\right)\right]u}{1+\left(sP\zeta^{NL}\left(\sqrt{u^2+L^2}\right)\right)^{-1}}du\right). \tag{5.15}$$

Proof   See Appendix A.   □

To ease the understanding on how to practically obtain this new coverage probability, p^{cov}, Figure 5.2 illustrates a flow chart that depicts the necessary enhancements to the traditional stochastic geometry framework presented in Chapter 3 – and updated in Chapter 4 – to compute the new results of Theorem 5.3.1.

Compared with the logic illustrated in Figure 4.2, and broadly speaking, three-dimensional distances instead of two-dimensional ones are considered in the channel model now. To avoid confusion, equation (5.1) is used in Theorem 5.3.1 to represent three-dimensional distances. It should be noted, however, that the outermost integration is still performed over two-dimensional distances, as it relates to the homogeneous Poisson point process (HPPP) modelling, and thus to the distances between the typical UE and the small cell BSs in the two-dimensional plane.

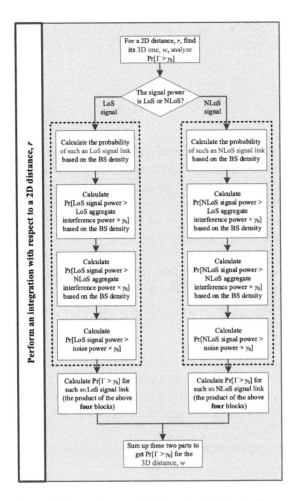

Figure 5.2 Logical steps within the standard stochastic geometry framework to obtain the results in Theorem 5.3.1, considering three-dimensional distances.

5.3.2 Area Spectral Efficiency

Similarly as in the previous chapters, plugging the coverage probability, p^{cov} – obtained from Theorem 5.3.1 – into equation (2.24) to compute the PDF, $f_\Gamma(\gamma)$, of the SINR, Γ, of the typical UE, we can obtain the ASE, A^{ASE}, solving equation (2.25). See Table 5.2 for further reference on the ASE formulation.

5.3.3 Computational Complexity

To calculate the coverage probability, p^{cov}, presented in Theorem 5.3.1, three folds of integrals are still required for the calculation of $\left\{f_{R,n}^{Path}(r)\right\}$, $\left\{\mathscr{L}_{I_{\text{agg}}}\left(\frac{\gamma_0}{P\zeta_n^{Path}(r)}\right)\right\}$ and $\left\{T_n^{Path}\right\}$, respectively, where the string variable, *Path*, takes the value of "L" (for the LoS case) or "NL" (for the NLoS case). Note that an additional fold of integral is needed for the calculation of the ASE, A^{ASE}, making it a four-fold integral computation.

5.4   Discussion

In this section, we use numerical results from static system-level simulations to evaluate the accuracy of the above theoretical performance analysis, and study the impact of the small cell BS antenna height on an ultra-dense network in terms of the coverage probability, p^{cov}, and the ASE, A^{ASE}.

As a reminder, it should be noted that to obtain the results on the coverage probability, p^{cov}, and the ASE, A^{ASE}, one should now plug equations (4.17) and (4.19), using three-dimensional distances, into Theorem 5.3.1, and follow the subsequent derivations. In this manner, the height of the antenna of the typical UE and those of the small cell BSs can be incorporated into the analysis.

5.4.1   Case Study

To assess the impact of the height of the antenna of the typical UE and those of the small cell BSs on an ultra-dense network, the same 3rd Generation Partnership Project (3GPP) case study as in Chapter 4 is used to allow an apple-to-apple comparison. For readers interested in a more detailed description of this 3GPP case study, please refer to Table 5.1 and Section 4.4.1.

For the sake of clarity, please recall that the following parameters are used in such a 3GPP case study:

- maximum antenna gain, $G_M = 0\,dB$,
- path loss exponents, $\alpha^L = 2.09$ and $\alpha^{NL} = 3.75$ – considering a carrier frequency of 2 GHz,
- reference path losses, $A^L = 10^{-10.38}$ and $A^{NL} = 10^{-14.54}$,
- transmit power, $P = 24\,dBm$, and
- noise power, $P^N = -95\,dBm$ – including a noise figure of 9 dB at the UE.

Antenna Configurations
As for the absolute antenna height difference, $L \geq 0$, we assume that the antenna height of the UE is 1.5 m, and that that of the small cell BSs varies between 1.5 m, 5 m, 10 m and 20 m. Accordingly, the antenna height difference, $L \geq 0$, takes the values, $L = \{0, 3.5, 8.5, 18.5\}$ m.

Small Cell BS Densities
As explained in Chapters 2–4, note that the selected path loss model based on [153] was constructed using measurements, and that due to the characteristics of the measurement campaign, the resulting path loss exponents and reference path losses apply

to transmitter-to-receiver distances no smaller than 10 m. Thus, given the introduction of the mentioned antenna height difference, $L \geq 0$, which results in larger minimum transmitter-to-receiver distances, we study in this chapter larger small cell BS densities than in the previous ones, of up to $\lambda = 10^5$ BSs/km^2.

Benchmark

In this performance evaluation, we use, as the benchmark,

- the results of the analysis with a single-slope path loss model, presented earlier in Section 3.4, as well as
- those with both LoS and NLoS transmissions, presented in Section 4.4.

Note that the results of the former were obtained with a reference path loss, $A^{NL} = 10^{-14.54}$, and a path loss exponent, $\alpha = 3.75$.

5.4.2   Coverage Probability Performance

Figure 5.3 shows the coverage probability, p^{cov}, for the SINR threshold, $\gamma_0 = 0$ dB, and the following six configurations:

- Conf. a): single-slope path loss model with no antenna height difference, $L = 0$ m (analytical results).
- Conf. b): single-slope path loss model with an antenna height difference, $L = 8.5$ m (analytical results).

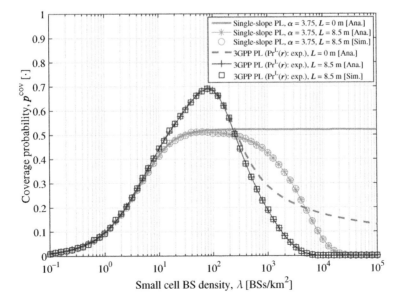

Figure 5.3 Coverage probability, p^{cov}, versus the small cell BS density, λ.

- Conf. c): single-slope path loss model with an antenna height difference, $L = 8.5$ m (simulated results).
- Conf. d): multi-slope path loss model with both LoS and NLoS transmissions and no antenna height difference, $L = 0$ m (analytical results).
- Conf. e): multi-slope path loss model with both LoS and NLoS transmissions and an antenna height difference, $L = 8.5$ m (analytical results).
- Conf. f): multi-slope path loss model with both LoS and NLoS transmissions and an antenna height difference, $L = 8.5$ m (simulated results).

Note that Conf. a) and d) were already presented and discussed in Chapters 3 and 4, respectively, and that they are incorporated for comparison purposes in Figure 5.3.

As can be observed from Figure 5.3, the analytical results, Conf. b) and Conf. e), match well the simulation results, Conf. c) and Conf. f), respectively. This corroborates the accuracy of Theorem 5.3.1 – and the derivations therein. Due to such significant accuracy, and since the results of the ASE, A^{ASE}, are computed solely based on the results of the coverage probability, p^{cov}, only analytical results are considered in our discussion hereafter in this chapter.

From Figure 5.3, we can also observe that

- the single-slope path loss model with no antenna height difference, $L = 0$ m – Conf. a) – results in an invariant SINR, Γ, for the typical UE, independent of the small cell BS density, λ, in the dense small cell BS density regimes, and in turn, the coverage probability, p^{cov}, remains constant, as such small cell BS density, λ, increases. See Chapter 3 for more details.
- The multi-slope path loss model with both LoS and NLoS transmissions and no antenna height difference, $L = 0$ m – Conf. d) – offers a significantly different behaviour, i.e.
  - ☐ when the small cell BS density, λ, is large enough, e.g. $\lambda > 10^2$ BSs/km^2, the coverage probability, p^{cov}, decreases with the small cell BS density, λ, due to the transition of a large number of interfering links from NLoS to LoS in dense networks.
  - ☐ When the small cell BS density, λ, is considerably large, e.g. $\lambda \geq 10^3$ BSs/km^2, the coverage probability, p^{cov}, decreases at a slower pace. This is because both the signal and the inter-cell interference power levels are increasingly LoS dominated, thus ultimately trending towards the LoS path loss exponent, and growing at the same pace.

  See Chapter 4 for more details.
- The coverage probabilities, p^{cov}, resulting from the models incorporating the multi-slope path loss model with both LoS and NLoS transmissions and an antenna height difference, $L = 8.5$ m – Conf. b), Conf. c), Conf. e) and Conf. f) – exhibit a significantly different behaviour than those resulting from the models with no antenna height difference, $L = 0$ m – Conf. a) and Conf. d). The new coverage probabilities, p^{cov}, resulting from the model incorporating an antenna height difference, $L = 8.5$ m, show a much more determined trajectory towards 0

when the network enters the ultra-dense regime. This is because the SINR, Γ, of the typical UE decreases at a faster pace with the small cell BS density, λ, when an antenna height difference, $L > 0$, exists. Since a UE cannot get infinitely close to its serving small cell BS due to such a finite and non-zero antenna height difference, $L > 0$, this introduces a cap on the maximum signal power that a UE can receive, which does not allow compensation for the increased inter-cell interference power, coming from the more and closer neighbouring small cell BSs in a densifying network. This translates into a dramatic degradation of the SINR, Γ, of the typical UE, and the observed damage of the coverage probability, p^{cov}.

In the following, we present Theorem 5.4.1 to formally explain this result, which not only impacts the coverage probability, p^{cov}, but also the ASE, A^{ASE}, as will be shown in the following section.

THEOREM 5.4.1 *If we have*

- *a finite and non-zero antenna height difference, $L > 0$,*
- *a finite SINR realization, γ, of the typical UE, $0 \leq \gamma < +\infty$, and*
- *a finite SINR threshold, $0 \leq \gamma_0 < +\infty$,*

then the asymptotic behaviours of both the coverage probability, p^{cov}, and the ASE, A^{ASE}, with the small cell BS density, λ, follow that $\lim_{\lambda \to +\infty} p^{\mathrm{cov}}(\lambda, \gamma_0) = 0$ *and* $\lim_{\lambda \to +\infty} A^{\mathrm{ASE}}(\lambda, \gamma_0) = 0$, *respectively.*

Proof See Appendix B.   □

In essence, Theorem 5.4.1 states that, under the analyzed scenario with a finite and non-zero antenna height difference, $L > 0$, both the coverage probability, p^{cov}, and the ASE, A^{ASE}, will decrease towards 0 when the small cell BS density, λ, is extremely large, and thus the typical UE will experience a total service outage in such a dense network. The key point of the proof reveals the reason for this phenomenon, which is sketched in the following.

* Let us consider the scenario with two small cell BSs presented in Figure 5.4, where
  □ r is the two-dimensional distance between the presented UE and its serving small cell BS and

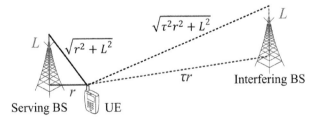

Figure 5.4 Illustration of a simple example with two small cell BSs.

☐ $\tau r, 1 < \tau < +\infty$, is that between the UE and an arbitrary interfering small cell BS.

In such a network, when the small cell BS density, λ, tends to infinity, i.e. $\lambda \to +\infty$, then

☐ the distance, r, between the presented UE and its serving small cell BS will tend to 0, i.e. $r \to 0$, and

☐ both the signal and the inter-cell interference power levels will be dominated by the first-piece LoS path loss function of equation (5.2), i.e.

$$\zeta_1^L(d) = A_1^L \left(\sqrt{r^2 + L^2}\right)^{-\alpha_1^L}.$$

With this in mind, the signal-to-interference ratio (SIR), $\bar{\gamma}$, of the presented UE can be calculated as

$$\bar{\gamma} = \frac{A_1^L \left(\sqrt{r^2 + L^2}\right)^{-\alpha_1^L}}{A_1^L \left(\sqrt{\tau^2 r^2 + L^2}\right)^{-\alpha_1^L}} = \left(\sqrt{\frac{1}{1 + \frac{\tau^2 - 1}{1 + \frac{L^2}{r^2}}}}\right)^{-\alpha_1^L}, \qquad (5.16)$$

where

☐ the noise power is not considered, as such an ultra-dense network is interference- and not noise-limited.

Based on this result, it is easy to demonstrate that, when the small cell BS density, λ, tends to infinity, the SIR, $\bar{\gamma}$, of the presented UE tends to 1, i.e. $\bar{\gamma} \to 1$ (0 dB). The intuition and consequences behind this conclusion are as follows:

☐ Due to the antenna height difference, $L > 0$, the presented UE cannot get arbitrarily close to the antenna of any small cell BS. Thus, this antenna height difference, $L > 0$, introduces a *cap* on the minimum UE to small cell BS antenna distance, and in turn, on the maximum signal and the maximum inter-cell interference power levels. These caps appear in the numerator and the denominator of the SIR, $\bar{\gamma}$, of the presented UE. However, this cap manifests earlier in the signal power than in the inter-cell interference one, i.e. the rate of increase of the numerator decreases earlier than that of the denominator with the densification level. This is because the presented UE is associated with the closest small cell BS, and thus the antenna height difference, $L > 0$, gets increasingly more significant in the computation of the distance between the presented UE and its serving small cell BS than with any other small cell BS.

☐ However, in the ultra-dense network of the example, where the small cell BS density, λ, tends to infinity, i.e. $\lambda \to +\infty$, both the serving and the interfering small cell BSs will be almost at the same distance, $\approx L$, from the presented UE, both directly overhead and equidistant above it. Thus, the signal and the interfering links will almost suffer from the same path loss, which leads to a SIR, $\bar{\gamma}$, of the presented UE of 1, i.e. $\bar{\gamma} \to 1$ (0 dB).

It is important to note that the intuition behind Theorem 5.4.1 is rooted in the geometry of the small cell BS deployment, and thus is valid regardless of the channel model.

Based on these conclusions, and extrapolating the logic of this example to a more general ultra-dense network with many more small cell BSs, we can infer that the aggregated inter-cell interference power will overwhelm the signal power due to the sheer number of strong interfering small cell BSs, approaching the typical UE from every direction in the ultra-dense regime.

This finalizes the sketch of the key point of the proof of Theorem 5.4.1.

Studying Various Antenna Height Differences

Figure 5.5 further explores the coverage probability, p^{cov}, paying special attention, this time, to the impact of various antenna height differences, $L = \{0, 3.5, 8.5, 18.5\}$ m. From Figure 5.5, we can observe that

- the larger the antenna height difference, $L \geq 0$, the faster the pace at which the coverage probability, p^{cov}, decreases towards 0. This is because a larger antenna height difference, $L \geq 0$, implies a larger minimum distance between the typical UE and its serving small cell BS, and in turn, a lower maximum received signal power. More importantly, this results in a tighter burden on the growth of the signal power with the small cell BS density, λ. As a result, the signal power faces a disadvantage earlier with respect to the inter-cell interference power. The latter continues to grow due to the sheer number of the rapidly approaching neighbouring small cell BS in a densifying network. Using as an example the simple case presented in Figure 5.4, the larger the antenna height difference, $L \geq 0$, the sooner the SINR, $\bar{\gamma}$, of the presented UE tends to 1 with the decrease of

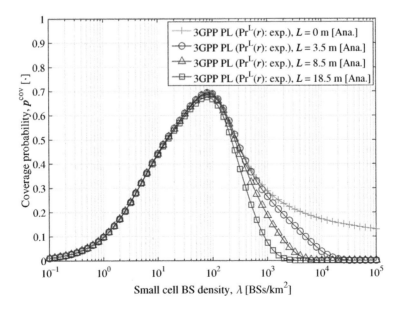

Figure 5.5 Coverage probability, p^{cov}, versus the small cell BS density, λ, for various antenna height differences, $L \geq 0$.

the scaling factor, τ, where a decrease of the scaling factor, τ, is equivalent to an increase of the small cell BS density, λ (see equation (5.16)).

- With respect to the baseline case with no antenna height difference, $L = 0$, and when considering a small cell BS density, $\lambda = 1 \times 10^4$ BSs/km^2, having an antenna height difference, $L = 8.5$ m, leads to a decrease of the coverage probability, p^{cov}, from 0.17 to 0.00023. This is a significant reduction of 99.86%.
- Looking at the results from a densification perspective, a reduction of the antenna height difference from $L = 8.5$ m to $L = 3.5$ m postpones the collapse of the coverage probability, p^{cov} – defined here by a coverage probability, $p^{\text{cov}} = 0.05$ – from a small cell BS density, $\lambda = 0.266 \times 10^4$ BSs/km^2, to a small cell BS density, $\lambda = 0.865 \times 10^4$ BSs/km^2.
- Importantly, it should be noted that with respect to the results presented in Chapter 4 one can also observe from Figure 5.5 that the collapse of the coverage probability, p^{cov}, due to the antenna height difference, $L > 0$, eclipsing – has a larger impact than – the decrease of the coverage probability, p^{cov}, due the transition of a large number of interfering links from NLoS to LoS in dense networks. The former is not even appreciated in the figure.

NetVisual Analysis

To visualize the fundamental behaviour of the coverage probability, p^{cov}, in a more intuitive manner, Figure 5.6 shows the coverage probability heat maps of three different scenarios with different small cell BS densities, i.e. 50, 250 and 2500 BSs/km^2, while considering the multi-slope path loss model with both LoS and NLoS transmissions with and without an antenna height difference, L. These heat maps are computed using NetVisual. As explained in Section 2.4, this tool is able to provide striking graphs to assess performance, which not only capture the mean, but also the standard deviation of the coverage probability, p^{cov}.

From Figure 5.6, we can see that, with respect to the case with no antenna height difference, $L = 0$ m,

- the larger antenna height difference, $L = 8.5$ m, has a negligible performance impact when the small cell BS density is not ultra-dense, i.e. when the small cell BS density, λ, is no larger than 250 BSs/km^2 in our example. The SINR maps for such small cell BS densities, λ, are basically the same in both Figures 5.6a and b. In more detail, the average and the standard deviation of the coverage probability, p^{cov}, do not change – with values of 0.71 and 0.22, respectively – for the case with 50 BSs/km^2, while they go from 0.51 and 0.23 to 0.48 and 0.21, respectively, for the case with 250 BSs/km^2, i.e. a 5.88% degradation in mean.
- Compared with Figure 5.6a, the SINR heat map in Figure 5.6b becomes much darker when the small cell BS density, λ, is around 2500 BSs/km^2, indicating a vast performance degradation due to the antenna height difference, $L > 0$. In more detail, the average and the standard deviation of the coverage probability, p^{cov}, go from 0.22 and 0.26 with no antenna height difference, $L = 0$ m to 0.05 and 0.05

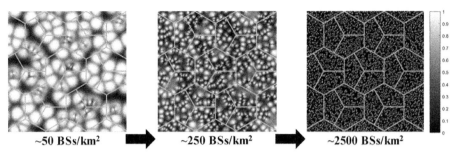

(a) Coverage probability, p^{cov}, under a multi-slope path loss model incorporating both LoS and NLoS transmissions and no antenna height difference, $L = 0\,\mathrm{m}$.

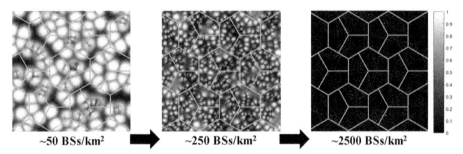

(b) Coverage probability, p^{cov}, under a multi-slope path loss model incorporating both LoS and NLoS transmissions and an antenna height difference, $L = 8.5\,\mathrm{m}$. Compared with Figure 5.6a, the SINR heat map becomes darker as the small cell BS density, λ, increases, showing a significant performance degradation because of the cap on the maximum signal power that a UE can receive imposed by the antenna height different, $L > 0$, in a dense network.

Figure 5.6 NetVisual plot of the coverage probability, p^{cov}, versus the small cell BS density, λ [Bright area: high probability, Dark area: low probability].

with the antenna height difference, $L = 8.5\,\mathrm{m}$, respectively – a 77.27% degradation in mean.

Summary of Findings: Coverage Probability Remark
Remark 5.1   Considering an antenna height difference, $L > 0$, the SINR, Γ, of the typical UE suffers in a dense network due to the cap on the maximum signal power that a UE can receive. As a result, the coverage probability, p^{cov}, does not remain constant as in Chapter 3, but significantly decreases – much more than in Chapter 4 – with the increase of the small cell BS density, λ, in the dense and the ultra-dense regimes.

5.4.3   Area Spectral Efficiency Performance

Let us now explore the behaviour of the ASE, A^{ASE}.

Figure 5.7 shows of the ASE, A^{ASE}, with an SINR threshold, $\gamma_0 = 0\,\mathrm{dB}$, for the following four configurations:

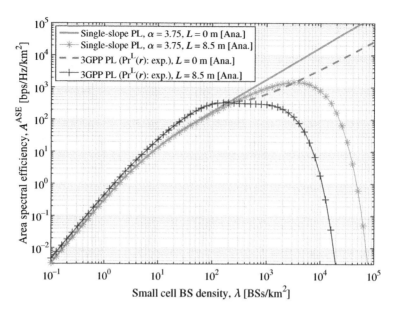

Figure 5.7 Area spectral efficiency, A^{ASE}, versus the small cell BS density, λ.

- Conf. a): single-slope path loss model with no antenna height difference, $L = 0$ m (analytical results).
- Conf. b): single-slope path loss model with an antenna height difference, $L = 8.5$ m (analytical results).
- Conf. c): multi-slope path loss model with both LoS and NLoS transmissions and no antenna height difference, $L = 0$ m (analytical results).
- Conf. d): multi-slope path loss model with both LoS and NLoS transmissions and an antenna height difference, $L = 8.5$ m (analytical results).

From Figure 5.7, we can observe that the ASEs, A^{ASE}, resulting from the models incorporating an antenna height difference, $L = 8.5$ m – Conf. b) and Conf. d) – also exhibit a totally different behaviour than those ASEs, A^{ASE}, resulting from the models with no antenna height difference, $L = 0$ m – Conf. a) and Conf. c). The new ASEs, A^{ASE}, resulting from the models incorporating an antenna height difference, $L > 0$, show how the network performance does not only stop growing linearly, but crashes and dramatically tends to 0, as the network enters the ultra-dense regime. This result is a direct consequence of the damage caused by the cap on the maximum signal power that a UE can receive imposed by the antenna height different, $L > 0$, in a dense network, and its impact on the coverage probability, p^{cov}. Theorem 5.4.1 formally explained the intuition behind this result.

Studying Various Antenna Height Differences
Similarly as in the discussion of the coverage probability, p^{cov}, Figure 5.8 further explores the ASE, A^{ASE}, paying special attention, this time, to the impact of various antenna height differences, $L = \{0, 3.5, 8.5, 18.5\}$ m.

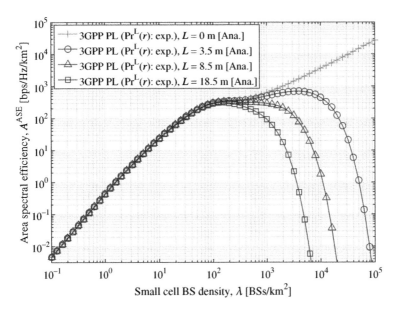

Figure 5.8 Area spectral efficiency, A^{ASE}, versus the small cell BS density, λ, for various antenna height differences, $L \geq 0$.

From Figure 5.8, we can observe that

- the larger the antenna height difference, $L > 0$, the earlier and more severe the crash of the ASE, A^{ASE}. This is because a larger antenna height difference, $L > 0$, imposes a tighter burden on the growth of the signal power with the small cell BS density, λ, which in turn, leads to an earlier degradation of the coverage probability, p^{cov}, as was shown in Figure 5.5, and thus of the ASE, A^{ASE}.
- With respect to the baseline case with no antenna height difference, $L = 0$, and when considering a small cell BS density, $\lambda = 1 \times 10^4$ BSs/km^2, having an antenna height difference, $L = 8.5$ m, leads to a decrease of the ASE, A^{ASE}, from 3624 to 1.80 bps/Hz/km^2. This is a significant reduction of 99.95%.
- Looking at the results from a densification perspective, a reduction of the antenna height difference from $L = 8.5$ m to $L = 3.5$ m postponed the crash of the ASE, A^{ASE} – defined here by an ASE, $A^{ASE} = 1$ bps/Hz/km^2 – from a small cell BS density, $\lambda = 1.1 \times 10^4$ BSs/km^2, to a small cell BS density, $\lambda = 5 \times 10^4$ BSs/km^2.
- It is also important to note that even in the most favourable case in the presented results – the configuration with an antenna height difference, $L = 3.5$ m – the ASE, A^{ASE}, which peaks at around a small cell BS density, $\lambda^* = 4000$ BSs/km^2, still suffers from a 60% loss with respect to that of the configuration where there is no antenna height difference, $L = 0$ m.
- As in the coverage probability analysis, it should also be noted that with respect to the results presented in Chapter 4 – the so-called ASE Crawl phenomenon – one can also observe from Figure 5.5 that the crash of the ASE, A^{ASE}, due to the antenna height difference, $L > 0$, eclipses – has a larger impact than – the crawl of

the ASE, A^{ASE}, due the transition of a large number of interfering links from NLoS to LoS in dense networks.

The ASE Crash

Overall, the fact that the ASE, A^{ASE}, will crash towards 0 as the small cell BS density, λ, marches into the ultra-dense regime shows the risk of deploying small cell BSs onto lampposts or facades of buildings, as has been traditionally done. This analysis shows that the antenna height difference, $L \geq 0$, matters, and should be considered in conjunction with the small cell BS density, λ. The fundamental reason for the tremendous negative impact that the antenna height difference, $L > 0$, may have on an ultra-dense network is the cap that it poses on the maximum signal power that a UE can receive. A way to address this issue, and prevent the crash of the ASE, A^{ASE}, is to set an antenna height difference, $L = 0$, which means lowering the height of the small cell BS antenna, not just by a few metres, but straight down to the height of the UE antenna. However, this requires a new approach to the architecture and the deployment of outdoor small cell BSs, to avoid tampering, vandalism and other undesirable effects.
    With this in mind, let us make the definition.

DEFINITION 5.4.2   The ASE Crash:
The existence of an antenna height difference, $L > 0$, between a UE and its serving small cell BS in an ultra-dense network imposes a minimum distance between them. Generally speaking, this introduces a cap on the maximum signal power that a UE can receive, which does not allow compensation for the increased inter-cell interference power, coming from the more and closer neighbouring small cell BSs in a densifying network. The ASE Crash is defined as the undesired – and significant – drop of the ASE, A^{ASE}, due to the cap on the maximum signal power that a UE can receive, imposed by the antenna height different, $L > 0$, in a dense network.

    It is worth noting that, even if the conclusion on the ASE Crash has been obtained from a particular model and set of parameters in this book, a number of studies using other antenna models and other sets of parameters corroborate the generality of these conclusions. We are going to show examples of this in the following section.

Summary of Findings: Area Spectral Efficiency Remark

Remark 5.2   Considering an antenna height difference, $L > 0$, the ASE, A^{ASE}, does not linearly increase with the increase of the small cell BS density, λ, as in Chapter 3, but crashes towards 0 as the small cell BS density, λ, increases. Due to the cap on the maximum signal power that a UE can receive, and the increasingly overwhelming inter-cell interference (from the more and closer neighbouring small cell BSs), the contribution of each small cell BS to the ASE, A^{ASE}, tends to 0 in a densifying network.

5.4.4   Impact of the Antenna Pattern and Down-Tilt – and Modelling – on the Main Results

It is important to note that the previous results have all been obtained assuming an omnidirectional antenna pattern at the small cell BSs. However, some antenna directivity can be gained, by using an antenna array comprized of multiple antenna elements, which can help to create a radiation beam, and focus the energy towards a given direction of – or point in – space. Obviously, such antenna directivity has an impact on the coverage probability, p^{cov}, and the ASE, A^{ASE}, and one may wonder whether it can alter the disastrous ASE Crash phenomenon, presented earlier.

To answer this question, in this section, we adopt a directional antenna pattern to the modelling of the small cell BS. In more detail, we incorporate the model of the four-element half-wave dipole antenna proposed in [150], and presented in Section 2.1.5. This model presents a good trade-off between accuracy and tractability.

Mathematically, we include the antenna gain, $\kappa \left(\varphi, \theta, \theta_{\text{tilt}} \right)$, modelled in equation (2.2) to the path loss function, $\zeta(d)$, presented in equation (5.2) as follows

$$\zeta(d) := \zeta(d) 10^{\frac{1}{10} \kappa(\varphi, \theta, \theta_{\text{tilt}})}, \tag{5.17}$$

where the following parameters have been selected,

- a maximum antenna gain, $\kappa_{\text{M}} = 8.15\,\text{dB}$,
- a horizontal attenuation offset, $\kappa_{\text{H}} \left(\varphi \right) [\text{dB}] = 0$,
- a fitting parameter, $n = 47.64$, and
- a secondary lobe level (SLL), $F_{\text{V}} = -12.0\,\text{dB}$.
- With respect to the down-tilt angle, θ_{tilt}, it should be noted that, for a given height of the small cell BS antenna, it should be increased as the small cell BS density, λ, increases to keep the same level of coverage. The down-tilt angle, θ_{tilt}, can be empirically calculated as [212]

$$\theta_{\text{tilt}} = \arctan \left(\frac{L}{r^{\text{cov}}} \right) + z B_{\text{V}}, \tag{5.18}$$

where
- $\square$ r^{cov} is the average two-dimensional distance from a cell-edge UE to its serving small cell BS, given in our analysis by, $r^{\text{cov}} = \sqrt{\frac{1}{\lambda \pi}}$,
- $\square$ B_{V} is the vertical half power beam width (HPBW), which is 19.5 degrees in our small cell BS antenna pattern and
- $\square$ z is an empirical parameter used to tune the trade-off between the signal and the inter-cell interference power levels, set to 0.7.

For more detail on this antenna model and its parameters, please refer to Section 2.1.5. For clarity, Figure 5.9 shows such antenna gain, $\kappa \left(\varphi, \theta, \theta_{\text{tilt}} \right)$, for the selected parameters and a number of different down-tilt angles, θ_{tilt}, which gradually increase from around 10 degrees to 90 degrees as the network densifies.

In the following, using the above antenna model, we investigate the impact of the antenna pattern and the down-tilt of the small cell BS on the ASE, A^{ASE}. Due to

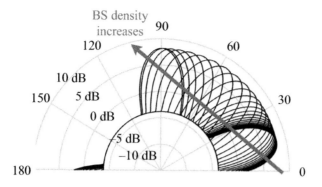

Figure 5.9 Antenna gain, $A\left(\varphi, \theta, \theta_{\text{tilt}}\right)$, versus the elevation angle, θ, for the four-element half-wave dipole antenna considered in this section.

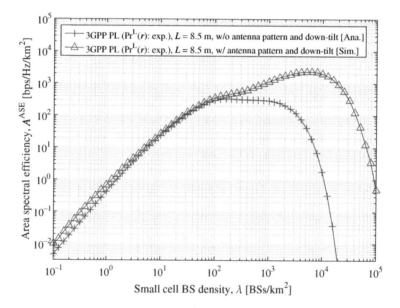

Figure 5.10 Area spectral efficiency, A^{ASE}, versus the small cell BS density, λ, for the practical antenna pattern and down-tilt shown in Figure 5.9.

the high mathematical complexity, we only present system-level simulation results in Figure 5.10, and concentrate on an antenna height difference, $L = 8.5$ m.

From Figure 5.10, we can draw the following conclusions:

- The practical antenna pattern and down-tilt of the small cell BS shown in Figure 5.9 help alleviate the ASE Crash because they constrain the small cell BS energy emission within certain geometrical areas, thus mitigating the inter-cell interference. However, the ASE Crash does not disappear – it is just postponed – and shows up at a larger small cell BS density, λ. This is because the cap on the signal power imposed by the antenna height difference, $L > 0$, in a dense network still persists, even if the small cell BS antenna faces downward with a down-tilt angle of 90 degrees. In this extreme case with a downwards facing antenna, the ASE Crash would occur at a very large small cell BS density, λ, when the (main or secondary) beam of a neighbouring small cell BS leaks into the intended coverage area of the small cell BS under consideration.

- Compared with the baseline ASE Crash performance, and looking at a level of ASE, $A^{ASE} = 1$ bps/Hz/km^2, the used antenna pattern and down-tilt for the small cell BS delay the ASE Crash from a small cell BS density, $\lambda = 10^4$ BSs/km^2, to a small cell BS density, $\lambda = 10^5$ BSs/km^2. Thus, antenna directionality and a proper down-tilt can be quite beneficial in ultra-dense networks and should be considered when deploying an ultra-dense network to mitigate the ASE Crash.

5.4.5 Impact of the Multi-Path Fading – and Modelling – on the Main Results

For completeness, and similarly to Chapter 4, in this section, we analyze the impact of the more accurate – but also less tractable – Rician multi-path fast fading model, presented in Section 2.1.8, on the coverage probability, p^{cov}, and the ASE, A^{ASE}.

Recall that

- the Rician multi-path fast fading model is applied to the LoS component only and that,
- in contrast to the Rayleigh multi-path fast fading model, it is able to capture the ratio between the power in the direct path and the power in the other scattered paths in the multi-path fast fading gain, h, as a function of the distance, r, between a UE and a small cell BS.

As discussed in Chapters 3 and 4, it should also be noted that

- the dynamic range of the multi-path fast fading gain, h, is smaller under the Rician multi-path fast fading model than that under the Rayleigh one – the former has larger minimum and smaller maximum values – and that importantly,
- the mean of the Rician multi-path fast fading gain, h, increases with the increase of the small cell BS density, λ (see Figure 2.4).

Our intention in this section is to check whether the conclusions obtained in this chapter about the ASE Crash hold when considering this more realistic – but less tractable – Rician-based multi-path fast fading model.

For the sake of mathematical complexity, we only present system-level simulation results with this alternative Rician-based multi-path fast fading model hereafter. Please refer to Section 2.1.8 for more details on both the Rayleigh and the Rician multi-path fast fading models.

Figure 5.11 illustrates the results on the coverage probability, p^{cov}, while Figure 5.12 shows those on the ASE, A^{ASE}. As a reminder, note that the assumptions and the parameters used to obtain the results in these two figures are the same as those used to

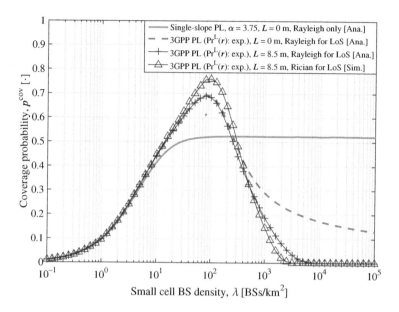

Figure 5.11 Coverage probability, p^{cov}, versus the small cell BS density, λ, for an alternative case study with multi-path Rician fading.

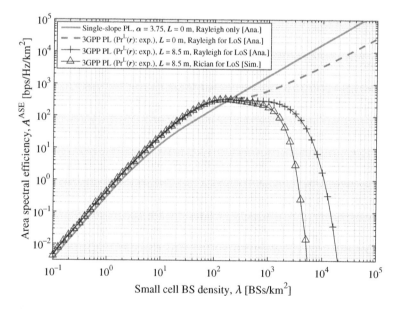

Figure 5.12 Area spectral efficiency, A^{ASE}, versus the small cell BS density, λ, for an alternative case study with multi-path Rician fading.

get the previous results shown in this chapter (see Table 5.1 and Section 5.4.1), except under the Rician multi-path fast fading model for the LoS component considered here.

As can be seen from Figures 5.11 and 5.12, all the observations with respect to the coverage probability, p^{cov}, and the ASE, A^{ASE}, obtained under the Rayleigh-based multi-path fast fading in Section 5.4 are qualitatively valid under the Rician-based multi-path fast fading model. The same trends are present.

Let us now analyze the coverage probability, p^{cov}, in more detail. The coverage probability, p^{cov}, first increases and then decreases with the small cell BS density, λ, and only a quantitative deviation exists. Specifically,

- in Figure 5.11, the maximum coverage probability, p^{cov}, is slightly larger under the Rician-based multi-path fast fading model, 0.77, than under the Rayleigh-based one, 0.7.
- Moreover, the small cell BS density, λ_0, for which such a maximum coverage probability, p^{cov}, is obtained, is also slightly larger under the Rician-based multi-path fast fading model, $100\,\mathrm{BSs/km}^2$, than under the Rayleigh-based one, $80\,\mathrm{BSs/km}^2$.
- These results are practically the same as those presented in Figure 4.8 of Chapter 4, and thus the same explanations apply. However, note that the coverage probability, p^{cov}, under the Rician-based multi-path fast fading model and that under the Rayleigh-based one cross at a small cell BS density, $\lambda = 500\,\mathrm{BSs/km}^2$, and it becomes worse for the former. This is because the phenomenon creating the ASE Crash – the antenna height difference, $L > 0$ – also plays a role at the multi-path fast fading level. It should be noted that, as illustrated in Figure 2.4, the multi-path fast fading gain, h, tends towards larger positive values as the distance, d, between a UE and a BS decreases, and in turn, the K-factor, K, increases. The antenna height difference, $L > 0$, thus prevents the enhancement of the multi-path fast fading gain, h, in the signal power when UEs are already quite close to their serving BSs in an ultra-dense network, while those in the interference links – of larger distance – continue to grow as the network keeps densifying.

As for the ASE, A^{ASE}, note that the behaviours under the Rician- and the Rayleigh-based multi-path fast fading models are almost identical, until the small cell BS density, $\lambda = 500\,\mathrm{BSs/km}^2$, is reached. From such a density on, the ASE, A^{ASE}, under the Rician-based multi-path fast fading model becomes worse, and tends faster towards 0. This is because of the effect of the antenna height difference, $L > 0$, on the multi-path fast fading, which was just explained when describing the coverage probability results.

Overall, and despite the quantitative differences, these results on the coverage probability, p^{cov}, and the ASE, A^{ASE}, confirm our statement, i.e. the main conclusions in this chapter are general, and do not qualitatively change due to the assumption on the multi-path fast fading model.[1]

In the remainder of this book, since the modelling of the multi-path fading did not have a profound impact in terms of qualitative results, we stick to Rayleigh multi-path fast fading modelling for mathematical tractability, unless otherwise stated.

---

[1] Note that this statement applies for the studied case with single-antenna UEs and small cell BSs, but may not apply to multi-antenna cases, where the fading correlation among the antennas of an array plays a key role.

5.5   Conclusions

In this chapter, we have brought to attention the importance of the modelling of the antenna characteristics in network performance analysis. In more detail, we have described the necessary upgrades that one has to perform to the system model presented so far in this book, in terms of antenna heights and patterns, to realize a more practical and close-to-reality study. We have also provided details on the new derivations conducted in the theoretical performance analysis, and presented the resulting expressions for the coverage probability, p^{cov}, and the ASE, A^{ASE}. Moreover, we have shared numerical results for small cell BS deployments with different small cell BS densities and characteristics. To finalize, we have also discussed the important conclusions drawn from this work, which are significantly different from those in Chapters 3 and 4.

Importantly, such conclusions, which are summarized in Remarks 5.1 and 5.2, have shown yet another new performance behaviour in the ultra-dense regime, referred to as the ASE Crash hereafter in this book. This new behaviour shows how a network operator or a service provider should not only carefully consider the densification level of its network, but also the heights of the antennas of the small cell BSs when planning its network. Otherwise, they may invest exponentially more money to densify their network to get an ultra-dense network unable to perform. This sends an important message:

The small cell BS antenna height matters!

In this chapter, we have also shown that this behaviour – the ASE Crash – is quantitatively – but not qualitatively – affected by (i) the antenna pattern and the downtilt of the small cell BS and/or (ii) the multi-path fast fading model. This showed the generality of the findings.

Appendix A Proof of Theorem 5.3.1

To ease the understanding of the readers, it should be noted that this proof shares a lot of similarities with the proof in Appendix 4A, where the main differences arise due to the need to use three-dimensional instead of two-dimensional distances in some derivations to capture the effect of the antenna height difference, L. The full proof is presented hereafter for completeness.

Having said that, let us first describe the main idea behind the proof of Theorem 5.3.1, and then proceed with a more detailed explanation.

In line with the guidelines provided in Section 2.3.2, and to evaluate the coverage probability, p^{cov}, we need proper expressions for

- the PDF, $f_R(r)$, of the random variable, R, characterizing the two-dimensional distance between the typical UE and its serving small cell BS for the event that the typical UE is associated to the strongest small cell BS with either an LoS or with an NLoS path and

- the conditional probability, $\Pr[\Gamma > \gamma_0 \,|\, r]$, where
  - $\Box$ $r = R(\omega)$ is a realization of the random variable of the two-dimensional distance, R, for both the LoS and the NLoS transmission.

Once such expressions are known, we can finally derive the coverage probability, p^{cov}, by performing the corresponding appropriate integrals, some of which will be shown in the following.

Before proceeding with the more detailed calculations, however, it is important to note that, from equations (2.22) and (3.3), we can derive the coverage probability, p^{cov}, as

$$
\begin{aligned}
p^{\mathrm{cov}}&(\lambda, \gamma_0) \\
&\overset{(a)}{=} \int_{r>0} \Pr\left[\mathrm{SINR} > \gamma_0 \,|\, r\right] f_R(r) dr \\
&= \int_{r>0} \Pr\left[\frac{P\zeta\left(\sqrt{r^2+L^2}\right)h}{I_{\mathrm{agg}}+P^{\mathrm{N}}} > \gamma_0\right] f_R(r) dr \\
&= \int_0^{d_1} \Pr\left[\frac{P\zeta_1^{\mathrm{L}}\left(\sqrt{r^2+L^2}\right)h}{I_{\mathrm{agg}}+P^{\mathrm{N}}} > \gamma_0\right] f_{R,1}^{\mathrm{L}}(r) dr + \int_0^{d_1} \Pr\left[\frac{P\zeta_1^{\mathrm{NL}}\left(\sqrt{r^2+L^2}\right)h}{I_{\mathrm{agg}}+P^{\mathrm{N}}} > \gamma_0\right] f_{R,1}^{\mathrm{NL}}(r) dr \\
&\quad + \cdots \\
&\quad + \int_{d_{N-1}}^{\infty} \Pr\left[\frac{P\zeta_N^{\mathrm{L}}\left(\sqrt{r^2+L^2}\right)h}{I_{\mathrm{agg}}+P^{\mathrm{N}}} > \gamma_0\right] f_{R,N}^{\mathrm{L}}(r) dr + \int_{d_{N-1}}^{\infty} \Pr\left[\frac{P\zeta_N^{\mathrm{NL}}\left(\sqrt{r^2+L^2}\right)h}{I_{\mathrm{agg}}+P^{\mathrm{N}}} > \gamma_0\right] f_{R,N}^{\mathrm{NL}}(r) dr \\
&\overset{\triangle}{=} \sum_{n=1}^{N} \left(T_n^{\mathrm{L}} + T_n^{\mathrm{NL}}\right),
\end{aligned}
\tag{5.19}
$$

where

- R_n^{L} and R_n^{NL} are the piecewise distributions of the two-dimensional distances over which the typical UE is associated to a small cell BS with an LoS or an NLoS path, respectively:
  - $\Box$ Note that these two events, i.e. the typical UE being associated with a small cell BS through an LoS or an NLoS, are disjoint, and thus the coverage probability, p^{cov}, is the sum of the corresponding probabilities of these two events.
- $f_{R,n}^{\mathrm{L}}(r)$ and $f_{R,n}^{\mathrm{NL}}(r)$ are the piecewise PDFs of the random variables of the two-dimensional distances, R_n^{L} and R_n^{NL}, respectively:
  - $\Box$ For clarity, both piecewise PDFs, $f_{R,n}^{\mathrm{L}}(r)$ and $f_{R,n}^{\mathrm{NL}}(r)$, are stacked into the PDF, $f_R(r)$, in step (a) of equation (5.19), where the stacked PDF, $f_R(r)$, takes a similar form as in equation (4.42) in Chapter 4.
  - $\Box$ Moreover, since the two events, i.e. the typical UE being associated with a small cell BS through an LoS or an NLoS, are disjoint, as mentioned before, we can rely on the following equality, $\sum_{n=1}^{N} \int_{d_{n-1}}^{d_n} f_{R,n}(r) dr = \sum_{n=1}^{N} \int_{d_{n-1}}^{d_n} f_{R,n}^{\mathrm{L}}(r) dr + \sum_{n=1}^{N} \int_{d_{n-1}}^{d_n} f_{R,n}^{\mathrm{NL}}(r) dr = 1$.

- T_n^{L} and T_n^{NL} are two piecewise functions, defined as

$$T_n^{\mathrm{L}} = \int_{d_{n-1}}^{d_n} \Pr\left[\frac{P\zeta_n^{\mathrm{L}}\left(\sqrt{r^2 + L^2}\right)h}{I_{\mathrm{agg}} + P^{\mathrm{N}}} > \gamma_0\right] f_{R,n}^{\mathrm{L}}(r)dr \qquad (5.20)$$

and

$$T_n^{\mathrm{NL}} = \int_{d_{n-1}}^{d_n} \Pr\left[\frac{P\zeta_n^{\mathrm{NL}}\left(\sqrt{r^2 + L^2}\right)h}{I_{\mathrm{agg}} + P^{\mathrm{N}}} > \gamma_0\right] f_{R,n}^{\mathrm{NL}}(r)dr, \qquad (5.21)$$

respectively, and

- d_0 and d_N are equal to 0 and $+\infty$, respectively.

$$f_R(r) = \begin{cases} f_{R,1}(r) = \begin{cases} f_{R,1}^{\mathrm{L}}(r), & \text{UE associated to an LoS BS} \\ f_{R,1}^{\mathrm{NL}}(r), & \text{UE associated to an NLoS BS}' \end{cases} & 0 \le r \le d_1 \\[2mm] f_{R,2}(r) = \begin{cases} f_{R,2}^{\mathrm{L}}(r), & \text{UE associated to an LoS BS} \\ f_{R,2}^{\mathrm{NL}}(r), & \text{UE associated to an NLoS BS}' \end{cases} & d_1 < r \le d_2 \\[2mm] \vdots & \vdots \\[2mm] f_{R,N}(r) = \begin{cases} f_{R,N}^{\mathrm{L}}(r), & \text{UE associated to an LoS BS} \\ f_{R,N}^{\mathrm{NL}}(r), & \text{UE associated to an NLoS BS}' \end{cases} & r > d_{N-1} \end{cases} \qquad (5.22)$$

Following this method, let us now dive into the more detailed derivations.

LoS-Related Calculations

Let us first look into the LoS transmissions, and show how to first calculate the PDF, $f_{R,n}^{\mathrm{L}}(r)$, and then the probability, $\Pr\left[\frac{P\zeta_n^{\mathrm{L}}\left(\sqrt{r^2+L^2}\right)h}{I_{\mathrm{agg}}+P^{\mathrm{N}}} > \gamma_0\right]$, in equation (5.19).

To that end, we define first the two following events:

- Event B^{L}: The nearest small cell BS to the typical UE with an LoS path is located at a two-dimensional distance, $x = X^{\mathrm{L}}(\omega)$, defined by the distance random variable, X^{L}. According to the results presented in Section 2.3.2, the PDF, $f_R(r)$, of the random variable of the two-dimensional distance, R, in Event B^{L} can be calculated by

$$f_X^{\mathrm{L}}(x) = \exp\left(-\int_0^x \Pr^{\mathrm{L}}\left(\sqrt{u^2 + L^2}\right) 2\pi u\lambda\, du\right) \Pr^{\mathrm{L}}\left(\sqrt{x^2 + L^2}\right) 2\pi x\lambda. \qquad (5.23)$$

This is because, according to [52], the complementary cumulative distribution function (CCDF), $\bar{F}_X^{\mathrm{L}}(x)$, of the random variable, X^{L}, in Event B^{L} is given by

$$\bar{F}_X^{\mathrm{L}}(x) = \exp\left(-\int_0^x \Pr^{\mathrm{L}}\left(\sqrt{u^2 + L^2}\right) 2\pi u\lambda\, du\right) \qquad (5.24)$$

and taking the derivative of the cumulative distribution function (CDF), $\left(1 - \bar{F}_X^{L}(x)\right)$, of the random variable, X^{L}, with respect to the two-dimensional distance, x, we can get the PDF, $f_X^{L}(x)$, of X^{L}, shown in equation (5.23). It is important to note that the derived equation (5.23) is more complex than equation (2.42) in Section 2.3.2. This is because the LoS small cell BS deployment is an inhomogeneous one due to the fact that the closer small cell BSs to the typical UE are more likely to establish LoS links than the further away ones. Compared with equation (2.42), we can see three important changes in equation (5.23).

☐ The probability of a two-dimensional disc of a radius, r, containing exactly 0 points, which took the form, $\exp\left(-\pi\lambda r^2\right)$, as presented in equation (2.38) in Section 2.3.2, has changed now, and is replaced by the expression,

$$\exp\left(-\int_0^x \Pr^{L}\left(\sqrt{u^2 + L^2}\right) 2\pi u\lambda\, du\right),$$

in equation (5.23). This is because the equivalent intensity of the LoS small cell BSs in an inhomogeneous Poisson point process (PPP) is distance-dependent, i.e. $\Pr^{L}\left(\sqrt{u^2 + L^2}\right)\lambda$. This adds a new integral to equation (5.23) with respect to the two-dimensional distance, u.

☐ It is important to note that, compared with equation (4.46) in Chapter 4, a three-dimensional distance, $\sqrt{u^2 + L^2}$, has been used in equation (5.23). This is because the LoS probability is defined on a three-dimensional distance, due to the existence of an antenna height difference, $L \geq 0$.

☐ The intensity, λ, of the HPPP in equation (2.42), has been replaced by the equivalent intensity, $\Pr^{L}\left(\sqrt{x^2 + L^2}\right)\lambda$, of the inhomogeneous PPP in equation (5.23).

• Event C^{NL} conditioned on the value, $x = X^{L}(\omega)$, of the random variable of the two-dimensional distance, X^{L}: The typical UE is associated to the nearest small cell BS located at a two-dimensional distance, $x = X^{L}(\omega)$, through an LoS path. If the typical UE is associated to the nearest LoS small cell BS located at a two-dimensional distance, $x = X^{L}(\omega)$, this LoS small cell BS must be at the smallest path loss, $\zeta^{L}\left(\sqrt{x^2 + L^2}\right)$. As a result, there must be no NLoS small cell BS inside the disc

☐ centred on the typical UE,
☐ with a two-dimensional radius, x_1,

where such a radius, x_1, satisfies the following condition,

$x_1 = \arg_{x_1}\left\{\zeta^{NL}\left(\sqrt{x_1^2 + L^2}\right) = \zeta^{L}\left(\sqrt{x^2 + L^2}\right)\right\}$. Otherwise, this NLoS small cell BS would outperform the LoS small cell BS at the two-dimensional distance, $x = X^{L}(\omega)$.

According to [52], the conditional probability, $\Pr\left[C^{NL}\,|\,X^{L} = x\right]$, of the Event, C^{NL}, conditioned on the realization, $x = X^{L}(\omega)$, of the random variable, X^{L}, is given by

$$\Pr\left[C^{\text{NL}} \,\middle|\, X^{\text{L}} = x\right] = \exp\left(-\int_0^{x_1}\left(1 - \Pr^{\text{L}}\left(\sqrt{u^2 + L^2}\right)\right)2\pi u\lambda\, du\right). \quad (5.25)$$

As a summary, note that Event B^{L} ensures that the path loss, $\zeta^{\text{L}}(x)$, associated with *an arbitrary LoS small cell BS* is always larger than that associated with *the considered LoS small cell BS* at the two-dimensional distance, $x = X^{\text{L}}(\omega)$. Besides, conditioned on such a two-dimensional distance, $x = X^{\text{L}}(\omega)$, Event C^{NL} guarantees that the path loss, $\zeta^{\text{NL}}(x)$, associated with *an arbitrary NLoS small cell BS* is also always larger than that associated with *the considered LoS small cell BS* at the two-dimensional distance, $x = X^{\text{L}}(\omega)$. With this, we can guarantee that the typical UE is associated with the strongest LoS small cell BS.

Let us thus now consider the resulting new event, in which the typical UE is associated with an LoS small cell BS, and such a small cell BS is located at the two-dimensional distance, $r = R^{\text{L}}(\omega)$. The CCDF, $\bar{F}_R^{\text{L}}(r)$, of such a random variable, R^{L}, can be derived as

$$
\begin{aligned}
\bar{F}_R^{\text{L}}(r) &= \Pr\left[R^{\text{L}} > r\right] \\
&\overset{(a)}{=} \mathbb{E}_{[X^{\text{L}}]}\left\{\Pr\left[R^{\text{L}} > r \,\middle|\, X^{\text{L}}\right]\right\} \\
&= \int_0^{+\infty} \Pr\left[R^{\text{L}} > r \,\middle|\, X^{\text{L}} = x\right] f_X^{\text{L}}(x)dx \\
&\overset{(b)}{=} \int_0^r 0 \times f_X^{\text{L}}(x)dx + \int_r^{+\infty} \Pr\left[C^{\text{NL}} \,\middle|\, X^{\text{L}} = x\right] f_X^{\text{L}}(x)dx \\
&= \int_r^{+\infty} \Pr\left[C^{\text{NL}} \,\middle|\, X^{\text{L}} = x\right] f_X^{\text{L}}(x)dx, \quad (5.26)
\end{aligned}
$$

where

- $\mathbb{E}_{[X]}\{\cdot\}$ in step (a) of equation (5.26) is the expectation operation over the random variable, X, and
- step (b) of equation (5.26) is valid because
  - □ $\Pr\left[R^{\text{L}} > r \,\middle|\, X^{\text{L}} = x\right] = 0$ for $0 < x \le r$, and
  - □ the conditional event, $\left[R^{\text{L}} > r \,\middle|\, X^{\text{L}} = x\right]$, is equivalent to the conditional event, $\left[C^{\text{NL}} \,\middle|\, X^{\text{L}} = x\right]$, in the range, $x > r$.

Now, given the CCDF, $\bar{F}_R^{\text{L}}(r)$, one can find its PDF, $f_R^{\text{L}}(r)$, by taking the derivative, $\frac{\partial\left(1 - \bar{F}_R^{\text{L}}(r)\right)}{\partial r}$, with respect to the two-dimensional distance, r, i.e.

$$f_R^{\text{L}}(r) = \Pr\left[C^{\text{NL}} \,\middle|\, X^{\text{L}} = r\right] f_X^{\text{L}}(r). \quad (5.27)$$

Considering the two-dimensional distance range, $(d_{n-1} < r \le d_n)$, we can find the PDF of such a segment, $f_{R,n}^{\text{L}}(r)$, from such a PDF, $f_R^{\text{L}}(r)$, as

$$f_{R,n}^{L}(r) = \exp\left(-\int_0^{r_1}\left(1 - \mathrm{Pr}^{L}\left(\sqrt{u^2 + L^2}\right)\right)2\pi u\lambda\,du\right)$$

$$\times \exp\left(-\int_0^{r}\mathrm{Pr}^{L}\left(\sqrt{u^2 + L^2}\right)2\pi u\lambda\,du\right)$$

$$\times \mathrm{Pr}_n^{L}\left(\sqrt{r^2 + L^2}\right)2\pi r\lambda,\ (d_{n-1} < r \le d_n),\quad (5.28)$$

where

- $r_1 = \underset{r_1}{\arg}\left\{\zeta^{\mathrm{NL}}\left(\sqrt{r_1^2 + L^2}\right) = \zeta_n^{L}\left(\sqrt{r^2 + L^2}\right)\right\}.$

Having obtained the PDF of a segment, $f_{R,n}^{L}(r)$, we can evaluate the probability, $\mathrm{Pr}\left[\frac{P\zeta_n^{L}\left(\sqrt{r^2+L^2}\right)h}{I_{\mathrm{agg}}+P^{N}} > \gamma_0\right]$, in equation (5.19) as

$$\mathrm{Pr}\left[\frac{P\zeta_n^{L}\left(\sqrt{r^2 + L^2}\right)h}{I_{\mathrm{agg}} + P^{N}} > \gamma_0\right] = \mathbb{E}_{[I_{\mathrm{agg}}]}\left\{\mathrm{Pr}\left[h > \frac{\gamma_0\left(I_{\mathrm{agg}} + P^{N}\right)}{P\zeta_n^{L}\left(\sqrt{r^2 + L^2}\right)}\right]\right\}$$

$$= \mathbb{E}_{[I_{\mathrm{agg}}]}\left\{\bar{F}_H\left(\frac{\gamma_0\left(I_{\mathrm{agg}} + P^{N}\right)}{P\zeta_n^{L}\left(\sqrt{r^2 + L^2}\right)}\right)\right\},\quad (5.29)$$

where

- $\bar{F}_H(h)$ is the CCDF of the multi-path fast fading channel gain, h, which is assumed to be drawn from a Rayleigh fading distribution (see Section 2.1.8).

Since the CCDF, $\bar{F}_H(h)$, of the multi-path fast fading channel gain, h, follows an exponential distribution with unitary mean given by the expression,

$$\bar{F}_H(h) = \exp(-h),$$

equation (5.29) can be further derived as

$$\mathrm{Pr}\left[\frac{P\zeta_n^{L}\left(\sqrt{r^2 + L^2}\right)h}{I_{\mathrm{agg}} + P^{N}} > \gamma_0\right]$$

$$= \mathbb{E}_{[I_{\mathrm{agg}}]}\left\{\exp\left(-\frac{\gamma_0\left(I_{\mathrm{agg}} + P^{N}\right)}{P\zeta_n^{L}\left(\sqrt{r^2 + L^2}\right)}\right)\right\}$$

$$\overset{(a)}{=} \exp\left(-\frac{\gamma_0 P^{N}}{P\zeta_n^{L}\left(\sqrt{r^2 + L^2}\right)}\right)\mathbb{E}_{[I_{\mathrm{agg}}]}\left\{\exp\left(-\frac{\gamma_0}{P\zeta_n^{L}\left(\sqrt{r^2 + L^2}\right)}I_{\mathrm{agg}}\right)\right\}$$

$$= \exp\left(-\frac{\gamma_0 P^{N}}{P\zeta_n^{L}\left(\sqrt{r^2 + L^2}\right)}\right)\mathscr{L}_{I_{\mathrm{agg}}}^{L}\left(\frac{\gamma_0}{P\zeta_n^{L}\left(\sqrt{r^2 + L^2}\right)}\right),\quad (5.30)$$

where

- $\mathscr{L}^{L}_{I_{agg}}(s)$ is the Laplace transform of the aggregated inter-cell interference random variable, I_{agg}, conditioned on LoS signal transmission, evaluated at the variable value, $s = \dfrac{\gamma_0}{P\zeta_n^L\left(\sqrt{r^2+L^2}\right)}$.

It should be noted that the use of the Laplace transform is to ease the mathematical representation. By definition, $\mathscr{L}^{L}_{I_{agg}}(s)$ is the Laplace transform of the PDF of the aggregated inter-cell interference random variable, I_{agg}, conditioned on LoS signal transmission, evaluated at the variable value, s, and the equality,

$$\mathscr{L}_X(s) = \mathbb{E}_{[X]}\left\{\exp\left(-sX\right)\right\},$$

follows from the definitions and derivations leading to equation (2.46) in Chapter 2.

Based on the condition of LoS signal transmission, the Laplace transform, $\mathscr{L}^{L}_{I_{agg}}(s)$, can be derived as

$$\mathscr{L}^{L}_{I_{agg}}(s) = \mathbb{E}_{[I_{agg}]}\left\{\exp\left(-sI_{agg}\right)\right\}$$

$$= \mathbb{E}_{[\Phi,\{\beta_i\},\{g_i\}]}\left\{\exp\left(-s\sum_{i\in\Phi/b_o}P\beta_ig_i\right)\right\}$$

$$\overset{(a)}{=} \exp\left(-2\pi\lambda\int\left(1 - \mathbb{E}_{[g]}\left\{\exp\left(-sP\beta\left(\sqrt{u^2+L^2}\right)g\right)\right\}\right)u\,du\right).$$

$$(5.31)$$

where

- Φ is the set of small cell BSs,
- b_o is the small cell BS serving the typical UE,
- b_i is the ith interfering small cell BS,
- β_i and g_i are the path loss and the multi-path fast fading gain between the typical UE and the ith interfering small cell BS and
- step (a) of equation (5.31) has been explained in detail in equation (3.18), and further derived in equation (3.19) of Chapter 3.

Compared with equation (3.19) which considers the inter-cell interference from a single-slope path loss model, the expression, $\mathbb{E}_{[g]}\left\{\exp\left(-sP\beta\left(\sqrt{u^2+L^2}\right)g\right)\right\}$, in equation (5.31) should consider the inter-cell interference from both the LoS and the NLoS paths. As a result, the Laplace transform, $\mathscr{L}^{L}_{I_{agg}}(s)$, can be further developed as

$$\mathscr{L}^{L}_{I_{agg}}(s) = \exp\left(-2\pi\lambda\int\left(1 - \mathbb{E}_{[g]}\left\{\exp\left(-sP\beta\left(\sqrt{u^2+L^2}\right)g\right)\right\}\right)u\,du\right)$$

$$\overset{(a)}{=} \exp\left(-2\pi\lambda\int\left[\mathrm{Pr}^L\left(\sqrt{u^2+L^2}\right)\left(1 - \mathbb{E}_{[g]}\left\{\exp\left(-sP\zeta^L\left(\sqrt{u^2+L^2}\right)g\right)\right\}\right)\right.\right.$$

$$\left.\left. + \left(1 - \mathrm{Pr}^L\left(\sqrt{u^2+L^2}\right)\right)\left(1 - \mathbb{E}_{[g]}\left\{\exp\left(-sP\zeta^{NL}\left(\sqrt{u^2+L^2}\right)g\right)\right\}\right)\right]u\,du\right)$$

$$\overset{(b)}{=} \exp\left(-2\pi\lambda\int_r^{+\infty}\mathrm{Pr}^L\left(\sqrt{u^2+L^2}\right)\left(1 - \mathbb{E}_{[g]}\left\{\exp\left(-sP\zeta^L\left(\sqrt{u^2+L^2}\right)g\right)\right\}\right)u\,du\right)$$

$$\times \exp\left(-2\pi\lambda \int_{r_1}^{+\infty} \left(1 - \Pr^{L}\left(\sqrt{u^2 + L^2}\right)\right)\left(1 - \mathbb{E}_{[g]}\left\{\exp\left(-sP\zeta^{NL}\left(\sqrt{u^2 + L^2}\right)g\right)\right\}\right) u\, du\right)$$

$$\overset{(c)}{\overset{}{=}} \exp\left(-2\pi\lambda \int_{r}^{+\infty} \frac{\Pr^{L}\left(\sqrt{u^2 + L^2}\right) u}{1 + \left(sP\zeta^{L}\left(\sqrt{u^2 + L^2}\right)\right)^{-1}}\, du\right)$$

$$\times \exp\left(-2\pi\lambda \int_{r_1}^{+\infty} \frac{\left[1 - \Pr^{L}\left(\sqrt{u^2 + L^2}\right)\right] u}{1 + \left(sP\zeta^{NL}\left(\sqrt{u^2 + L^2}\right)\right)^{-1}}\, du\right), \tag{5.32}$$

where

- in step (a), the integration is probabilistically divided into two parts considering the LoS and the NLoS inter-cell interference,
- in step (b), the LoS and the NLoS inter-cell interference come from a distance larger than the distances, r and r_1, respectively, and
- in step (c), the PDF, $f_G(g) = 1 - \exp(-g)$, of the Rayleigh random variable, g, has been used to calculate the expectations, $\mathbb{E}_{[g]}\left\{\exp\left(-sP\zeta^{L}\left(\sqrt{u^2 + L^2}\right)g\right)\right\}$ and $\mathbb{E}_{[g]}\left\{\exp\left(-sP\zeta^{NL}\left(\sqrt{u^2 + L^2}\right)g\right)\right\}$.

To give some intuition for equation (5.30), it should be noted that

- the exponential factor, $\exp\left(-\dfrac{\gamma_0 P^{N}}{P\zeta_n^{L}\left(\sqrt{r^2 + L^2}\right)}\right)$, measures the probability that the signal power exceeds the noise power by at least a factor, γ_0, while
- the Laplace transform, $\mathscr{L}_{I_{agg}}\left(\dfrac{\gamma_0}{P\zeta_n^{L}\left(\sqrt{r^2 + L^2}\right)}\right)$, measures the probability that the signal power exceeds the aggregated inter-cell interference power by at least a factor, γ_0.

As a result, and because the multi-path fast fading channel gain, h, follows an exponential distribution, the product of the above probabilities, shown in step (a) of equation (5.30), yields the probability that the signal power exceeds the sum power of the noise and the aggregated inter-cell interference by at least a factor, γ_0.

NLoS-Related Calculations

Let us now look into the NLoS transmissions, and show how to first calculate the PDF, $f_{R,n}^{NL}(r)$, and then the probability, $\Pr\left[\dfrac{P\zeta_n^{NL}\left(\sqrt{r^2 + L^2}\right)h}{I_{agg} + P^{N}} > \gamma_0\right]$, in equation (5.19).

To that end, we define the two following events:

- Event B^{NL}: The nearest small cell BS to the typical UE with an NLoS path is located at a two-dimensional distance, $x = X^{NL}(\omega)$, defined by the two-dimensional distance random variable, X^{NL}. Similarly to equation (5.23), one can find the PDF, $f_X^{NL}(x)$, as

$$f_X^{\text{NL}}(x) = \exp\left(-\int_0^x \left(1 - \text{Pr}^{\text{L}}\left(\sqrt{u^2 + L^2}\right)\right) 2\pi u \lambda \, du\right) \left(1 - \text{Pr}^{\text{L}}\left(\sqrt{x^2 + L^2}\right)\right) 2\pi x \lambda.$$

(5.33)

It is important to note that the derived equation (5.33) is more complex than equation (2.42) of Section 2.3.2. This is because the NLoS small cell BS deployment is an inhomogeneous one due to the fact that the further away small cell BSs to the typical UE are more likely to establish NLoS links than the closer ones. Compared with equation (2.42), we can see three important changes in equation (5.33):

☐ The probability of a two-dimensional disc of a radius, r, containing exactly 0 points, which took the form, $\exp\left(-\pi \lambda r^2\right)$, as presented in equation (2.38) of Section 2.3.2, has changed now, and is replaced by the expression,

$$\exp\left(-\int_0^x \left(1 - \text{Pr}^{\text{L}}\left(\sqrt{u^2 + L^2}\right)\right) 2\pi u \lambda \, du\right),$$

in equation (5.33). This is because the equivalent intensity of the NLoS small cell BSs in an inhomogeneous PPP is distance-dependent, i.e. $\left(1 - \text{Pr}^{\text{L}}\left(\sqrt{u^2 + L^2}\right)\right) \lambda$. This adds a new integral to equation (5.33) with respect to the two-dimensional distance, u.

☐ It is important to note that, compared with equation (4.56) in Chapter 4, a three-dimensional distance, $\sqrt{u^2 + L^2}$, has been used in equation (5.33). This is because the LoS probability is defined on a three-dimensional distance, due to the existence of an antenna height difference, $L \geq 0$.

☐ The intensity, λ, of the HPPP in equation (2.42), has been replaced by the equivalent intensity, $\left(1 - \text{Pr}^{\text{L}}\left(\sqrt{u^2 + L^2}\right)\right) \lambda$, of the inhomogeneous PPP in equation (5.33).

- Event C^{L} conditioned on the value, $x = X^{\text{NL}}(\omega)$, of the random variable of two-dimensional distance, X^{NL}: The typical UE is associated to the nearest small cell BS located at a two-dimensional distance, $x = X^{\text{NL}}(\omega)$, through an NLoS path. If the typical UE is associated to the nearest NLoS small cell BS located at a two-dimensional distance, $x = X^{\text{NL}}(\omega)$, this NLoS small cell BS must be at the smallest path loss. As a result, there must be no LoS small cell BS inside the disc
  ☐ centred on the typical UE,
  ☐ with a radius, x_2,
  where such a radius, x_2, satisfies the following condition,

$$x_2 = \arg_{x_2}\left\{\zeta^{\text{L}}\left(\sqrt{x_2^2 + L^2}\right) = \zeta_n^{\text{NL}}\left(\sqrt{x^2 + L^2}\right)\right\}.$$ Otherwise, this LoS small cell BS would outperform the NLoS small cell BS at the two-dimensional distance, $x = X^{\text{NL}}(\omega)$.

Similarly to equation (5.25), the conditional probability, $\text{Pr}\left[C^{\text{L}} \mid X^{\text{NL}} = x\right]$, of Event C^{L} conditioned on the realization, $x = X^{\text{NL}}(\omega)$, of the random variable, X^{NL}, is given by

$$\Pr\left[C^L \mid X^{NL} = x\right] = \exp\left(-\int_0^{x_2} \Pr^L\left(\sqrt{u^2 + L^2}\right) 2\pi u \lambda \, du\right). \tag{5.34}$$

Let us thus now consider the resulting new event, in which the typical UE is associated with an NLoS small cell BS, and such a small cell BS is located at the two-dimensional distance, $r = R^{NL}(\omega)$. The CCDF, $\bar{F}_R^{NL}(r)$, of such a random variable, R^{NL}, can be derived as

$$\bar{F}_R^{NL}(r) = \Pr\left[R^{NL} > r\right]$$

$$= \int_r^{+\infty} \Pr\left[C^L \mid X^{NL} = x\right] f_X^{NL}(x) dx. \tag{5.35}$$

Now, given the CCDF, $\bar{F}_R^{NL}(r)$, one can find its PDF, $f_R^{NL}(r)$, by taking the derivative, $\frac{\partial(1 - \bar{F}_R^{NL}(r))}{\partial r}$, with respect to the two-dimensional distance, r, i.e.

$$f_R^{NL}(r) = \Pr\left[C^L \mid X^{NL} = r\right] f_X^{NL}(r). \tag{5.36}$$

Considering the two-dimensional distance range, $(d_{n-1} < r \leq d_n)$, we can find the PDF of such a segment, $f_{R,n}^{NL}(r)$, from such a PDF, $f_R^{NL}(r)$, as

$$f_{R,n}^{NL}(r) = \exp\left(-\int_0^{r_2} \Pr^L\left(\sqrt{u^2 + L^2}\right) 2\pi u \lambda \, du\right)$$

$$\times \exp\left(-\int_0^r \left(1 - \Pr^L\left(\sqrt{u^2 + L^2}\right)\right) 2\pi u \lambda \, du\right)$$

$$\times \left(1 - \Pr_n^L\left(\sqrt{r^2 + L^2}\right)\right) 2\pi r \lambda, \; (d_{n-1} < r \leq d_n), \tag{5.37}$$

where

- $r_2 = \underset{r_2}{\arg}\left\{\zeta^L\left(\sqrt{r_2^2 + L^2}\right) = \zeta_n^{NL}\left(\sqrt{r^2 + L^2}\right)\right\}.$

Having obtained the PDF of a segment, $f_{R,n}^{NL}(r)$, we can evaluate the probability, $\Pr\left[\frac{P\zeta_n^{NL}\left(\sqrt{r^2 + L^2}\right)h}{I_{agg} + P^N} > \gamma_0\right]$, in equation (5.19) as

$$\Pr\left[\frac{P\zeta_n^{NL}\left(\sqrt{r^2 + L^2}\right)h}{I_{agg} + P^N} > \gamma_0\right] = \mathbb{E}_{[I_{agg}]}\left\{\Pr\left[h > \frac{\gamma_0\left(I_{agg} + P^N\right)}{P\zeta_n^{NL}\left(\sqrt{r^2 + L^2}\right)}\right]\right\}$$

$$= \mathbb{E}_{[I_{agg}]}\left\{\bar{F}_H\left(\frac{\gamma_0\left(I_{agg} + P^N\right)}{P\zeta_n^{NL}\left(\sqrt{r^2 + L^2}\right)}\right)\right\}. \tag{5.38}$$

Since the CCDF, $\bar{F}_H(h)$, of the multi-path fast fading channel gain, h, follows an exponential distribution with unitary mean given by

$$\bar{F}_H(h) = \exp(-h),$$

equation (5.38) can be further derived as

$$\Pr\left[\frac{P\zeta_n^{NL}\left(\sqrt{r^2+L^2}\right)h}{I_{agg} + P^N} > \gamma_0\right]$$

$$= \mathbb{E}_{[I_{agg}]}\left\{\exp\left(-\frac{\gamma_0\left(I_{agg} + P^N\right)}{P\zeta_n^{NL}\left(\sqrt{r^2+L^2}\right)}\right)\right\}$$

$$= \exp\left(-\frac{\gamma_0 P^N}{P\zeta_n^{NL}\left(\sqrt{r^2+L^2}\right)}\right)\mathbb{E}_{[I_{agg}]}\left\{\exp\left(-\frac{\gamma_0}{P\zeta_n^{NL}\left(\sqrt{r^2+L^2}\right)}I_{agg}\right)\right\}$$

$$= \exp\left(-\frac{\gamma_0 P^N}{P\zeta_n^{NL}\left(\sqrt{r^2+L^2}\right)}\right)\mathscr{L}_{I_{agg}}^{NL}\left(\frac{\gamma_0}{P\zeta_n^{NL}\left(\sqrt{r^2+L^2}\right)}\right). \tag{5.39}$$

Based on the condition of NLoS signal transmission, the Laplace transform, $\mathscr{L}_{I_{agg}}^{NL}(s)$, can be derived as

$$\mathscr{L}_{I_{agg}}^{NL}(s) = \mathbb{E}_{[I_{agg}]}\left\{\exp\left(-sI_{agg}\right)\right\}$$

$$= \mathbb{E}_{[\Phi,\{\beta_i\},\{g_i\}]}\left\{\exp\left(-s\sum_{i\in\Phi/b_o}P\beta_i g_i\right)\right\}$$

$$\overset{(a)}{=} \exp\left(-2\pi\lambda\int\left(1 - \mathbb{E}_{[g]}\left\{\exp\left(-sP\beta\left(\sqrt{u^2+L^2}\right)g\right)\right\}\right)u\,du\right), \tag{5.40}$$

where

- step (a) of equation (5.40) has been explained in detail in equation (3.18), and further derived in equation (3.19) of Chapter 3.

Compared with equation (3.19), which considers the inter-cell interference from a single-slope path loss model, the expression, $\mathbb{E}_{[g]}\left\{\exp\left(-sP\beta\left(\sqrt{u^2+L^2}\right)g\right)\right\}$, in equation (5.40) should consider the inter-cell interference from both the LoS and the NLoS paths. As a result, similar to equation (5.41), the Laplace transform, $\mathscr{L}_{I_{agg}}^{NL}(s)$, can be further developed as

$$\mathscr{L}_{I_{agg}}^{NL}(s) = \exp\left(-2\pi\lambda\int\left(1 - \mathbb{E}_{[g]}\left\{\exp\left(-sP\beta\left(\sqrt{u^2+L^2}\right)g\right)\right\}\right)u\,du\right)$$

$$\overset{(a)}{=} \exp\left(-2\pi\lambda\int_{r_2}^{+\infty}\Pr^L\left(\sqrt{u^2+L^2}\right)\left(1 - \mathbb{E}_{[g]}\left\{\exp\left(-sP\zeta^L\left(\sqrt{u^2+L^2}\right)g\right)\right\}\right)u\,du\right.$$

$$\times \exp\left(-2\pi\lambda \int_{r}^{+\infty} \left(1 - \Pr^{L}\left(\sqrt{u^2 + L^2}\right)\right) \left(1 - \mathbb{E}_{[g]}\left\{\exp\left(-sP\zeta^{NL}\left(\sqrt{u^2 + L^2}\right)g\right)\right\}\right) u\, du\right)$$

$$\overset{(b)}{=} \exp\left(-2\pi\lambda \int_{r_2}^{+\infty} \frac{\Pr^{L}\left(\sqrt{u^2 + L^2}\right) u}{1 + \left(sP\zeta^{L}\left(\sqrt{u^2 + L^2}\right)\right)^{-1}} du\right)$$

$$\times \exp\left(-2\pi\lambda \int_{r}^{+\infty} \frac{\left[1 - \Pr^{L}\left(\sqrt{u^2 + L^2}\right)\right] u}{1 + \left(sP\zeta^{NL}\left(\sqrt{u^2 + L^2}\right)\right)^{-1}} du\right), \tag{5.41}$$

where

- in step (a), the LoS and the NLoS inter-cell interference come from a distance larger than the distances, r_2 and r, respectively, and
- in step (b), the PDF, $f_G(g) = 1 - \exp(-g)$, of the Rayleigh random variable, g, has been used to calculate the expectations, $\mathbb{E}_{[g]}\left\{\exp\left(-sP\zeta^{L}\left(\sqrt{u^2 + L^2}\right)g\right)\right\}$ and $\mathbb{E}_{[g]}\left\{\exp\left(-sP\zeta^{NL}\left(\sqrt{u^2 + L^2}\right)g\right)\right\}$.

Our proof of Theorem 5.3.1 is completed by plugging equations (5.28), (5.30), (5.37) and (5.39) into equation (5.19).

Appendix B Proof of Theorem 5.4.1

Let us consider the scenario proposed, and apply the theory of limits on the coverage probability, p^{cov}, derived in Theorem 5.3.1. As a result, we obtain that

$$\lim_{\lambda \to +\infty} p^{\text{cov}} = \lim_{\lambda \to +\infty} T_1^{L} + \lim_{\lambda \to +\infty} T_1^{NL}, \tag{5.42}$$

where

- T_1^{L} and T_1^{NL} are the coverage probabilities of the first-piece LoS and the first-piece NLoS paths, respectively.

In other words, the two-dimensional distance, r, from the typical UE to its serving small cell BS tends to 0, $r \to 0$, when the small cell BS density, λ, tends to infinity, $\lambda \to +\infty$, and as a consequence, the propagation between them is dominated by either the first-piece LoS path loss or the first-piece NLoS path loss function in such an ultra-dense network. Note that, in this case, we are assuming that both the two-dimensional distance, r, and the antenna height difference, $L \geq 0$, are smaller than distance, d_1, as defined in equation (4.1), where, as a reminder, note that d_1 is the three-dimensional distance at which the multi-slope path loss model transits from the first to the second piece.

NLoS-Related Calculations

Let us focus now on the component, T_1^{NL}, where we can find the following results.

The component, T_1^{NL}, tends to 0, when the small cell BS density, λ, tends to infinity, $\lambda \to +\infty$, i.e. $\lim\limits_{\lambda \to +\infty} T_1^{\text{NL}} = 0$.

This is because,

- according to equation (5.11), $r_2^{\min} \triangleq \lim\limits_{r \to 0} r_2 = \arg\limits_{r_2} \left\{ \zeta^{\text{L}} \left(\sqrt{r_2^2 + L^2} \right) = \zeta_1^{\text{NL}}(L) \right\}$,

  where if the antenna height difference, $L > 0$, is larger than 0, the term, $r_2^{\min}$, is also larger than 0.

- As a consequence, the term, $\exp\left(- \int_0^{r_2} \text{Pr}^{\text{L}} \left(\sqrt{u^2 + L^2} \right) 2\pi u \lambda \, du \right)$, in equation (5.9) can be upper-bounded by the term,

  $\exp\left(-\text{Pr}^{\text{L}} \left(\sqrt{\left(r_2^{\min}\right)^2 + L^2} \right) \pi \lambda \left(r_2^{\min}\right)^2 \right)$, and thus this term also tends to 0, when the small cell BS density, λ, tends to infinity, $\lambda \to +\infty$, i.e.

  $\lim\limits_{\lambda \to +\infty} \exp\left(-\text{Pr}^{\text{L}} \left(\sqrt{\left(r_2^{\min}\right)^2 + L^2} \right) \pi \lambda \left(r_2^{\min}\right)^2 \right) = 0$. As introduced in Section 5.2, we are assuming here that the probability of LoS function, $\text{Pr}^{\text{L}}(d)$, is a monotonically decreasing function with respect to the three-dimensional distance, d, between the typical UE and the arbitrary small cell BS.

- Using the above findings, and considering the definition of the PDF, $f_{R,1}^{\text{NL}}(r)$, in equation (5.9), we can show that $\lim\limits_{\lambda \to +\infty} f_{R,1}^{\text{NL}}(r) = 0$.

- As a result, we can conclude that

$$\lim\limits_{\lambda \to +\infty} T_1^{\text{NL}} = 0. \tag{5.43}$$

LoS-Related Calculations

Let us focus now on the component, T_1^{L}, where we can find the following results.

The component, T_1^{L}, tends to 0, when the small cell BS density, λ, tends to infinity, $\lambda \to +\infty$, i.e. $\lim\limits_{\lambda \to +\infty} T_1^{\text{L}} = 0$.

This is due to the following three reasons:

- Using equations (5.12) and (5.13), we can demonstrate that

$$\text{Pr}\left[\frac{P\zeta_1^{\text{L}}\left(\sqrt{r^2 + L^2}\right)h}{I_{\text{agg}} + N_0} > \gamma_0 \right]$$

$$< \exp\left(-2\pi\lambda \int_r^{+\infty} \frac{\text{Pr}^{\text{L}}\left(\sqrt{u^2 + L^2}\right)u}{1 + \left(\frac{\gamma_0 P\zeta^{\text{L}}\left(\sqrt{u^2 + L^2}\right)}{P\zeta_1^{\text{L}}\left(\sqrt{r^2 + L^2}\right)} \right)^{-1}} du \right) \tag{5.44}$$

$$< \exp\left(-2\pi\lambda \int_r^{\tau r} \frac{\text{Pr}^{\text{L}}\left(\sqrt{u^2 + L^2}\right)u}{1 + \frac{1}{\gamma_0} \left(\frac{\sqrt{r^2 + L^2}}{\sqrt{u^2 + L^2}} \right)^{-\alpha_1^{\text{L}}}} du \right) \tag{5.45}$$

$$< \exp \left(-2\pi\lambda \frac{\Pr^{\mathrm{L}}\left(\sqrt{\tau^2 r^2 + L^2}\right) \int_r^{\tau r} u\, du}{1 + \frac{1}{\gamma_0}\left(\frac{\sqrt{r^2+L^2}}{\sqrt{\tau^2 r^2 + L^2}}\right)^{-\alpha_1^{\mathrm{L}}}} \right) \tag{5.46}$$

$$\overset{r\to 0}{<} \exp \left(-\frac{\Pr^{\mathrm{L}}(L)\left(\tau^2 - 1\right)}{1 + \frac{1}{\gamma_0}} \right), \tag{5.47}$$

where

- ☐ in inequality (5.44), we made the change, $s = \frac{\gamma_0}{P\zeta_1^{\mathrm{L}}\left(\sqrt{r^2+L^2}\right)}$, and disregarded the NLoS inter-cell interference. Thus, we only consider the probability that the signal power is larger than the aggregated inter-cell interference power from all LoS small cell BSs by at least a factor, γ_0, which results in an overestimation of the coverage probability, p^{cov}, in our example.
- ☐ In inequality (5.45), we further concentrated on the LoS inter-cell interference, and assumed that it is only contributed by the first-piece LoS path loss function, i.e. $u \in (r, \tau r]$, $(1 < \tau < +\infty)$.
- ☐ To get inequality (5.46), we used the following inequalities in step (5.45), which follow from the fact that the LoS probability function, $\Pr^{\mathrm{L}}(d)$, is a monotonically decreasing function with respect to the three-dimensional distance, d, between the typical UE and an arbitrary small cell BS:
  - ○ $\left(\frac{\sqrt{r^2+L^2}}{\sqrt{u^2+L^2}}\right)^{-\alpha_1^{\mathrm{L}}} < \left(\frac{\sqrt{r^2+L^2}}{\sqrt{\tau^2 r^2 + L^2}}\right)^{-\alpha_1^{\mathrm{L}}}$, $u \in (r, \tau r]$, and
  - ○ $\Pr^{\mathrm{L}}\left(\sqrt{u^2 + L^2}\right) > \Pr^{\mathrm{L}}\left(\sqrt{\tau^2 r^2 + L^2}\right)$, $u \in (r, \tau r]$.
- ☐ Finally, to obtain the result in inequality (5.47) we plugged the following equations into step (5.46):
  - ○ $\lim_{\lambda\to +\infty} \Pr^{\mathrm{L}}\left(\sqrt{\tau^2 r^2 + L^2}\right) = \Pr^{\mathrm{L}}(L)$,
  - ○ $\lim_{\lambda\to +\infty} \frac{\sqrt{r^2+L^2}}{\sqrt{\tau^2 r^2 + L^2}} = 1$,
  - ○ $\int_r^{\tau r} u\, du = \frac{1}{2}\left(\tau^2 - 1\right) r^2$ and
  - ○ $\lim_{\lambda\to +\infty} \pi r^2 \lambda = 1$. This last result on the limit follows because the typical coverage area, πr^2, in an HPPP is in the order of $\frac{1}{\lambda}$, and thus we made the above conclusion, such limit is equal to 1.

- Since the parameter, τ, can take any arbitrary finite value larger than 1, $1 < \tau < +\infty$, in our example, it can be shown that $\lim_{\lambda\to +\infty} \Pr\left[\frac{P\zeta_1^{\mathrm{L}}\left(\sqrt{r^2+L^2}\right)h}{I_{\mathrm{agg}}+N_0} > \gamma_0\right] = 0$. This is because, with a sufficiently large parameter, τ, we can reduce the left-hand side of the inequality in equation (5.47) to any arbitrarily small value.

- As a result, we can conclude that

$$\lim_{\lambda \to +\infty} T_1^L = \lim_{\lambda \to +\infty} \Pr \left[\frac{P \zeta_1^L \left(\sqrt{r^2 + L^2} \right) h}{I_{agg} + N_0} > \gamma_0 \right] = 0. \qquad (5.48)$$

Plugging equations (5.43) and (5.48) into equation (5.42), we can finally show that the coverage probability, p^{cov} – and in turn, the ASE, A^{ASE} – tend to 0, when the small cell BS density, λ, tends to infinity, $\lambda \to +\infty$, i.e. $\lim_{\lambda \to +\infty} A^{ASE} = \lim_{\lambda \to +\infty} p^{cov} = 0$.

This completes our proof.

6 The Impact of a Finite User Density on Ultra-Dense Wireless Networks

6.1 Introduction

In traditional outdoor macrocell deployments, such as in the widely deployed 3rd Generation Partnership Project (3GPP) ones, based on universal mobile telecommunication system (UMTS) and long-term evolution (LTE), cells are continuously transmitting cell-specific reference signals and broadcasting system information, regardless of the traffic activity in the cell. Such technologies are cell-centric, aimed to ease the tasks of the user equipment (UE). An important reason for this is to reduce the complexity of UE procedures, and facilitate those UEs with no data to transmit, i.e. in idle state, the detection of the presence of a cell. In other words, if there were no transmissions of reference signals and system information from a cell, there would be nothing for the UE to measure upon, and thus the UE would not be able to detect the presence of – and connect to – such cell [13]. This design principle makes sense in a large macrocell deployment, where there is a relatively high probability of at least one UE being active in a cell. Thus, there is no loss resulting from a continuous transmission of all these control signals, also known as "always-on" or pilot signals. Such always-on signals include, for example, signals for cell detection and selection, demodulation purposes and channel estimation, as well as the broadcast of system information. For more details on the basics of these always-on signals for UMTS and LTE, please refer to [213] and [214], respectively.

In a dense or ultra-dense deployment with many small cells, however, things are different. The probability of a small cell not serving any UE at a given time can be significantly high in some scenarios, and thus the always-on signals have two negative impacts [13]:

- they impose an upper limit on the achievable network energy performance; and
- they cause inter-cell interference, thereby reducing the achievable data rates.

Particularly, it should be highlighted that the downlink inter-cell interference scenario experienced by some UEs in dense deployments may be severe, even if the neighbouring cells are empty. These UEs may experience very low signal-to-interference-plus-noise ratios (SINRs) due to the inter-cell interference coming from the always-on signals of those neighbouring and potentially empty small cells, especially if there is a large amount of line-of-sight (LoS) propagation, as already shown in Chapters 4 and 5.

To address this important issue, in the past few years, a number of mechanisms have been introduced for turning on/off individual small cells – or at least their always-on signals – as a function of the traffic load to reduce the power consumption and the inter-cell interference. Some of these mechanisms are proprietary ones and others led to new standardized features. However, it is fair to say that they are not widely used as of today in the currently deployed small cell base station (BS) hardware, and it is expected that this feature will be more widely implemented in the fifth generation of radio technology.

One such proprietary mechanism is the idle mode capability for small cells presented in [215], which relies on the existence of macrocell coverage, and detects active UEs within the range of the small cell BS based on a rise in the measured uplink noise level. This procedure allows disabling the always-on signals and most processing operations at the small cell BS, when they are not needed to support active transmissions. This is achieved through a low-power sniffer capability in the small cell BS, which allows the detection of an active transmission from a UE to the underlying macrocell BS. When a UE located inside the coverage range of the small cell BS transmits to – or receives from – the macrocell BS, the sniffer detects a rise in the received power on the uplink frequency band. If this noise rise exceeds a predetermined threshold, the detected UE is deemed close enough to be potentially covered by the small cell BS, and thus the small cell BS is powered on. Otherwise, it remains inactive. Once the small cell BS is in active mode, the UE reports the received pilot signal strength from the small cell BS to the macrocell BS to which it is connected, and the UE is then handed over from the macrocell BS to the small cell BS. In a conceptually simple manner, while compatible with any standard due to its over-the-top implementation, this idle mode capability allows the small cell BS to switch off all pilot transmissions as well as the processing associated with the wireless reception when no UE is involved in an active transmission. However, it should be noted that this procedure requires macrocell coverage, since it relies on detecting transmissions from a UE to a macrocell BS. Therefore, the small cell BS needs to identify if sufficient macrocell coverage is available. This can be detected through measuring the macrocell BS pilot signals at the small cell BS, and/or by UE measurement reports. The idle mode capability must be deactivated in the small cell BS, if there is no sufficient macrocell overlay coverage. Please refer to [146] for more details on this idle mode capability.

Switching on/off a small cell BS may be simplified if other cells can provide basic coverage in the area handled by the small cell BS. However, another important drawback of this type of idle mode capability is that it may take many hundreds of milliseconds to transition from a dormant state to a fully active one. It also takes some time until the UE discovers a small cell BS that has just been turned on. Bearing in mind these delays, mechanisms for a significantly more rapid small cell on/off operation in a dense deployment, even at a millisecond level, were extensively discussed during the development of 3GPP LTE Release 12 [87]. Based on these discussions, it was decided to base this small cell on/off operation on the activation and deactivation mechanism in the carrier aggregation framework. As a result, this turning on/off operation is restricted to secondary cells in active mode, and the primary carrier is always on. When a secondary component carrier is turned off, there is no transmission of always-on signals, i.e. synchronization signals, cell-specific reference signals, channel state information (CSI) reference signals or system information from such deactivated cells. It should be noticed that although a carrier being completely silent leads to the best energy savings and the lowest inter-cell interference, it also implies that the UE cannot maintain synchronization to that carrier or perform any measurements, for example, mobility-related ones. To address these aspects, a new reference signal – the discovery reference signal – was introduced. This signal is transmitted less frequently and used by the UE to maintain synchronization and perform mobility measurements. Please refer to [216] for more details on these standardized small cell on/off enhancements.

To address the much more challenging performance requirements of today, the 3GPP new radio (NR) Release 15 breaks with the design principles of its predecessor, LTE, and has no backward compatibility requirements with it. Among such new design principles, the ultra-lean network design to reduce the energy consumption and mitigate the inter-cell interference from empty cells should be highlighted here. This ultra-lean design principle targets at minimizing as much as possible the always-on transmissions. By moving to a user-centric design, and changing the signalling and procedures required for the UE to reckon the presence of – and connect to – a cell, NR tackles the energy efficiency and inter-cell interference problem posed by always-on signals from the root [13]. As mentioned earlier, LTE follows a cell-centric design, which is heavily based on always-on signals, i.e. signals that are always present and use for cell detection and selection, demodulation purposes and channel estimation. In NR, however, many of these procedures have been revisited, and modified to mitigate the burden of always-on signals. For example, the 20 ms periodicity of the NR synchronization signal-block is four times longer than the corresponding 5 ms periodicity of the LTE primary synchronization channel (PSS) and secondary synchronization channel secondary synchronization channel (SSS). Importantly, and unlike LTE, NR does not include cell-specific reference signals. Reference signals for both demodulation and channel estimation are user-specific, and they are only transmitted when there is data to transmit [13].

With all these advancements on small cell technology in mind, and particularly considering that the model used in Chapters 3–5 did not consider either a finite UE density, $\rho < +\infty$, or the important benefits brought about by these idle mode capabilities in terms of inter-cell interference mitigation, it is fair to wonder about the impact of such realistic and soon-to-come features on the performance of a future ultra-dense network. In more detail, one may hope that the pessimistic conclusions of Chapters 4 and 5, i.e.

- the area spectral efficiency (ASE) Crawl, presented in Remarks 4.1 and 4.2, as well as
- the ASE Crash, summarized in Remarks 5.1 and 5.2,

respectively, can be addressed through this idle mode capabilities.

This gives rise to a fundamental question:

What is the impact of a simple idle mode capability at the small cell BSs on the performance of a practical ultra-dense network with a finite UE density?

In this chapter, we answer this fundamental question by means of in-depth theoretical analyses.

The rest of this chapter is organized as follows:

- Section 6.2 introduces the system model and the assumptions taken in the theoretical performance analysis framework presented in this chapter, considering a finite UE density and a simple idle mode capability at the small cell BSs, while embracing the channel model presented in Chapter 4, with both LoS and non-line-of-sight (NLoS) transmissions.
- Section 6.3 presents the theoretical expressions for the coverage probability and the ASE under the new assumptions.
- Section 6.4 provides results for a number of small cell BS deployments with different densities and characteristics. It also presents the conclusions drawn by this theoretical performance analysis, shedding new light on the importance of the larger number of small cell BSs with respect to that of UEs and their idle mode capabilities in a dense network. Importantly, and for completeness, this section also studies via system-level simulations the impact of a Rician multi-path fast fading – instead of a Rayleigh one – on the derived results and conclusions.
- Section 6.5, finally, summarizes the key takeaways of this chapter.

6.2   Updated System Model

To assess the impact of the finite UE density and the idle mode capabilities described in Section 6.1 on an ultra-dense network, in this section, we upgrade the system model described in Section 4.2, which has accounted for

- both LoS and NLoS transmissions and
- a practical user association strategy, based on the strongest received signal strength,

by incorporating the following features,

- a finite UE density and
- a simple idle mode capability at the small cell BSs.

To facilitate the reader understanding the complete system model used in this chapter, Table 6.1 provides a concise and updated summary of it.

In the following, we present how the system model is upgraded by including the finite UE density and the simple idle mode capability at the small cell BSs in more detail. Before that, however, it is important to note that the height difference between the antenna of the typical UE and that of an arbitrary small cell BS, considered in Chapter 5, is not taken into account in this one. This is to better understand – in

Table 6.1. System model

Model	Description	Reference
Transmission link		
Link direction	Downlink only	Transmissions from the small cell BSs to the UEs
Deployment		
Small cell BS deployment	HPPP with a finite density, $\lambda < +\infty$	Section 3.2.2
UE deployment	HPPP with full load, resulting in at least one UE per cell	Section 3.2.3
UE to small cell BS association		
Strongest small cell BS	UEs connect to the small cell BS providing the strongest received signal strength	Section 3.2.8 and references therein, equation (2.14)
Path loss		
3GPP UMi [153]	Multi-slope path loss with probabilistic LoS and NLoS transmissions	Section 4.2.1, equation (4.1)
	☐ LoS component	equation (4.2)
	☐ NLoS component	equation (4.3)
	☐ Exponential probability of LoS	equation (4.19)
Multi-path fast fading		
Rayleigh*	Highly scattered scenario	Section 3.2.7 and references therein equation (2.15)
Shadow fading		
Not modelled	For tractability reasons, shadowing is not modelled since it has no qualitative impact on the results, see Section 3.4.5	Section 3.2.6 references therein
Antenna		
Small cell BS antenna	Isotropic single-antenna element with 0 dBi gain	Section 3.2.4
UE antenna	Isotropic single-antenna element with 0 dBi gain	Section 3.2.4
Small cell BS antenna height	Not considered	—
UE antenna height	Not considered	—
Idle mode capability at small cell BSs		
Connection-aware	Small cell BSs with no UE in their coverage areas are switched off	Section 6.2.2
Scheduler at small cell BSs		
Round robin	UEs take turns in accessing the radio channel	Section 7.1

* Section 6.4.5 presents simulated results with Rician multi-path fast fading to demonstrate its impact on the obtained results.

isolation – the impact of the finite UE density and the idle mode capability at the small cell BSs on an ultra-dense network. As a result, all distances are two-dimensional in this chapter.

6.2.1   Finite User Equipment Density

To model the finite UE density, we consider that all UEs within the network scenario are active UEs,[1] and that such UEs are distributed according to a stationary homogeneous Poisson point process (HPPP), Φ^{UE}, of intensity, $\rho < +\infty$, in UEs/km^2 on the two-dimensional plane. Contrary to the description in Section 3.2, the UE density, $\rho < +\infty$, adopts finite reasonable values in this chapter, and thus we cannot assume that every small cell BS has at least one UE in its coverage area.

6.2.2   Idle Mode Capability at the Small Cell Base Stations

To model the idle mode capability at the small cell BSs, we consider a simple approach, in which a small cell BS is active if it has at least a UE within its coverage area. Otherwise, it is in the idle mode.

Given the previous assumptions on the small cell BS and the UE distributions, and embracing the above idle mode capability at the small cell BSs, we consider that the resulting active small cell BSs follow another stationary HPPP, $\tilde{\Phi}$, of intensity, $\tilde{\lambda}$, in BSs/km^2.

With regard to this assumption on the modelling of the active small cell BSs, it is important to note that, up to now, there is no theoretical proof showing that the set of active small cell BSs follows an HPPP, even if the set of small cell BSs follows an HPPP. Theoretically speaking, the activation of each small cell BS depends on the activity of the UEs in its vicinity, and thus both processes should be correlated. Having said that, this HPPP assumption has been widely used in the literature, such as in seminal works [160, 217–219]. To further back up this assumption, both theoretical and simulated results will be presented in Section 6.4, which even if they have been obtained through different methodologies,[2] they obtain the very same results. The intuition behind such equal results is that since UEs are distributed according to a stationary HPPP, and no channel correlations are considered in this framework, the activation and deactivation of each small cell BS – which depends on the activity of the UEs, and thus on their distribution – is uniformly and randomly distributed across the entire network, which leads to the assumed HPPP.

With this in mind, it is important to note the following properties of the model:

[1] A UE is deemed as an active UE in this downlink-based model, if its serving small cell BS has packets to transmit to it. In other words, an active UE has packets to receive.
[2] While no assumption was made on the active small cell BS distribution in the simulations, and small cell BSs were activated based on the activity of the UEs, the HPPP assumption was taken to obtain the analytical results.

Table 6.2. Key performance indicators

Metric	Formulation	Reference
Coverage probability, p^{cov}	$p^{\text{cov}}(\lambda, \gamma_0) = \text{Pr}\left[\Gamma > \gamma_0\right] = \bar{F}_{\Gamma}\left(\gamma_0\right)$	Section 2.1.12, equation (2.22)
ASE, $A^{\text{ASE}*}$	$A^{\text{ASE}}(\lambda, \gamma_0) = \frac{\lambda}{\ln 2} \int_{\gamma_0}^{+\infty} \frac{p^{\text{cov}}(\lambda, \gamma)}{1+\gamma} d\gamma$ $+ \lambda \log_2\left(1 + \gamma_0\right) p^{\text{cov}}(\lambda, \gamma_0)$	Section 2.1.12, equation (2.25)

* Note that the idle mode capability at the small cell BSs will change the ASE definition, as will be shown in Section 6.3.3.

- The active small cell BS density, $\tilde{\lambda}$, is no larger than the small cell BS density, λ, i.e. $\tilde{\lambda} \leq \lambda$.
- The active small cell BS density, $\tilde{\lambda}$, is no larger than the finite UE density, $\rho < +\infty$, i.e. $\tilde{\lambda} \leq \rho$, since one UE is served by at most one small cell BS in our analysis.[3]

From the above two points, it follows that a larger finite UE density, $\rho < +\infty$, leads to a larger or equal, but never smaller, active small cell BS density, $\tilde{\lambda}$.

6.3 Theoretical Performance Analysis and Main Results

In this section, given the definition of the coverage probability, p^{cov}, and the ASE, A^{ASE}, also summarized for convenience in Table 6.2, we derive expressions for these two key performance indicators, while considering the finite UE density, $\rho < +\infty$, and the simple idle mode capability at the small cell BSs presented in Section 6.2. These new expressions will allow one to assess the performance of an ultra-dense network under these much more realistic network assumptions.

6.3.1 Coverage Probability

In the following, we present the new main result on the coverage probability, p^{cov}, through Theorem 6.3.1, considering the above introduced system model, where

- the finite UE density, $\rho < +\infty$, and
- the simple idle mode capability at the small cell BSs

should be highlighted. Readers interested in the research article originally presenting these results are referred to [88].

THEOREM 6.3.1   *Considering the path loss model in equation (4.1), and the strongest small cell BS association presented in Section 4.2, the coverage probability, p^{cov}, can be derived as*

---

[3] No coordinated inter-small cell BS transmissions/receptions are considered in this chapter.

$$p^{\text{cov}}(\lambda, \gamma_0) = \sum_{n=1}^{N} \left(T_n^{\text{L}} + T_n^{\text{NL}} \right), \tag{6.1}$$

where

-

$$T_n^{\text{L}} = \int_{d_{n-1}}^{d_n} \Pr \left[\frac{P \zeta_n^{\text{L}}(r) h}{I_{\text{agg}} + P^{\text{N}}} > \gamma_0 \right] f_{R,n}^{\text{L}}(r) dr, \tag{6.2}$$

-

$$T_n^{\text{NL}} = \int_{d_{n-1}}^{d_n} \Pr \left[\frac{P \zeta_n^{\text{NL}}(r) h}{I_{\text{agg}} + P^{\text{N}}} > \gamma_0 \right] f_{R,n}^{\text{NL}}(r) dr \tag{6.3}$$

and
- d_0 and d_N are defined as 0 and $+\infty$, respectively.

Moreover, the probability density function (PDF), $f_{R,n}^{\text{L}}(r)$, and the PDF, $f_{R,n}^{\text{NL}}(r)$, within the range, $d_{n-1} < r \le d_n$, are given by

$$f_{R,n}^{\text{L}}(r) = \exp\left(-\int_0^{r_1} \left(1 - \Pr^{\text{L}}(u)\right) 2\pi u \lambda \, du\right)$$

$$\times \exp\left(-\int_0^{r} \Pr^{\text{L}}(u) 2\pi u \lambda \, du\right) \Pr_n^{\text{L}}(r) 2\pi r \lambda, \tag{6.4}$$

and

$$f_{R,n}^{\text{NL}}(r) = \exp\left(-\int_0^{r_2} \Pr^{\text{L}}(u) 2\pi u \lambda \, du\right)$$

$$\times \exp\left(-\int_0^{r} \left(1 - \Pr^{\text{L}}(u)\right) 2\pi u \lambda \, du\right) \left(1 - \Pr_n^{\text{L}}(r)\right) 2\pi r \lambda, \tag{6.5}$$

where

- r_1 *and* r_2 *are determined by*

$$r_1 = \arg_{r_1} \left\{ \zeta^{\text{NL}}(r_1) = \zeta_n^{\text{L}}(r) \right\} \tag{6.6}$$

and

$$r_2 = \arg_{r_2} \left\{ \zeta^{\text{L}}(r_2) = \zeta_n^{\text{NL}}(r) \right\}, \tag{6.7}$$

respectively.

The probability, $\Pr\left[\frac{P\zeta_n^{\mathrm{L}}(r)h}{I_{\mathrm{agg}}+P^{\mathrm{N}}} > \gamma_0\right]$, is computed as

$$\Pr\left[\frac{P\zeta_n^{\mathrm{L}}(r)h}{I_{\mathrm{agg}} + P^{\mathrm{N}}} > \gamma_0\right] = \exp\left(-\frac{\gamma_0 P^{\mathrm{N}}}{P\zeta_n^{\mathrm{L}}(r)}\right) \mathscr{L}_{I_{\mathrm{agg}}}^{\mathrm{L}}\left(\frac{\gamma_0}{P\zeta_n^{\mathrm{L}}(r)}\right), \qquad (6.8)$$

where

- $\mathscr{L}_{I_{\mathrm{agg}}}^{\mathrm{L}}(s)$ is the Laplace transform of the aggregated inter-cell interference random variable, I_{agg}, for the LoS signal transmission evaluated at point, $s = \frac{\gamma_0}{P\zeta_n^{\mathrm{L}}(r)}$, which can be expressed as

$$\mathscr{L}_{I_{\mathrm{agg}}}^{\mathrm{L}}(s) = \exp\left(-2\pi\tilde\lambda \int_r^{+\infty} \frac{\Pr^{\mathrm{L}}(u)u}{1 + \left(sP\zeta^{\mathrm{L}}(u)\right)^{-1}} du\right)$$

$$\times \exp\left(-2\pi\tilde\lambda \int_{r_1}^{+\infty} \frac{\left[1 - \Pr^{\mathrm{L}}(u)\right]u}{1 + \left(sP\zeta^{\mathrm{NL}}(u)\right)^{-1}} du\right). \qquad (6.9)$$

Moreover, the probability, $\Pr\left[\frac{P\zeta_n^{\mathrm{NL}}(r)h}{I_{\mathrm{agg}}+P^{\mathrm{N}}} > \gamma_0\right]$, is computed as

$$\Pr\left[\frac{P\zeta_n^{\mathrm{NL}}(r)h}{I_{\mathrm{agg}} + P^{\mathrm{N}}} > \gamma_0\right] = \exp\left(-\frac{\gamma_0 P^{\mathrm{N}}}{P\zeta_n^{\mathrm{NL}}(r)}\right) \mathscr{L}_{I_{\mathrm{agg}}}^{\mathrm{NL}}\left(\frac{\gamma_0}{P\zeta_n^{\mathrm{NL}}(r)}\right), \qquad (6.10)$$

where

- $\mathscr{L}_{I_{\mathrm{agg}}}^{\mathrm{NL}}(s)$ is the Laplace transform of the aggregated inter-cell interference random variable, I_{agg}, for the NLoS signal transmission evaluated at point, $s = \frac{\gamma_0}{P\zeta_n^{\mathrm{NL}}(r)}$, which can be written as

$$\mathscr{L}_{I_{\mathrm{agg}}}^{\mathrm{NL}}(s) = \exp\left(-2\pi\tilde\lambda \int_{r_2}^{+\infty} \frac{\Pr^{\mathrm{L}}(u)u}{1 + \left(sP\zeta^{\mathrm{L}}(u)\right)^{-1}} du\right)$$

$$\times \exp\left(-2\pi\tilde\lambda \int_r^{+\infty} \frac{\left[1 - \Pr^{\mathrm{L}}(u)\right]u}{1 + \left(sP\zeta^{\mathrm{NL}}(u)\right)^{-1}} du\right). \qquad (6.11)$$

Proof See Appendix A $\square$

Analyzing Theorem 6.3.1, the following important conclusion, summarized in Lemma (6.3.2), can be drawn:

LEMMA 6.3.2   *The coverage probability, p^{cov}, when considering the finite UE density, $\rho < +\infty$, and the simple idle mode capability at the small cell BSs presented earlier, is always no smaller than when all small cell BSs are active and have no idle model capability.*

Proof See Appendix B. $\square$

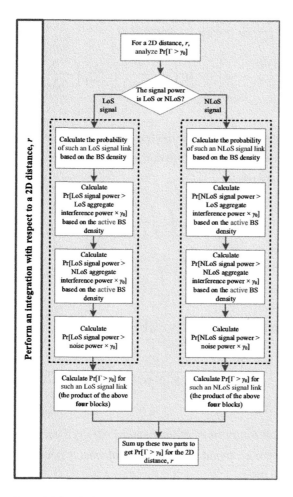

Figure 6.1 Logical steps within the standard stochastic geometry framework to obtain the results in Theorem 6.3.1, considering a finite UE density, $\rho < +\infty$, and a simple idle mode capability at the small cell BSs.

To facilitate the understanding on how to practically obtain this new coverage probability, p^{cov}, Figure 6.1 illustrates the pseudo code that depicts the necessary enhancements to the traditional stochastic geometry framework presented in Chapter 3 – and updated in Chapter 4 – to compute the new results of Theorem 6.3.1.

Compared with the logic illustrated in Figure 4.2, it is important to note that the inter-cell interference does not come from all the small cell BSs, but only from the active ones.

Diving a little bit further into the details, and comparing Theorem 6.3.1 with Theorem 4.3.1, where all small cell BSs are active with at least one UE per small cell BS, it is important to note that:

- The impact of the serving small cell BS selection on the coverage probability, p^{cov}, is measured by equations (6.4) and (6.5). These expressions are based on the small cell BS density, λ, and not on the active small cell BS density, $\tilde{\lambda}$. This is the same in both theorems.

- The impact of the aggregated inter-cell interference, I_{agg}, on the coverage probability, p^{cov}, is measured by equations (6.9) and (6.11). Since only the active small cell BS generate inter-cell interference, these expressions are thus based on the active small cell BS density, $\tilde{\lambda}$, and not on the small cell BS density, λ. This is different from Theorem 4.3.1.
- The derivation of the active small cell BS density, $\tilde{\lambda}$, is not trivial, and it is presented in Section 6.3.2.

6.3.2 Active Small Cell Base Station Density

In this section, we elaborate on the theoretical computation of the active small cell BS density, $\tilde{\lambda}$.

To carry out this task, we build on previous contributions. Particularly, we use the result in [217], where the authors derived an approximate expression for the active small cell BS density, $\tilde{\lambda}$, based on the distribution of the Voronoi cell size, assuming that each UE associates, not with the strongest, but with the nearest small cell BS. The following expression for the active small cell BS density, $\tilde{\lambda}$, is the main result obtained in [217]:

$$\tilde{\lambda}^{\mathrm{minDis}} \approx \lambda \left[1 - \frac{1}{\left(1 + \frac{\rho}{q\lambda}\right)^q} \right] \triangleq \lambda_0 \left(\lambda, \rho, q\right), \qquad (6.12)$$

where

- $\tilde{\lambda}^{\mathrm{minDis}}$ is the active small cell BS density based on the nearest small cell BS association,
- λ is the small cell BS density,
- ρ is the finite UE density and
- q is a fitting parameter, which takes an empirical value of 3.5.

It is important to note that, up to now, the exact expression for the active small cell BS density, $\tilde{\lambda}^{\mathrm{minDis}}$, is still unknown, but the above approximation has been shown to be quite accurate in a number of recent works [217, 219, 220], where the nearest small cell BS association is embraced. This approximation, however, cannot be used in the more realistic modelling targeted in this book, where each UE connects, not to the nearest, but to the strongest small cell BS. Thus, a new approximate expression for the active small cell BS density, $\tilde{\lambda}$, considering both probabilistic LoS and NLoS transmissions and the strongest small cell BS association, is thus needed to carry on with the analysis.

In the following, a lower and an upper bound as well as a tight approximation of the active small cell BS density, $\tilde{\lambda}$, embracing the assumptions made in this chapter, are presented.

A Lower Bound of the Active Small Cell BS Density

The following Theorem 6.3.3 proposes that the active small cell BS density, $\tilde{\lambda}^{\text{minDis}}$, based on the nearest small cell BS association, presented in equation (6.12), is a lower bound of the active small cell BS density, $\tilde{\lambda}$, based on the strongest small cell BS association.

THEOREM 6.3.3   *Based on the path loss model in equation (4.1), and the strongest small cell BS association presented in Section 4.2, the active small cell BS density, $\tilde{\lambda}$, can be lower-bounded by*

$$\tilde{\lambda} \geq \tilde{\lambda}^{\text{minDis}} \triangleq \tilde{\lambda}^{\text{LB}}. \tag{6.13}$$

Proof   See Appendix C.   □

Intuitively speaking, the proof of Theorem 6.3.3 states that from a typical UE's point of view, the equivalent small cell BS density, $\tilde{\lambda}$, based on the strongest small cell BS association is no less than the equivalent small cell BS density, $\tilde{\lambda}^{\text{minDis}}$, based on the nearest small cell BS association, i.e. $\tilde{\lambda} \geq \tilde{\lambda}^{\text{minDis}}$.

The tightness of this lower bound, $\tilde{\lambda}^{\text{LB}}$, will be verified using numerical results in Section 6.4.2.

An Upper Bound of the Active Small Cell BS Density

The following Theorem 6.3.4 proposes an upper bound for the active small cell BS density, $\tilde{\lambda}$, based on the strongest small cell BS association.

THEOREM 6.3.4   *Based on the path loss model in equation (4.1), and the strongest small cell BS association presented in Section 4.2, the active small cell BS density, $\tilde{\lambda}$, can be upper-bounded by*

$$\tilde{\lambda} \leq \lambda \left(1 - Q^{\text{off}}\right) \triangleq \tilde{\lambda}^{\text{UB}}, \tag{6.14}$$

where

•

$$Q^{\text{off}} = \lim_{r_{\text{max}} \to +\infty} \sum_{k=0}^{+\infty} \{\Pr[w \not\sim b]\}^k \frac{\lambda_\Omega^k e^{-\lambda_\Omega}}{k!}, \tag{6.15}$$

where
□ $\lambda_\Omega = \rho \pi r_{\text{max}}^2$ *and*
□ $\Pr[w \not\sim b]$ *is the probability of a UE, u, not being associated with small cell BS, b, and it can be computed as*

$$\Pr[w \not\sim b] = \int_0^{r_{\text{max}}} \Pr[w \not\sim b \mid r] \frac{2r}{r_{\text{max}}^2} dr, \tag{6.16}$$

where

$$\Pr[w \sim b|\, r] = \left[F_R^{\mathrm{L}}(r) + F_R^{\mathrm{NL}}(r_1) \right] \Pr^{\mathrm{L}}(r)$$
$$+ \left[F_R^{\mathrm{L}}(r_2) + F_R^{\mathrm{NL}}(r) \right] \left[1 - \Pr^{\mathrm{L}}(r) \right], \qquad (6.17)$$

where

$$F_R^{\mathrm{L}}(r) = \int_0^r f_R^{\mathrm{L}}(u)du, \qquad (6.18)$$

$$F_R^{\mathrm{NL}}(r) = \int_0^r f_R^{\mathrm{NL}}(u)du \qquad (6.19)$$

and

◇ r_1 *and* r_2 *are defined in equations (6.6) and (6.7), respectively.*

Proof   See Appendix D.   □

Intuitively speaking, the proof of Theorem 6.3.4 takes a disc area, $\mathcal{D}$,

- centred at the typical small cell BS,
- with radius, r_{max},

and calculates the probability, Q^{off}, of such a typical small cell BS entering in idle mode, i.e. having no UE associated with it, inside the disc area, $\mathcal{D}$.

It is important to note that the correlation between the UE association of two nearby UEs is not considered in the calculation of the probability, Q^{off}. In reality, if a UE, k, is not associated with the small cell BS, b, this may imply that a nearby UE, k', may also not be associated with the small cell BS, b, with a large probability. This correlation is not considered in the analysis above, which leads to an underestimation of the probability, Q^{off}, and in turn, to an overestimation of the active small cell BS density, $\tilde{\lambda}$ – see equation (6.14).

The tightness of this upper bound, $\tilde{\lambda}^{\mathrm{UB}}$, will be verified using numerical results later in Section 6.4.2.

An Approximation of the Active Small Cell BS Density

Considering the tightness of the lower bound, $\tilde{\lambda}^{\mathrm{LB}}$, which will be shown in Section 6.4.2, and the fact that such a lower bound, $\tilde{\lambda}^{\mathrm{LB}}$, is defined by an increasing function of the fitting parameter, q, the following Corollary 6.3.5 is used to obtain a tight approximation of the active small cell BS density, $\tilde{\lambda}$.

COROLLARY 6.3.5   *Based on the path loss model in equation (4.1), and the strongest small cell BS association presented in Section 4.2, the active small cell BS density, $\tilde{\lambda}$, can be approximated by*

$$\tilde{\lambda} \approx \lambda_0\left(\lambda, \rho, q^*\right),\tag{6.20}$$

where

- q^* *is the optimal fitting parameter, which takes an empirical value within the range,*

$$3.5 \leq q^* \leq \arg_x \left\{\lambda_0\left(\lambda, \rho, x\right) = \tilde{\lambda}^{\mathrm{UB}}\right\},\tag{6.21}$$

where

☐ $\tilde{\lambda}^{\mathrm{UB}}$ *is the upper bound for the active small cell BS density, $\tilde{\lambda}$, based on the strongest small cell BS association, computed from equation (6.14).*

The optimal fitting parameter, q^*, in Corollary 6.3.5 can be obtained according to the lower bound, $\tilde{\lambda}^{\mathrm{LB}}$, and the upper bound, $\tilde{\lambda}^{\mathrm{UB}}$, presented in Theorems 6.3.3 and 6.3.4, respectively. It is important to note that such an optimal value, q^*, depends on a number of parameters, among others, the small cell BS density, λ, the UE density, ρ, and the specifics of the path loss model, which in this case are given by equations (4.4) and (4.5). Generally, the optimal fitting parameter, q^*, can be calculated using offline solvers together with the deterministic bounds found for it in Corollary 6.3.5. To achieve a balance between accuracy and tractability, we assume that the optimal fitting parameter, q^*, is only a function of the UE density, ρ. This assumption avoids the complication of having the optimal fitting parameter, q^*, dependent on the small cell BS density, λ, inside the function of the active small cell BS density, $\tilde{\lambda}$, which is already a function of the small cell BS density, λ. With this in mind, the average difference between the approximate results and the simulated ones for all possible values of the small cell BS density, λ, can be measured by using, for example, the mean square error (MSE), and the optimal fitting parameter, q^*, can thus be searched using the minimum mean square error (MMSE) criteria. As an example, the bisection method may be used for this task [209], minimizing the difference between the approximate results of the active small cell BS, $\tilde{\lambda}$, whose range is characterized by equation (6.14), and the simulated ones. In more detail, we can set a left point and right point for the fitting parameter, q^*, as 3.5 and $\arg\left\{\lambda_0\left(\lambda, \rho, x\right) = \tilde{\lambda}^{\mathrm{UB}}\right\}$, respectively. Then, a middle point can be found as the average of the left and right points. By comparing with the simulation results, if the middle point represents a better estimation of the optimal fitting parameter, q^*, than the left point according to the MMSE criteria, then the middle point will be used as the new left point. Otherwise, it will be used as the new right point. The process continues until the MSE is small enough to yield a satisfactory estimation of the optimal fitting parameter, q^*.

For the sake of complexity, however, it should be noted that, while computing the optimal fitting parameter, q^*, as a function of the small cell BS density, λ, and the UE density, ρ, and the specifics of the path loss model ensures the most accurate approximation for the active small cell BS density, $\tilde{\lambda}$, it also leads to the computationally intensive process. To avoid such a complication, the authors in [217] proposed to use a constant fitting parameter, q, in the approximated expression across all small cell

BS and UE densities. In this chapter, however, due to the introduction of the LoS and NLoS transmissions, such an approach may not be suitable.

The tightness of the approximation, $\lambda_0(q^*)$, will be verified using numerical results later in Section 6.4.2.

Asymptotic Behaviour of the Active Small Cell BS Density

For the sake of completeness, and because it will be useful in the subsequent chapters, let us derive in the following lemma the limit of the active small cell BS density, $\tilde{\lambda}$, when the small cell BS density, $\tilde{\lambda}$, tends to infinity, i.e. $\lim_{\lambda \to +\infty} \tilde{\lambda}$.

As a disclaimer, before proceeding with the result, it is important to note that one UE is served by at most one small cell BS in this book. In other words, no inter-small cell BS coordination techniques are considered.

LEMMA 6.3.6   *For a given finite UE density, $\rho < +\infty$, the limit, $\lim_{\lambda \to +\infty} \tilde{\lambda}$, tends to a finite UE density, $\rho < +\infty$, i.e.*

$$\lim_{\lambda \to +\infty} \tilde{\lambda} = \rho. \tag{6.22}$$

Proof   The proof is straightforward since an infinite small cell BSs density, λ, serving a finite UE density, $\rho < +\infty$, would lead to the extreme case of the one-UE-per-BS limit. Thus, the limit of the active small cell BS density, $\tilde{\lambda}$, equals to the UE density, ρ. $\square$

6.3.3   Area Spectral Efficiency

In the following, we present the theoretical results on the ASE, A^{ASE}, considering the above introduced system model. Importantly, it should be noted that not all the small cell BSs – but only the active ones – contribute to the ASE, A^{ASE}. As a result, the definition of the ASE, A^{ASE}, in Section 2.1.12 does not apply, and equation (2.23) has to be reformulated as

$$A^{\text{ASE}}(\lambda, \gamma_0) = \tilde{\lambda} \int_{\gamma_0}^{\infty} \log_2 (1 + \gamma) f_{\Gamma}(\gamma) d\gamma, \tag{6.23}$$

where

- the small cell BS density, λ, in the original formulation has been replaced by the active small cell BS density, $\tilde{\lambda}$, for the reasons stated above.

For the sake of generality, and because the active small cell BS density, $\tilde{\lambda}$, is intimately related to the small cell BS density, λ, we still keep the nomenclature of the ASE variable as $A^{\text{ASE}}(\lambda, \gamma_0)$, and not as $A^{\text{ASE}}(\tilde{\lambda}, \gamma_0)$.

Following the same steps as in the original formulation, using the partial integration theorem, the ASE, A^{ASE}, can be obtained as

$$A^{ASE}(\lambda, \gamma_0) = \frac{\tilde{\lambda}}{\ln 2} \int_{\gamma_0}^{+\infty} \frac{p^{cov}(\lambda, \gamma)}{1 + \gamma} d\gamma$$

$$+ \tilde{\lambda} \log_2 (1 + \gamma_0) p^{cov}(\lambda, \gamma_0). \qquad (6.24)$$

Table 6.2 captures this result.

Using this new ASE expression, and similarly as in Chapters 4 and 5, plugging the coverage probability, p^{cov} – obtained from Theorem 6.3.1 – into equation (2.24) to compute the PDF, $f_\Gamma(\gamma)$, of the SINR, Γ, of the typical UE, we can obtain the ASE, A^{ASE}.

6.3.4   Computational Complexity

To calculate the coverage probability, p^{cov}, presented in Theorem 6.3.1, three folds of integrals are still required for the calculation of $\left\{f_{R,n}^{Path}(r)\right\}$, $\left\{\mathscr{L}_{I_{agg}}\left(\frac{\gamma_0}{P\zeta_n^{Path}(r)}\right)\right\}$ and $\left\{T_n^{Path}\right\}$, respectively, where the string variable, *Path*, takes the value of "L" (for the LoS case) or "NL" (for the NLoS case). Note that an additional fold of integral is needed for the calculation of the ASE, A^{ASE}, making it a four-fold integral computation.

6.4   Discussion

In this section, we use numerical results from static system-level simulations to evaluate the accuracy of the above theoretical performance analysis, and study the impact of a finite UE density, $\rho < +\infty$, and a simple idle mode capability at the small cell BSs on an ultra-dense network in terms of the coverage probability, p^{cov}, and the ASE, A^{ASE}.

As a reminder, it should be noted that to obtain the results on the coverage probability, p^{cov}, and the ASE, A^{ASE}, one should now plug equations (4.17) and (4.19), using two-dimensional distances, into Theorem 6.3.1, and follow the subsequent derivations. In this manner, a finite UE density, $\rho < +\infty$, and a simple idle mode capability at the small cell BSs can be incorporated in the analysis.

6.4.1   Case Study

To assess the impact of a finite UE density, $\rho < +\infty$, and a simple idle mode capability at the small cell BSs on an ultra-dense network, the same 3GPP case study as in Chapter 4 is used to allow an apple-to-apple comparison. For readers interested in a more detailed description of this 3GPP case study, please refer to Table 6.1 and Section 4.4.1.

For the sake of presentation, please recall that the following parameters are used in our 3GPP case study:

- maximum antenna gain, $G_M = 0\,dB$,
- path loss exponents, $\alpha^L = 2.09$ and $\alpha^{NL} = 3.75$ – considering a carrier frequency of 2 GHz,
- reference path losses, $A^L = 10^{-10.38}$ and $A^{NL} = 10^{-14.54}$,
- transmit power, $P = 24\,dBm$, and
- noise power, $P^N = -95\,dBm$ – including a noise figure of 9 dB at the UE.

Antenna Configurations

In this chapter, the height difference between the antenna height of the typical UE and that of an arbitrary small cell BS is not considered, as we want to isolate the performance impact of the antenna height from that of the proportional fair (PF) scheduler at the small cell BSs.

UE Densities

As for the finite UE density, $\rho < +\infty$, we study the following three cases in this performance evaluation, $\rho = \{100, 300, 600\}$ UEs/km^2.

Small Cell BS Densities

Since antenna heights are not considered in this chapter, to embrace the minimum transmitter-to-receiver distance of the selected path loss model of 10 m, while still using a simple system model based on HPPP deployments, we analyze small cell BS densities up to $\lambda = 10^4$ BSs/km^2 in our studies hereafter.

Benchmark

In this performance evaluation, we use, as the benchmark,

- the results of the analysis with a single-slope path loss model, presented earlier in Section 3.4, as well as
- those with both LoS and NLoS transmissions, presented in Section 4.4.

Note that the results of the former were obtained with a reference path loss, $A^{NL} = 10^{-14.54}$, and a path loss exponent, $\alpha = 3.75$.

6.4.2   Active Small Cell Base Station Density

In this section, we show the accuracy of the theoretical analysis presented in Section 6.3.2 for the calculation of the active small cell BS density, $\tilde{\lambda}$, as a function of both the small cell BS, λ, and the finite UE density, $\rho < +\infty$.

To this end, Figure 6.2 shows the simulated results on the active small cell BS density, $\tilde{\lambda}$, for a wide range of small cell BS densities, λ, and for the following three UE densities, $\rho = \{100, 300, 600\}$ UEs/km^2. These simulated results will serve as ground truth for our theoretical evaluation.

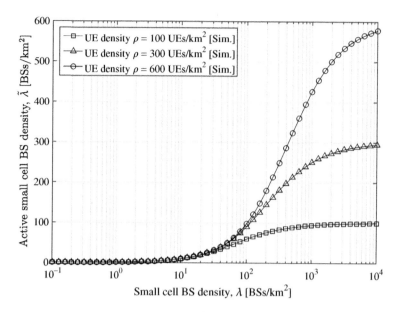

Figure 6.2 Active small cell BS density, $\tilde{\lambda}$, versus the small cell BS density, λ.

From this figure, one can observe that

- the active small cell BS density, $\tilde{\lambda}$, increases as the small cell BS density, λ, increases and that
- the active small cell BS density, $\tilde{\lambda}$, is upper bounded by the finite UE density, $\rho < +\infty$, as one UE activates at most one small cell BS for its service – no multi-cell coordination schemes are considered here.

Before diving into the details of these results, it should be mentioned that, considering Corollary 6.3.5, and as explained earlier, a bisection search was conducted to numerically find the optimal fitting parameter, q^*, for computing the approximated active small cell BS density, $\tilde{\lambda}$. Based on the MMSE criteria, and for the various finite UE densities under analysis, $\rho = \{100, 300, 600\}$ UEs/km^2, such optimal fitting parameters are $q^* = \{4.73, 4.18, 3.97\}$, respectively.

To assess the accuracy of the theoretical analysis of the active small cell BS density, $\tilde{\lambda}$, Figures 6.3–6.5 show the average error between the simulated values – ground truth – and the derived lower bound, $\tilde{\lambda}^{LB}$, the derived upper bound, $\tilde{\lambda}^{UB}$, and the approximation, $\lambda_0(q^*)$, in Section 6.3.2. Note that each of the mentioned figures deals with one of the different UE densities under study, $\rho = \{100, 300, 600\}$ UEs/km^2, and that a fitting parameter, $q = 3.5$, is used for the lower bound, $\tilde{\lambda}^{LB}$.

From these three figures, one can clearly see that the analyzed average errors follow the same trend for the three different UE densities under study, $\rho = \{100, 300, 600\}$ UEs/km^2, i.e. the same qualitative results with some quantitative deviations. With this in mind, we focus on Figure 6.4, which has a finite UE density, $\rho = 300$ UEs/km^2, to analyze the results. The following conclusions can be drawn:

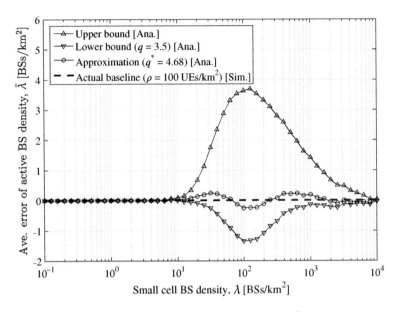

Figure 6.3 Average error of the active small cell BS density, $\tilde{\lambda}$, with a finite UE density, $\rho = 100\,\text{UEs/km}^2$, versus the small cell BS density, λ.

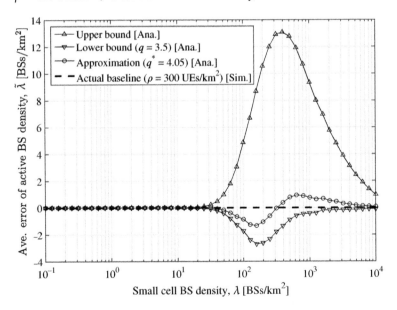

Figure 6.4 Average error of the active small cell BS density, $\tilde{\lambda}$, with a finite UE density, $\rho = 300\,\text{UEs/km}^2$, versus the small cell BS density, λ.

- The lower bound, $\tilde{\lambda}^{\text{LB}}$, and upper bound, $\tilde{\lambda}^{\text{LB}}$, derived in this chapter behave as such, i.e. the lower bound is always smaller (showing negative errors) and the upper bound is always lager (showing positive errors) than the simulation baseline, respectively.

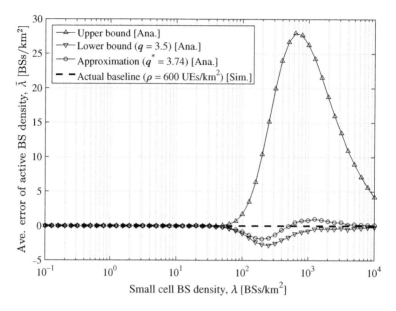

Figure 6.5 Average error of the active small cell BS density, $\tilde{\lambda}$, with a finite UE density, $\rho = 600$ UEs/km^2, versus the small cell BS density, λ.

- The upper bound, $\tilde{\lambda}^{\mathrm{UB}}$, is tighter than the lower bound, $\tilde{\lambda}^{\mathrm{LB}}$, when the small cell BS density, λ, is sparse, e.g. $\lambda < 30$ BSs/km^2.
- The lower bound, $\tilde{\lambda}^{\mathrm{LB}}$, is much tighter than the upper bound, $\tilde{\lambda}^{\mathrm{UB}}$, when the small cell BS density, λ, is dense and ultra-dense, e.g. $\lambda > 100$ BSs/km^2.
- The presented approximation, $\lambda_0(q^*)$, performs better than both the lower bound, $\tilde{\lambda}^{\mathrm{LB}}$, and the upper bound, $\tilde{\lambda}^{\mathrm{LB}}$. For example, for a finite UE density, $\rho = 300$ UEs/km^2, and a fitting parameter, $q^* = 4.18$, the maximum error resulting from the approximation, $\lambda_0(q^*)$, is around ± 0.5 BSs/km^2, while those given by the upper bound, $\tilde{\lambda}^{\mathrm{UB}}$, and the lower bound, $\tilde{\lambda}^{\mathrm{LB}}$, are around 12 BSs/km^2 and -2 BSs/km^2, respectively.

Based on the above analysis and results, it can be concluded that the approximation, $\lambda_0(q^*)$, is tight, and useful for the analysis of the coverage probability, p^{cov}, and the ASE, A^{ASE}.

6.4.3 Coverage Probability Performance

Figure 6.6 shows the coverage probability, p^{cov}, for an SINR threshold, $\gamma_0 = 0$ dB, an optimal fitting parameter, $q^* = 4.18$, and the following four configurations:

- Conf. a): single-slope path loss model with an infinite number of UEs – or at least one UE per cell – thus all small cell BSs are active (analytical results).

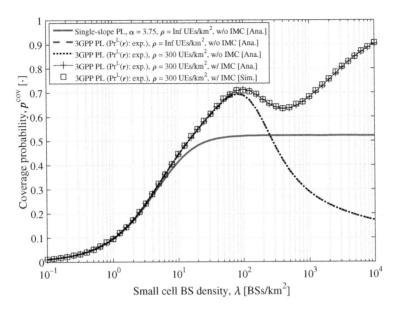

Figure 6.6 Coverage probability, p^{cov}, versus the small cell BS density, λ.

- Conf. b): multi-slope path loss model with both LoS and NLoS transmissions, and an infinite number of UEs – or at least one UE per cell – thus all small cell BSs are active (analytical results).
- Conf. c): multi-slope path loss model with both LoS and NLoS transmissions, a finite UE density, $\rho < +\infty$, but with no simple idle mode capability at the small cell BSs, thus all small cell BSs are active (analytical results).
- Conf. d): multi-slope path loss model with both LoS and NLoS transmissions, a finite UE density, $\rho < +\infty$, and a simple idle mode capability at the small cell BSs, thus not all small cell BSs are active (analytical results).
- Conf. e): multi-slope path loss model with both LoS and NLoS transmissions, a finite UE density, $\rho < +\infty$, and a simple idle mode capability at the small cell BSs, thus not all small cell BSs are active (simulated results).

In this figure, a finite UE density, $\rho = 300 \, \mathrm{UEs/km}^2$, is selected for the analysis. Note that Conf. a) and Conf. b) were already presented and discussed in Chapters 3 and 4, respectively, and they are incorporated for comparison purposes in Figure 6.6.

As can be observed from Figure 6.6, the analytical results, Conf. d), match well the simulation results, Conf. e). This corroborates the accuracy of Theorem 6.3.1 – and the derivations therein. Due to such significant accuracy, and since the results of the ASE, A^{ASE}, are computed based on the results of the coverage probability, p^{cov}, only analytical results are considered in our discussion hereafter in this chapter.

From Figure 6.6, we can also observe that:

- The single-slope path loss model with all small cell BSs activated – Conf. a) – results in an invariant SINR, Γ, for the typical UE, independent of the small cell BS density, λ, in the dense small cell BS density regimes, and in turn, the coverage probability, p^{cov}, remains constant, as such a small cell BS density, λ, increases. See Chapter 3 for more details.
- The multi-slope path loss model with both LoS and NLoS transmissions with all small cell BSs activated – Conf. b) and Conf. c) – offers a different behaviour, i.e.
  - ☐ when the small cell BS density, λ, is large enough, e.g. $\lambda > 10^2$ BSs/km^2, the coverage probability, p^{cov}, decreases with the small cell BS density, λ, due to the transition of a large number of interfering links from NLoS to LoS in dense networks.
  - ☐ When the small cell BS density, λ, is considerably large, e.g. $\lambda \geq 10^3$ BSs/km^2, the coverage probability, p^{cov}, decreases at a slower pace. This is because both the signal and the inter-cell interference power levels are increasingly LoS dominated, thus ultimately trending towards the LoS path loss exponent, and growing at the same pace.

  See Chapter 4 for more details. Also note that the presence of an infinite – Conf. b) – or a finite – Conf. c) – number of UEs does not bring any change in the coverage probability, p^{cov}, as in the downlink the inter-cell interference originates from the small cell BSs, and thus the SINR, Γ, of the typical UE does not depend on the neighbouring UEs. That is not the case in the uplink, where the interference sources are the UEs.
- The coverage probability, p^{cov}, resulting from the model incorporating the multi-slope path loss model with both LoS and NLoS transmissions, a finite UE density, $\rho < +\infty$, and a simple idle mode capability at the small cell BSs – Conf. d) and Conf. e) – exhibits a significantly different behaviour than those resulting from the models with no such features – Conf. a) and Conf. b). The new coverage probabilities, p^{cov}, resulting from the models incorporating a finite UE density, $\rho < +\infty$, and a simple idle mode capability at the small cell BSs, show a much more optimistic behaviour, significantly increasing towards 1 – or values close to 1 – when the network enters the ultra-dense regime. Since the UE density, $\rho < +\infty$, is finite, the active small cell BS density, $\tilde{\lambda}$, is also finite, as every UE is served by at most one small cell BS – no multi-cell coordination schemes are considered here – and thus there cannot be more active small cell BSs than UEs in the network. As a result, the inter-cell interference becomes automatically bounded, as the number of interfering sources – active small cell BSs – reaches a constant value in the ultra-dense regime, at most equal to the number of UEs in the network. This leads to the increase of the SINR, Γ, of the typical UE, in such a regime. This is because the signal power continues to increase due to the smaller distance between the typical UE and its serving small cell BS in a densifying network, while the inter-cell interference power is constant and bounded, as just explained.

Despite this good news, it should be noted that the new coverage probabilities, p^{cov}, still show a decreasing behaviour in the small cell BS density range, $\lambda = [100-500]$ BSs/km^2. This is due to the transition of a large number of interfering links from NLoS to LoS in dense networks.

Studying Various UE Densities

Figure 6.7 further explores the coverage probability, p^{cov}, paying special attention, this time, to the impact of various UE densities, $\rho = \{100, 300, 600\}$ UEs/km^2. Since Conf. b) and Conf. c) offer the same performance in terms of coverage probability, p^{cov}, only Conf. b) with the mentioned various UE densities is considered here.

From Figure 6.7, we can observe that

- the smaller the finite UE density, $\rho < +\infty$, the faster the pace at which the coverage probability, p^{cov}, increases towards 1 – or values close to 1. This is because a smaller finite UE density, $\rho < +\infty$, implies a smaller number of interfering small cell BSs, and in turn, a lower maximum received inter-cell interference power, and more importantly, a tighter burden on the growth of the inter-cell interference power with the small cell BS density, λ. As a result, the inter-cell interference power is at a disadvantage earlier with respect to the signal power. The latter continues to grow due to the smaller distance between the typical UE and its serving small cell BS in a densifying network.
- With respect to the baseline case with all small cell BSs active, and when considering a small cell BS density, $\lambda = 1 \times 10^4$ BSs/km^2, having a finite UE

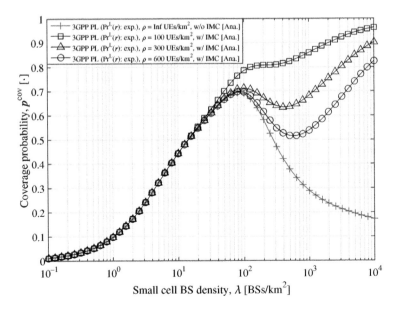

Figure 6.7 Coverage probability p^{cov} versus the small cell BS density, λ, for various UE densities, $\rho < +\infty$.

density, $\rho = 300$ UEs/km^2, leads to an increase of the coverage probability, p^{cov}, from 0.173 to 0.902. This is a significant boost of $4.21\times$.

- Looking at the results from a densification perspective, a reduction from a finite UE density, $\rho = 600$ UEs/km^2, to a finite UE density, $\rho = 100$ UEs/km^2, anticipates the targeted coverage probability, p^{cov} – defined here by a coverage probability, $p^{\mathrm{cov}} = 0.8$ – from a small cell BS density, $\lambda = 0.8 \times 10^4$ BSs/km^2, to a small cell BS density, $\lambda = 0.0126 \times 10^4$ BSs/km^2.
- Importantly, it should be noted that with respect to the results presented in Chapter 4, the smaller the finite UE density, $\rho < +\infty$, the earlier the inter-cell interference power becomes bounded, which results in an earlier increase of the coverage probability, p^{cov}, with the increase of the small cell BS, λ. This helps to alleviate the coverage probability decrease – even almost to remove it in some cases – due to the transition of a large number of interfering links from NLoS to LoS in dense networks.

NetVisual Analysis

To visualize the fundamental behaviour of the coverage probability, p^{cov}, in a more intuitive manner, Figure 6.8 shows the coverage probability heat map of three different scenarios with different small cell BS densities, i.e. 50, 250 and 2500 BSs/km^2, while considering the multi-slope path loss model with both LoS and NLoS transmissions with and without a finite UE density, $\rho < +\infty$, and a simple idle mode capability at the small cell BSs. A finite UE density, $\rho = 300$ UEs/km^2, is selected for the analysis. As in Chapters 3–5, these heat maps are computed using NetVisual – a tool introduced in Section 2.4, which is able to capture not only the mean, but also the standard deviation of the coverage probability, p^{cov}.

From Figure 6.8, we can see that, with respect to the case with an infinite UE density, $\rho = \infty$, and no idle mode capability at the small cell BSs,

- the finite UE density, $\rho < +\infty$, and the simple idle mode capability at the small cell BSs have a negligible performance impact when the small cell BS density, λ, is sparse, i.e. when the small cell BS density, λ, is no larger than 50 BSs/km^2 in our example. The SINR maps for such a small cell BS density, λ, are basically the same in both Figure 6.8a and b. Since the number of small cell BSs is much smaller than that of the UEs in this case, most of the small cell BSs have to be activated, as they are likely to have at least a UE within their coverage area. As a result, the idle mode capability does not play a role. The average and the standard deviation of the coverage probability, p^{cov}, for these two cases are approximately 0.71 and 0.22, respectively.
- Compared with Figure 6.8a, the SINR heat map in Figure 6.8b becomes much brighter when the small cell BS density, λ, is around 250 BSs/km^2, and even more for that when $\lambda = 2500$ BSs/km^2. This indicates a vast performance improvement due to the finite UE density, $\rho < +\infty$, and the simple idle mode capability at the small cell BSs. As an example, for the case with a small cell BS density, $\lambda = 2500$ BSs/km^2, the average and the standard deviation of the coverage

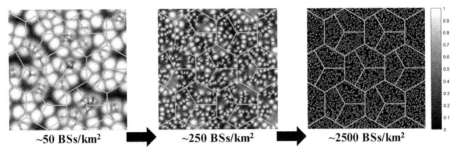

(a) Coverage probability, p^{cov}, under a multi-slope path loss model incorporating both LoS and NLoS transmissions, an infinite UE density and no idle mode capability at the small cell BSs.

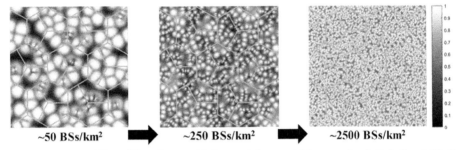

(b) Coverage probability, p^{cov}, under a multi-slope path loss model incorporating both LoS and NLoS transmissions, a finite UE density, $\rho = 300\,\mathrm{UEs/km^2}$, and an idle mode capability at the small cell BSs. Compared with Figure 6.8a, the SINR heat map becomes lighter as the small cell BS density, λ, increases, showing a significant performance enhancement because of the reduced inter-cell interference power caused by the subset of small cell BSs that have gone into idle mode.

Figure 6.8 NetVisual plot of the coverage probability, p^{cov}, versus the small cell BS density, λ [Bright area: high probability, Dark area: low probability].

probability, p^{cov}, go from 0.22 and 0.26 with an infinite UE density, $\rho = \infty$, and no idle mode capability at the small cell BSs to 0.79 and 0.13 with the finite UE density, $\rho < +\infty$, and the simple idle mode capability at the small cell BSs, respectively – a 259.09% improvement in mean.

Summary of Findings: Coverage Probability Remarks

Remark 6.1 Considering a finite UE density, $\rho < +\infty$, and a simple idle mode capability at the small cell BSs, the SINR, Γ, of the typical UE improves in a dense network due to the cap on the maximum inter-cell interference power that a UE can receive. As a result, the coverage probability, p^{cov}, neither stays constant – nor crashes towards 0 – with the increase of the small cell BS density, λ, as in Chapters 3 and 5, respectively. Instead, it increases towards 1 – or a value very close to 1 – with the increase of the small cell BS density, λ, in the dense and the ultra-dense regimes.

6.4.4 Area Spectral Efficiency Performance

Let us now explore the behaviour of the ASE, A^{ASE}.

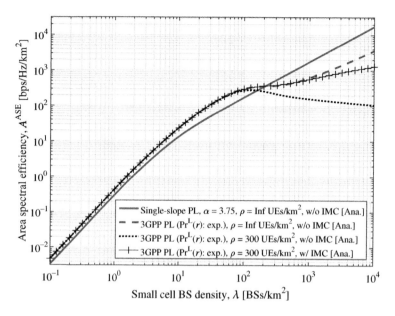

Figure 6.9 Area spectral efficiency, A^{ASE}, versus the small cell BS density, λ.

Figure 6.9 shows the ASE, A^{ASE}, for an SINR threshold, $\gamma = 0\,\mathrm{dB}$, and an optimal fitting parameter, $q^* = 4.18$, for the following three configurations:

- Conf. a): single-slope path loss model with an infinite number of UEs – or at least one UE per cell – thus all small cell BSs are active (analytical results).
- Conf. b): multi-slope path loss model with both LoS and NLoS transmissions, and an infinite number of UEs – or at least one UE per cell – thus all small cell BSs are active (analytical results).
- Conf. c): multi-slope path loss model with both LoS and NLoS transmissions, a finite UE density, $\rho < +\infty$, but with no simple idle mode capability at the small cell BSs, thus all small cell BSs are active (analytical results).
- Conf. d): multi-slope path loss model with both LoS and NLoS transmissions, a finite UE density, $\rho < +\infty$, and a simple idle mode capability at the small cell BSs, thus not all small cell BSs are active (analytical results).

From Figure 6.9, we can observe that the ASEs, A^{ASE}, resulting from the models incorporating the multi-slope path loss model with both LoS and NLoS transmissions, and a finite UE density, $\rho < +\infty$, without and with a simple idle mode capability at the small cell BSs – Conf. c) and Conf. d) – exhibit a different behaviour than those resulting from the models with no such features – Conf. a) and Conf. b). Two important aspects should be highlighted from Figure 6.9:

- The finite UE density, $\rho < +\infty$, plays a key role in the ASEs, A^{ASE}, where a trade-off should be noticed. The larger the finite UE density, $\rho < +\infty$, the larger the active small cell BS density, $\tilde{\lambda}$, because more small cell BSs may be needed to serve the more UEs. As a result of the larger finite UE density, $\rho < +\infty$, and the larger active small cell BS density, $\tilde{\lambda}$,

☐ the spatial reuse is larger in the network, which results into more bits per second transmitted per unit of frequency and area – a positive impact. However,

☐ the inter-cell interference is also larger in the system, which results in a lower SINR, Γ, of the typical UE, and in turn, in a lower coverage probability, p^{cov}, as was shown in Figure 6.7 – a negative impact.

With this in mind, the results of Figure 6.9 show that the configuration with an infinite UE density, $\rho = \infty$ – Conf. a) and Conf. b) – outperform those with a finite one – Conf. c) and Conf. d). This indicates that the effect of the spatial reuse outweighs that of the inter-cell interference – at least – in this scenario. This can be broadly explained by equation (1.1), which indicates that the former has a linear scaling effect on capacity, while the latter only a logarithmic one. As a result, the models considering a finite UE density, $\rho < +\infty$, show a worse performance in terms of ASE, A^{ASE}, regardless of whether the idle mode capability is active or not. In other words, the more UEs the network serves, the better the ASE, A^{ASE}. Importantly, it should be remembered that the performance of each individual UE is not better but worse with the larger active small cell BS density, $\tilde{\lambda}$.

• However, considering the more realistic networks with a finite UE density, $\rho < +\infty$, we should note that the configuration with a simple idle mode capability – Conf. d) – outperforms that without it – Conf. c). With a finite UE density, $\rho < +\infty$, the spatial reuse in the network is upper-bounded, and does not grow with the small cell BSs, λ. When considering no idle mode capability, however, the inter-cell interference in the network continues to grow with the increase of the small cell BSs, λ, due to the "always-on" signals transmitted by the always-on small cell BSs. This has a negative effect in the SINR, Γ, of the typical UE, as shown in the previous coverage probability analysis. When a simple mode capability is considered, instead, small cell BSs with no UE in their coverage areas are switched off, and thus they do not transmit "always-on" signals. As a result, the inter-cell interference becomes automatically bounded, as the number of interfering sources – active small cell BSs – reaches a constant value once all the UE are served in the ultra-dense regime. This leads to an increase of the SINR, Γ, of the typical UE, which in turn, increases the ASE, A^{ASE}. This also follows from the previous coverage probability results. The existing UEs have better signal quality and transmit at a higher rate, thus increasing the bits per unit of frequency, time and area.

Studying Various UE Densities

Figure 6.10 further explores the ASE, A^{ASE}, paying special attention, this time, to the impact of various UE densities, $\rho = \{100, 300, 600\}$ UEs/km^2. In contrast to the previous coverage probability results, Figure 6.10 considers both Conf. b) and Conf. c), as they have a different impact on the spatial reuse, and thus on the ASE, A^{ASE}.

From Figure 6.10, we can observe the same performance trend as shown earlier in Figure 6.9, and thus the same core explanations apply.

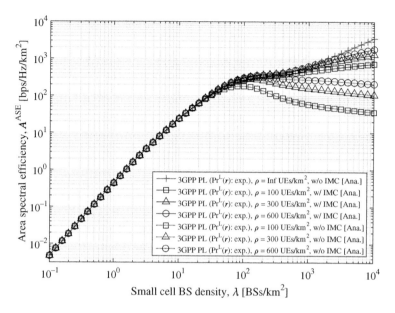

Figure 6.10 Area spectral efficiency, A^{ASE}, versus the small cell BS density, λ, for various UE densities, $\rho < +\infty$.

- The larger the finite UE density, $\rho < +\infty$, the better the performance in terms of the ASE, A^{ASE}, due to the better spatial reuse. In other words, the linear scaling effect due to the spatial reuse outweighs the logarithmic one resulting from the increase of the SINR, Γ, of the typical UE, originated by the mitigated inter-cell interference.
- For a small cell BS density of $\lambda = 1 \times 10^4$ BSs/km^2, and using as the benchmark the configuration with a finite UE density, $\rho = 100$ UEs/km^2, when considering the idle mode capability at the small cell BSs, the ASE, A^{ASE}, increases from 732.8 bps/Hz/km^2 to 1289.7 and 1822.6 bps/Hz/km^2 for the UE densities, $\rho = \{300, 600\}$ UEs/km^2, respectively. This means a 76.00% and a 2.49× performance increase, respectively. When the UE density is infinity, $\rho = \infty$, and there is a UE in a every small cell BS, the upper-bound ASE, A^{ASE}, is 3624.1 bps/Hz/km^2.
- When considering a finite UE density, $\rho < +\infty$, the idle mode capability at the small cell BSs is able to effectively mitigate inter-cell interference by switching off "always-on" signals, and thus Conf. d) always outperforms Conf. c). Importantly, the smaller the finite UE density, $\rho < +\infty$, the larger the benefit of the idle mode capability at the small cell BSs, as more small cell BSs can be switched off. As a result, more inter-cell interference can be removed, with the corresponding increase of the SINR, Γ, of the typical UE.
- For a small cell BS density of $\lambda = 1 \times 10^4$ BSs/km^2, and using as the benchmark the configuration without the idle mode capability at the small cell BSs, when considering the idle mode capability at the small cell BSs, the ASE, A^{ASE},

increases from 36.0, 106.7 and 209.4 bps/Hz/km^2 to 732.8, 1289.7 and 1822.6 bps/Hz/km^2 for the UE densities, $\rho = \{100, 300, 600\}$ UEs/km^2, respectively. This means a 1935.6%, 1108.7% and 770.4% performance increase, respectively.

The ASE Climb

Overall, the fact that the ASE, A^{ASE}, will climb towards a large value as the small cell BS density, λ, marches into the ultra-dense regime shows the benefit of deploying a larger number of small cell BSs equipped with an idle mode capability. This analysis shows that both the finite UE density, $\rho < +\infty$, and the simple idle mode capability at the small BSs matter, and should be considered when planning the small cell BS density, λ. The fundamental reason for the large positive impact that a simple idle mode capability at the small BSs may have on an ultra-dense network in the presence of a finite UE density, $\rho < +\infty$, is the cap that it poses on the maximum inter-cell interference power that a UE can receive, which such an idle mode capability at the small cell BSs can impose. As indicated in the introduction of this chapter, the realization of effective idle mode capabilities at the small cell BSs has recently been a hot topic in the telecommunications industry, where new features have been specified to allow small cell BSs to go into idle mode as often as possible. We should see effective implementations of such idle mode capabilities in the near future, which even allow microsleep periods, permitting the switch off a small cell BS, not only when there is no UE connect to it at a macroscopic time level, but also when connected UEs do not have traffic to transmit/receive at a microscopic one. Equally important is the role played by the finite UE density, $\rho < +\infty$. Note that even in scenarios with large UE densities, the network could adjust the number of UEs that have access to the network in a given time period through UE access control and/or medium access control (MAC) scheduling decisions. This can allow the regulation of the finite UE density, $\rho < +\infty$, across the network at a given scheduling period, and harvest the gains shown in this chapter. This concept is further elaborated in Chapter 9.

   With this in mind, let us make the definition.

DEFINITION 6.4.1   The ASE Climb:
In an uncoordinated network, where each UE is connected to a single small cell BS, the existence of a finite UE density, $\rho < +\infty$, and a simple idle mode capability at the small cell BSs limits the number of small cell BSs that activate to serve UEs in an ultra-dense network. There cannot be more active small cell BSs than UEs in such a network. As a result, this limit on the number of active small cell BSs poses a cap on the maximum inter-cell interference power that a UE can receive. The ASE Climb is defined as the desired – and significant – increase of the ASE, A^{ASE}, due to the cap on the maximum inter-cell interference power that a UE can receive, which can be imposed through an idle mode capability at the small cell BSs taking advantage of the finite UE density, $\rho < +\infty$, or more generally, other network and/or traffic circumstances.

It is worth noting that, even if the conclusion on the ASE Climb has been obtained from a particular model and set of parameters in this book, a number of studies using other antenna models and other sets of parameters corroborate the generality of these conclusions. We are going to show examples of this in the following section.

Summary of Findings: Area Spectral Efficiency Remarks
Remark 6.2   Considering a finite UE density, $\rho < +\infty$, and a simple idle mode capability at the small cell BSs, the ASE, A^{ASE}, neither linearly increases – nor crashes towards 0 – with the increase of the small cell BS density, λ, as in Chapter 3 and Chapter 5, respectively. Instead, it climbs towards a larger positive value at a good pace as the small cell BS density, λ, increases. Due to the cap on the maximum inter-cell interference power that a UE can receive, and the increasing signal power (from the shorter UE-to-serving small cell BSs distances), the contribution of each small cell BS to the ASE, A^{ASE}, increases in a densifying ultra-dense network.

6.4.5   Impact of the Multi-Path Fading – and Modelling – on the Main Results

For completeness, and similarly to Chapters 3–5, in this section, we also analyze the impact of the more accurate – but also less tractable – Rician multi-path fast fading model, presented in Section 2.1.8, on the coverage probability, p^{cov}, and the ASE, A^{ASE}.

Recall that

- the Rician multi-path fast fading model is applied to the LoS component only and that
- in contrast to the Rayleigh multi-path fast fading model, it is able to capture the ratio between the power in the direct path and the power in the other scattered paths in the multi-path fast fading gain, h, as a function of the distance, r, between a UE and a small cell BS.

As discussed in Chapters 3–5, it should also be noted that

- the dynamic range of the multi-path fast fading gain, h, is smaller under the Rician multi-path fast fading model than that under the Rayleigh one – the former has larger minimum and smaller maximum values – and that importantly,
- the mean of the Rician multi-path fast fading gain, h, increases with the increase of the small cell BS density, λ (see Figure 2.4).

Our intention in this section is to check whether the conclusions obtained in this chapter about the ASE Climb hold when considering this more realistic – but less tractable – Rician-based multi-path fast fading model.

For the sake of mathematical complexity, we only present system-level simulation results with this alternative Rician-based multi-path fast fading model hereafter. Please refer to Section 2.1.8 for more details on both the Rayleigh and the Rician multi-path fast fading models.

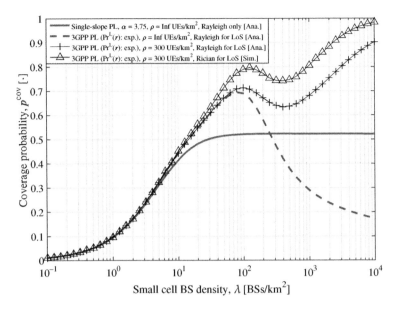

Figure 6.11 Coverage probability, p^{cov}, versus the small cell BS density, λ, for an alternative case study with multi-path Rician fading.

Figure 6.11 illustrates the results on the coverage probability, p^{cov}, while Figure 6.12 shows those on the ASE, A^{ASE}. The assumptions and the parameters used to obtain the results in these two figures are the same as those used to get the previous results shown in this chapter, except under the Rician multi-path fast fading model for the LoS component considered here.

As can be seen from Figures 6.11 and 6.12, all the observations with respect to the coverage probability, p^{cov}, and the ASE, A^{ASE}, obtained under the Rayleigh-based multi-path fast fading model in Section 6.4 are qualitatively valid for those obtained under the Rician-based multi-path fast fading model. The same trends are present.

Let us now analyze the coverage probability, p^{cov}, in more detail. The coverage probability, p^{cov}, first increases, then decreases and finally increases again with the small cell BS density, λ, and only a quantitative deviation exists when considering these different multi-path fast fading models. Specifically,

- in Figure 6.11, the first peak of the coverage probability, p^{cov}, is larger under the Rician-based multi-path fast fading model, 0.80, than under the Rayleigh-based one, 0.71.
- Moreover, the small cell BS density, λ_0, for which such a first peak of the coverage probability, p^{cov}, is obtained, is also slightly larger under the Rician-based multi-path fast fading model, 126 BSs/km^2, than under the Rayleigh-based one, 100 BSs/km^2.
- These results are practically the same as those presented in Figure 4.8 of Chapter 4, and the same explanations apply. However, it should be noted that such a first peak of the coverage probability, p^{cov}, for both multi-path fast fading models

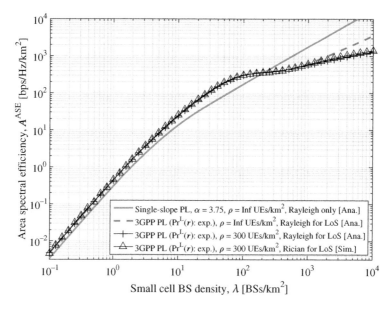

Figure 6.12 Area spectral efficiency, A^{ASE}, versus the small cell BS density, λ, for an alternative case study with multi-path Rician fading.

presents a slightly better performance. This is because the inter-cell interference mitigation introduced by the idle mode capability at the small cell BSs is already acting at these small cell BSs densities, increasing and prolonging such coverage probability peaks in terms of absolute value and densification level, respectively. In more detail, the idle mode capability at the small cell BSs switches off progressively more and more small cell BSs with the increase of the small cell BS density, λ, thus

☐ increasing the average distance between a UE and its neighbouring interfering small cell BSs, and in turn,

☐ delaying the NLoS to LoS inter-cell interference transition.

This latter fact exacerbates the difference in performance between the Rician- and the Rayleigh-based multi-path fast fading models.[4]

• Another important observation is that, once the coverage probability, p^{cov}, of the Rician-based multi-path fast fading model and that of the Rayleigh-based one reach their first peak, the coverage probability gap between them is apparently maintained with the increase of the small cell BS density, λ, and does not decrease with it as quickly as in Chapter 5. This is again because of the idle mode capability at the small cell BSs, which switches off progressively more and more small cell BSs and delays the NLoS to LoS inter-cell interference transition. This implicitly delays the Rayleigh to Rician multi-path fast fading transition, and as explained

---

[4] Remember, going from NLoS to LoS transmissions results in a change from Rayleigh to Rician modelling in the multi-path fast fading.

earlier, prolongs the difference in performance between the Rician- and the Rayleigh-based multi-path fast fading models to larger small cell BS densities.

• Such a coverage probability gap only decreases at very large small cell BSs densities. For a small cell BS density, $\lambda = \{10^3, 5 \times 10^3, 10^4\}\,\text{BSs/km}^2$, the coverage probability, p^{cov}, of the Rician-based multi-path fast fading model and that of the Rayleigh-based one are $\{0.81, 0.96, 0.98\}$ and $\{0.68, 0.85, 0.90\}$, respectively – a $\{19.12, 12.94, 8.89\}\%$ difference. Following the same line of thought, this closing coverage probability gap at larger small cell BS densities is due to the transition of a good number of neighbouring interfering small cell BSs from Rayleigh to Rician multi-path fast fading. To give some intuition, let us look at the extreme case, where both the signal and the stronger inter-cell interference power levels are fully LoS dominated. As shown in Figure 4.8 and discussed in Chapter 4, in this case, it makes no difference whether the Rician- or the Rayleigh-based multi-path fast fading model is used in terms of the coverage probability, p^{cov}. This is because we have independent and identical distributed (i.i.d.) multi-path fast fading channel gains, h – whether Rician or Rayleigh – in the signal and the stronger inter-cell interference links, which statistically cancel each other out in the calculation of the SINRs, Γ, of the UEs when averaging across the entire network. This is the reason for the closing gap.

As for the ASE, A^{ASE}, note that the behaviours under the Rayleigh- and the Rician-based multi-path fast fading models in Figure 6.12 are almost identical. This is because the relatively small change in the coverage probability, p^{cov}, results in an even smaller change in the ASE, A^{ASE}.

Overall, and despite the quantitative differences, these results on the coverage probability, p^{cov}, and the ASE, A^{ASE}, confirm our statement, i.e. the main conclusions in this chapter are general, and do not qualitatively change due to the assumption on the multi-path fast fading model.[5]

In the remainder of this book, since the modelling of the multi-path fading did not have a profound impact in terms of qualitative results, we stick to Rayleigh multi-path fast fading modelling for mathematical tractability, unless otherwise stated.

6.5   Conclusions

In this chapter, we have brought to attention the importance of the modelling of the inter-cell interference in network performance analysis. In more detail, we have described the necessary upgrades that one has to perform to the system model presented so far in this book to capture a finite UE density, $\rho < +\infty$, as well as a simple idle mode capability at the small cell BSs. These two features significantly impact the inter-cell interference and are essential for a more practical and close-to-reality

---

[5] Note that this statement applies for the studied case with single-antenna UEs and small cell BSs, but may not apply to multi-antenna cases, where the fading correlation among the antennas of an array plays a key role.

study in the ultra-dense regime. We have also provided details on the new derivations conducted in the theoretical performance analysis, and presented the resulting expressions for the coverage probability, p^{cov}, and the ASE, A^{ASE}. Moreover, we have shared numerical results for small cell BS deployments with different small cell BS densities and characteristics. To finalize, we have also discussed the important conclusions drawn from this work, which are significantly different from those in the previous three chapters.

Importantly, such conclusions, which are summarized in Remarks 6.1 and 6.2, have shown yet another new performance behaviour in the ultra-dense regime, referred to as the ASE Climb hereafter in this book. This new behaviour shows how a network operator or a service provider may significantly improve their network performance by not only carefully considering the densification level of its network, but also embracing the finite nature of the UE density, $\rho < +\infty$, and exploiting it through a simple idle mode capability at the small cell BSs. Contrary to the results of Chapters 4 and 5, this analysis sends an important optimistic message:

The increased inter-cell interference power in an ultra-dense network can be fought through the idle mode capability at the small cell BSs!

In this chapter, we have also shown that this behaviour – the ASE Climb – is quantitatively – but not qualitatively – affected by the multi-path fast fading model. This verified the generality of the findings.

Appendix A Proof of Theorem 6.3.1

To ease the understanding of the readers, it should be noted that this proof shares a lot of similarities with the proof in Appendix A, where the main differences arise due to the need of using the active small cell BSs density, $\tilde{\lambda}$, instead of the small cell BSs density, λ, in some derivations to capture the effect of the idle mode capability at the small cell BSs. The full proof is presented hereafter for completeness.

Having said that, let us first describe the main idea behind the proof of Theorem 6.3.1, and then proceed with a more detailed explanation.

In line with the guidelines provided in Section 2.3.2, to evaluate the coverage probability, p^{cov}, we need proper expressions for

- the PDF, $f_R(r)$, of the random variable, R, characterizing the distance between the typical UE and its serving small cell BS for the event that the typical UE is associated to the strongest small cell BS through either an LoS or an NLoS path and
- the conditional probability, $\Pr\left[\Gamma > \gamma_0 \middle| r\right]$, where
  - □ $r = R(\omega)$ is a realization of the random variable, R, for both the LoS and the NLoS transmission.

Once such expressions are known, we can finally derive, p^{cov}, by performing the corresponding appropriate integrals, some of which will be shown in the following.

Before proceeding with the more detailed calculations, however, it is important to note that from equations (2.22) and (3.3), we can derive the coverage probability, p^{cov}, as

$$
\begin{aligned}
p^{\mathrm{cov}}(\lambda, \gamma_0) &\overset{(a)}{=} \int_{r>0} \Pr\left[\mathrm{SINR} > \gamma_0 \,\middle|\, r\right] f_R(r) dr \\
&= \int_{r>0} \Pr\left[\frac{P\zeta(r)h}{I_{\mathrm{agg}} + P^{\mathrm{N}}} > \gamma_0\right] f_R(r) dr \\
&= \int_0^{d_1} \Pr\left[\frac{P\zeta_1^{\mathrm{L}}(r)h}{I_{\mathrm{agg}} + P^{\mathrm{N}}} > \gamma_0\right] f_{R,1}^{\mathrm{L}}(r) dr + \int_0^{d_1} \Pr\left[\frac{P\zeta_1^{\mathrm{NL}}(r)h}{I_{\mathrm{agg}} + P^{\mathrm{N}}} > \gamma_0\right] f_{R,1}^{\mathrm{NL}}(r) dr \\
&\quad + \cdots \\
&\quad + \int_{d_{N-1}}^{\infty} \Pr\left[\frac{P\zeta_N^{\mathrm{L}}(r)h}{I_{\mathrm{agg}} + P^{\mathrm{N}}} > \gamma_0\right] f_{R,N}^{\mathrm{L}}(r) dr + \int_{d_{N-1}}^{\infty} \Pr\left[\frac{P\zeta_N^{\mathrm{NL}}(r)h}{I_{\mathrm{agg}} + P^{\mathrm{N}}} > \gamma_0\right] f_{R,N}^{\mathrm{NL}}(r) dr \\
&\overset{\Delta}{=} \sum_{n=1}^{N} \left(T_n^{\mathrm{L}} + T_n^{\mathrm{NL}}\right),
\end{aligned}
\tag{6.25}
$$

where

- R_n^{L} and R_n^{NL} are the piecewise distributions of the distances over which the typical UE is associated to a small cell BS through an LoS and an NLoS path, respectively:
  - ☐ Note that these two events, i.e. the typical UE being associated with a small cell BS through an LoS or an NLoS path, are disjoint, and thus the coverage probability, p^{cov}, is the sum of the corresponding probabilities of these two events.
- $f_{R,n}^{\mathrm{L}}(r)$ and $f_{R,n}^{\mathrm{NL}}(r)$ are the piecewise PDFs of the random variables, R_n^{L} and R_n^{NL}, respectively:
  - ☐ For clarity, both piecewise PDFs, $f_{R,n}^{\mathrm{L}}(r)$ and $f_{R,n}^{\mathrm{NL}}(r)$, are stacked into the PDF, $f_R(r)$, in step (a) of equation (6.25), where the stacked PDF, $f_R(r)$, takes a similar form as in equation (4.1), and is defined in equation (4.45).
  - ☐ Moreover, since the two events, i.e. the typical UE being associated with a small cell BS through an LoS or an NLoS path, are disjoint, as mentioned before, we can rely on the following equality,
  $$\sum_{n=1}^{N} \int_{d_{n-1}}^{d_n} f_{R,n}(r) dr = \sum_{n=1}^{N} \int_{d_{n-1}}^{d_n} f_{R,n}^{\mathrm{L}}(r) dr + \sum_{n=1}^{N} \int_{d_{n-1}}^{d_n} f_{R,n}^{\mathrm{NL}}(r) dr = 1.$$
- T_n^{L} and T_n^{NL} are two piecewise functions, defined as

$$
T_n^{\mathrm{L}} = \int_{d_{n-1}}^{d_n} \Pr\left[\frac{P\zeta_n^{\mathrm{L}}(r)h}{I_{\mathrm{agg}} + P^{\mathrm{N}}} > \gamma_0\right] f_{R,n}^{\mathrm{L}}(r) dr
\tag{6.26}
$$

and

$$
T_n^{\mathrm{NL}} = \int_{d_{n-1}}^{d_n} \Pr\left[\frac{P\zeta_n^{\mathrm{NL}}(r)h}{I_{\mathrm{agg}} + P^{\mathrm{N}}} > \gamma_0\right] f_{R,n}^{\mathrm{NL}}(r) dr,
\tag{6.27}
$$

respectively, and

- d_0 and d_N are equal to 0 and $+\infty$, respectively.

$$
f_R(r) = \begin{cases}
f_{R,1}(r) = \begin{cases} f_{R,1}^{L}(r), & \text{UE associated to an LoS BS} \\ f_{R,1}^{NL}(r), & \text{UE associated to an NLoS BS} \end{cases} & 0 \le r \le d_1 \\[2ex]
f_{R,2}(r) = \begin{cases} f_{R,2}^{L}(r), & \text{UE associated to an LoS BS} \\ f_{R,2}^{NL}(r), & \text{UE associated to an NLoS BS} \end{cases} & d_1 < r \le d_2 \\[2ex]
\vdots & \vdots \\[2ex]
f_{R,N}(r) = \begin{cases} f_{R,N}^{L}(r), & \text{UE associated to an LoS BS} \\ f_{R,N}^{NL}(r), & \text{UE associated to an NLoS BS} \end{cases} & r > d_{N-1}
\end{cases}
\tag{6.28}
$$

Following this method, let us now dive into the more detailed derivations.

LoS-Related Calculations

Let us first look into the LoS transmissions, and show how to first calculate the PDF, $f_{R,n}^{L}(r)$, and then the probability, $\Pr\left[\frac{P\zeta_n^{L}(r)h}{I_{\mathrm{agg}}+P^{N}} > \gamma_0\right]$, in equation (6.25).
To that end, we define first the two following events:

- Event B^{L}: The nearest small cell BS to the typical UE with an LoS path is located at a distance, $x = X^{L}(\omega)$, defined by the distance random variable, X^{L}. According to the results presented in Section 2.3.2, the PDF, $f_R(r)$, of the random variable, R, in Event B^{L} can be calculated by

$$
f_X^{L}(x) = \exp\left(-\int_0^x \Pr^{L}(u)2\pi u\lambda \, du\right) \Pr^{L}(x)2\pi x\lambda.
\tag{6.29}
$$

This is because, according to [52], the complementary cumulative distribution function (CCDF), $\bar{F}_X^{L}(x)$, of the random variable, X^{L}, in Event B^{L} is given by

$$
\bar{F}_X^{L}(x) = \exp\left(-\int_0^x \Pr^{L}(u)2\pi u\lambda \, du\right),
\tag{6.30}
$$

and taking the derivative of the cumulative distribution function (CDF), $\left(1 - \bar{F}_X^{L}(x)\right)$, of the random variable, X^{L}, with respect to the distance, x, we can get the PDF, $f_X^{L}(x)$, of the random variable, X^{L}, shown in equation (6.29). It is important to note that the derived equation (6.29) is more complex than equation (2.42) of Section 2.3.2. This is because the LoS small cell BS deployment is an inhomogeneous one due to the fact that the closer small cell BSs to the typical UE are more likely to establish LoS links than the further away ones. Compared with equation (2.42), we can see two important changes in equation (6.29).

☐ The probability of a disc of a radius, r, containing exactly 0 points, which takes the form, $\exp\left(-\pi\lambda r^2\right)$, as presented in equation (2.38) in Section 2.3.2, has changed now, and is replaced by the expression,

$$\exp\left(-\int_0^x \text{Pr}^L(u)2\pi u\lambda\, du\right),$$

in equation (6.29). This is because the equivalent intensity of the LoS small cell BSs in an inhomogeneous Poisson point process (PPP) is distance-dependent, i.e. $\text{Pr}^L(u)\lambda$. This adds a new integral to equation (6.29) with respect to the distance, u.

☐ The intensity, λ, of the homogeneous PPP in equation (2.42) has been replaced by the equivalent intensity, $\text{Pr}^L(x)\lambda$, of the inhomogeneous PPP in equation (6.29).

• Event C^{NL} conditioned on the value, $x = X^L(\omega)$, of the random variable, X^L: The typical UE is associated to the nearest small cell BS located at a distance, $x = X^L(\omega)$, through an LoS path. If the typical UE is associated to the nearest LoS small cell BS located at a distance, $x = X^L(\omega)$, this LoS small cell BS must be at the smallest path loss, $\zeta(r)$. As a result, there must be no NLoS small cell BS inside the disc

☐ centred on the typical UE,

☐ with a radius, x_1,

where such a radius, x_1, satisfies the following condition, $x_1 = \arg_{x_1}\left\{\zeta^{NL}(x_1) = \zeta^L(x)\right\}$. Otherwise, this NLoS small cell BS would outperform the LoS small cell BS at the distance, $x = X^L(\omega)$.

According to [52], the conditional probability, $\text{Pr}\left[C^{NL}\,\middle|\,X^L = x\right]$, of the Event, C^{NL}, conditioned on the realization, $x = X^L(\omega)$, of the random variable, X^L, is given by

$$\text{Pr}\left[C^{NL}\,\middle|\,X^L = x\right] = \exp\left(-\int_0^{x_1}\left(1 - \text{Pr}^L(u)\right)2\pi u\lambda\, du\right). \tag{6.31}$$

As a summary, note that Event B^L ensures that the path loss, $\zeta^L(x)$, associated with an arbitrary LoS small cell BS is always larger than that associated with the considered LoS small cell BS at the distance, $x = X^L(\omega)$. Besides, conditioned on such a distance, $x = X^L(\omega)$, Event C^{NL} guarantees that the path loss, $\zeta^{NL}(x)$, associated with an arbitrary NLoS small cell BS is also always larger than that associated with the considered LoS small cell BS at the distance, $x = X^L(\omega)$. With this, we can guarantee that the typical UE is associated with the strongest LoS small cell BS.

Let us thus now consider the resulting new event, in which the typical UE is associated with an LoS small cell BS, and such a small cell BS is located at the distance, $r = R^L(\omega)$. The CCDF, $\bar{F}_R^L(r)$, of such a random variable, R^L, can be derived as

$$\bar{F}_R^L(r) = \Pr\left[R^L > r\right]$$

$$\overset{(a)}{=} \mathbb{E}_{[X^L]}\left\{\Pr\left[R^L > r \,\middle|\, X^L\right]\right\}$$

$$= \int_0^{+\infty} \Pr\left[R^L > r \,\middle|\, X^L = x\right] f_X^L(x)\,dx$$

$$\overset{(b)}{=} \int_0^r 0 \times f_X^L(x)\,dx + \int_r^{+\infty} \Pr\left[C^{NL} \,\middle|\, X^L = x\right] f_X^L(x)\,dx$$

$$= \int_r^{+\infty} \Pr\left[C^{NL} \,\middle|\, X^L = x\right] f_X^L(x)\,dx, \tag{6.32}$$

where

- $\mathbb{E}_{[X]}\{\cdot\}$ in step (a) of equation (6.32) is the expectation operation over the random variable, X, and
- step (b) of equation (6.32) is valid because
  - $\Pr\left[R^L > r \,\middle|\, X^L = x\right] = 0$ for $0 < x \le r$ and
  - the conditional event, $\left[R^L > r \,\middle|\, X^L = x\right]$, is equivalent to the conditional event, $\left[C^{NL} \,\middle|\, X^L = x\right]$, in the range, $x > r$.

Now, given the CCDF, $\bar{F}_R^L(r)$, one can find its PDF, $f_R^L(r)$, by taking the derivative, $\frac{\partial(1 - \bar{F}_R^L(r))}{\partial r}$, with respect to the distance, r, i.e.

$$f_R^L(r) = \Pr\left[C^{NL} \,\middle|\, X^L = r\right] f_X^L(r). \tag{6.33}$$

Considering the distance range, $(d_{n-1} < r \le d_n)$, we can find the PDF of such a segment, $f_{R,n}^L(r)$, from such a PDF, $f_R^L(r)$, as

$$f_{R,n}^L(r) = \exp\left(-\int_0^{r_1}\left(1 - \Pr^L(u)\right) 2\pi u\lambda\,du\right)$$

$$\times \exp\left(-\int_0^r \Pr^L(u) 2\pi u\lambda\,du\right) \Pr_n^L(r) 2\pi r\lambda, \ (d_{n-1} < r \le d_n),$$

$$\tag{6.34}$$

where

- $r_1 = \underset{r_1}{\arg}\left\{\zeta^{NL}(r_1) = \zeta_n^L(r)\right\}.$

Having obtained the PDF of a segment, $f_{R,n}^L(r)$, we can evaluate the probability, $\Pr\left[\frac{P\zeta_n^L(r)h}{I_{agg}+P^N} > \gamma_0\right]$, in equation (6.25) as

$$\Pr\left[\frac{P\zeta_n^L(r)h}{I_{\text{agg}} + P^N} > \gamma_0\right] = \mathbb{E}_{[I_{\text{agg}}]}\left\{\Pr\left[h > \frac{\gamma_0\left(I_{\text{agg}} + P^N\right)}{P\zeta_n^L(r)}\right]\right\}$$

$$= \mathbb{E}_{[I_{\text{agg}}]}\left\{\bar{F}_H\left(\frac{\gamma_0\left(I_{\text{agg}} + P^N\right)}{P\zeta_n^L(r)}\right)\right\}, \tag{6.35}$$

where

- $\bar{F}_H(h)$ is the CCDF of the multi-path fast fading channel gain, h, which is assumed to be drawn from a Rayleigh fading distribution (see Section 2.1.8).

Since the CCDF, $\bar{F}_H(h)$, of the multi-path fast fading channel gain, h, follows an exponential distribution with unitary mean given by

$$\bar{F}_H(h) = \exp(-h),$$

equation (6.35) can be further derived as

$$\Pr\left[\frac{P\zeta_n^L(r)h}{I_{\text{agg}} + P^N} > \gamma_0\right] = \mathbb{E}_{[I_{\text{agg}}]}\left\{\exp\left(-\frac{\gamma_0\left(I_{\text{agg}} + P^N\right)}{P\zeta_n^L(r)}\right)\right\}$$

$$\overset{(a)}{=} \exp\left(-\frac{\gamma_0 P^N}{P\zeta_n^L(r)}\right)\mathbb{E}_{[I_{\text{agg}}]}\left\{\exp\left(-\frac{\gamma_0}{P\zeta_n^L(r)}I_{\text{agg}}\right)\right\}$$

$$= \exp\left(-\frac{\gamma_0 P^N}{P\zeta_n^L(r)}\right)\mathscr{L}_{I_{\text{agg}}}\left(\frac{\gamma_0}{P\zeta_n^L(r)}\right), \tag{6.36}$$

where

- $\mathscr{L}_{I_{\text{agg}}}(s)$ is the Laplace transform of the aggregated inter-cell interference random variable, I_{agg}, evaluated at the variable value, $s = \frac{\gamma_0}{P\zeta_n^L(r)}$.

It should be noted that the use of the Laplace transform is to ease the mathematical representation. By definition, $\mathscr{L}_{I_{\text{agg}}}(s)$ is the Laplace transform of the PDF of the aggregated inter-cell interference random variable, I_{agg}, evaluated at the variable value, s, and the equality,

$$\mathscr{L}_X(s) = \mathbb{E}_{[X]}\left\{\exp(-sX)\right\},$$

follows from the definitions and derivations leading to equation (2.46) in Chapter 2.

Based on the condition of the LoS signal transmission, the Laplace transform, $\mathscr{L}_{I_{\text{agg}}}^L(s)$, can be derived as

$$\mathscr{L}_{I_{\text{agg}}}^L(s) = \mathbb{E}_{[I_{\text{agg}}]}\left\{\exp(-sI_{\text{agg}})\right\}$$

$$= \mathbb{E}_{[\Phi, \{\beta_i\}, \{g_i\}]}\left\{\exp\left(-s\sum_{i\in\Phi/b_o} P\beta_i g_i\right)\right\}$$

$$\overset{(a)}{=} \exp\left(-2\pi\tilde{\lambda}\int\left(1 - \mathbb{E}_{[g]}\left\{\exp\left(-sP\beta(u)g\right)\right\}\right)u\,du\right), \tag{6.37}$$

where

- Φ is the set of small cell BSs,
- b_o is the small cell BS serving the typical UE,
- b_i is the ith interfering small cell BS,
- β_i and g_i are the path loss and the multi-path fast fading gain between the typical UE and the ith interfering small cell BS and
- step (a) of equation (6.37) has been explained in detail in equation (3.18), and further derived in equation (3.19) of Chapter 3.

Importantly, note that unlike equation (3.18), the detailed calculation of the Laplace transform, $\mathscr{L}_{I_{\text{agg}}}^{L}(s)$, given by equation (6.37) only involves active small cell BSs with a density, $\tilde{\lambda}$, because only the active small cell BS generate inter-cell interference.

Compared with equation (3.19), which considers the inter-cell interference from a single-slope path loss model, the expression, $\mathbb{E}_{[g]}\{\exp(-sP\beta(u)g)\}$, in equation (6.37) should consider the inter-cell interference from both the LoS and the NLoS paths. As a result, the Laplace transform, $\mathscr{L}_{I_{\text{agg}}}^{L}(s)$, can be further developed as

$$\mathscr{L}_{I_{\text{agg}}}^{L}(s) = \exp\left(-2\pi\tilde{\lambda}\int\left(1 - \mathbb{E}_{[g]}\{\exp(-sP\beta(u)g)\}\right)u\,du\right)$$

$$\overset{(a)}{=} \exp\left(-2\pi\tilde{\lambda}\int\left[\Pr^{L}(u)\left(1 - \mathbb{E}_{[g]}\{\exp(-sP\zeta^{L}(u)g)\}\right)\right.\right.$$
$$\left.\left. + \left(1 - \Pr^{L}(u)\right)\left(1 - \mathbb{E}_{[g]}\{\exp(-sP\zeta^{NL}(u)g)\}\right)\right]u\,du\right)$$

$$\overset{(b)}{=} \exp\left(-2\pi\tilde{\lambda}\int_{r}^{+\infty}\Pr^{L}(u)\left(1 - \mathbb{E}_{[g]}\{\exp(-sP\zeta^{L}(u)g)\}\right)u\,du\right)$$

$$\times \exp\left(-2\pi\tilde{\lambda}\int_{r_1}^{+\infty}\left(1 - \Pr^{L}(u)\right)\left(1 - \mathbb{E}_{[g]}\{\exp(-sP\zeta^{NL}(u)g)\}\right)u\,du\right)$$

$$\overset{(c)}{=} \exp\left(-2\pi\tilde{\lambda}\int_{r}^{+\infty}\frac{\Pr^{L}(u)u}{1 + \left(sP\zeta^{L}(u)\right)^{-1}}du\right)$$

$$\times \exp\left(-2\pi\tilde{\lambda}\int_{r_1}^{+\infty}\frac{\left[1 - \Pr^{L}(u)\right]u}{1 + \left(sP\zeta^{NL}(u)\right)^{-1}}du\right), \tag{6.38}$$

where

- in step (a), the integration is probabilistically divided into two parts considering the LoS and the NLoS inter-cell interference,
- in step (b), the LoS and the NLoS inter-cell interference come from a distance larger than distances, r and r_1, respectively, and
- in step (c), the PDF, $f_G(g) = 1 - \exp(-g)$, of the Rayleigh random variable, g, has been used to calculate the expectations, $\mathbb{E}_{[g]}\{\exp(-sP\zeta^{L}(u)g)\}$ and $\mathbb{E}_{[g]}\{\exp(-sP\zeta^{NL}(u)g)\}$.

To give some intuition for equation (6.36), it should be noted that

- the exponential factor, $\exp\left(-\frac{\gamma_0 P^{N}}{P\zeta_n^{L}(r)}\right)$, measures the probability that the signal power exceeds the noise power by at least a factor, γ_0, while

- the Laplace transform, $\mathscr{L}_{I_{\text{agg}}}\left(\frac{\gamma_0}{P\zeta_n^L(r)}\right)$, measures the probability that the signal power exceeds the aggregated inter-cell interference power by at least a factor, γ_0.

As a result, and because the multi-path fast fading channel gain, h, follows an exponential distribution, the product of the above probabilities, shown in step (a) of equation (6.36), yields the probability that the signal power exceeds the sum power of the noise and the aggregated inter-cell interference by at least a factor, γ_0.

NLoS-Related Calculations

Let us now look into the NLoS transmissions, and show how to first calculate the PDF, $f_{R,n}^{NL}(r)$, and then the probability, $\Pr\left[\frac{P\zeta_n^{NL}(r)h}{I_{\text{agg}}+P^N} > \gamma_0\right]$, in equation (6.25).
   To that end, we define the two following events:

- Event B^{NL}: The nearest small cell BS to the typical UE with an NLoS path is located at a distance, $x = X^{NL}(\omega)$, defined by the distance random variable, X^{NL}. Similarly to equation (6.29), one can find the PDF, $f_X^{NL}(x)$, as

$$f_X^{NL}(x) = \exp\left(-\int_0^x \left(1 - \Pr^L(u)\right) 2\pi u\lambda\, du\right)\left(1 - \Pr^L(x)\right) 2\pi x\lambda. \qquad (6.39)$$

It is important to note that the derived equation (6.39) is more complex that the equation (2.42) of Section 2.3.2. This is because the NLoS small cell BS deployment is an inhomogeneous one due to the fact that the further away small cell BSs to the typical UE are more likely to establish NLoS links than the closer ones. Compared with equation (2.42), we can see two important changes in equation (6.39):

☐ The probability of a disc of a radius, r, containing exactly 0 points, which take the form, $\exp\left(-\pi\lambda r^2\right)$, as presented in equation (2.38) in Section 2.3.2, has changed now, and is replaced by the expression,

$$\exp\left(-\int_0^x \left(1 - \Pr^L(u)\right) 2\pi u\lambda\, du\right),$$

in equation (6.39). This is because the equivalent intensity of the NLoS small cell BSs in an inhomogeneous PPP is distance-dependent, i.e. $\left(1 - \Pr^L(u)\right)\lambda$. This adds a new integral to equation (6.29) with respect to the distance, u.

☐ The intensity, λ, of the HPPP in equation (2.42), has been replaced by the equivalent intensity, $\left(1 - \Pr^L(u)\right)\lambda$, of the inhomogeneous PPP in equation (6.39).

- Event C^L conditioned on the value, $x = X^{NL}(\omega)$, of the random variable, X^{NL}: The typical UE is associated to the nearest small cell BS located at a distance, $x = X^{NL}(\omega)$, through an NLoS path. If the typical UE is associated to the nearest NLoS small cell BS located at a distance, $x = X^{NL}(\omega)$, this NLoS small cell BS must be at the smallest path loss, $\zeta(r)$. As a result, there must be no LoS small cell BS inside the disc

☐ centred on the typical UE,

☐ with a radius, x_2,

where such a radius, x_2, satisfies the following condition,

$x_2 = \underset{x_2}{\arg}\left\{\zeta^L(x_2) = \zeta^{NL}(x)\right\}$. Otherwise, this LoS small cell BS would outperform

the NLoS small cell BS at the distance, $x = X^{NL}(\omega)$.

Similarly to equation (6.31), the conditional probability, $\Pr\left[C^L\middle|X^{NL} = x\right]$, of Event C^L conditioned on the realization, $x = X^{NL}(\omega)$, of the random variable, X^{NL}, is given by

$$\Pr\left[C^L\middle|X^{NL} = x\right] = \exp\left(-\int_0^{x_2} \Pr^L(u)2\pi u\lambda\, du\right). \tag{6.40}$$

Let us thus now consider the resulting new event, in which the typical UE is associated with an NLoS small cell BS, and such a small cell BS is located at the distance, $r = R^{NL}(\omega)$. The CCDF, $\bar{F}_R^{NL}(r)$, of such a random variable, R^{NL}, can be derived as

$$\bar{F}_R^{NL}(r) = \Pr\left[R^{NL} > r\right]$$

$$= \int_r^{+\infty} \Pr\left[C^L\middle|X^{NL} = x\right]f_X^{NL}(x)dx. \tag{6.41}$$

Now, given the CCDF, $\bar{F}_R^{NL}(r)$, one can find its PDF, $f_R^{NL}(r)$, by taking the derivative, $\frac{\partial\left(1-\bar{F}_R^{NL}(r)\right)}{\partial r}$, with respect to the distance, r, i.e.

$$f_R^{NL}(r) = \Pr\left[C^L\middle|X^{NL} = r\right]f_X^{NL}(r). \tag{6.42}$$

Considering the distance range, $(d_{n-1} < r \le d_n)$, we can find the PDF of such a segment, $f_{R,n}^{NL}(r)$, from such a PDF, $f_R^{NL}(r)$, as

$$f_{R,n}^{NL}(r) = \exp\left(-\int_0^{r_2} \Pr^L(u)2\pi u\lambda\, du\right)$$

$$\times \exp\left(-\int_0^r \left(1 - \Pr^L(u)\right)2\pi u\lambda\, du\right)\left(1 - \Pr_n^L(r)\right)2\pi r\lambda,\ (d_{n-1} < r \le d_n),$$

$$\tag{6.43}$$

where

• $r_2 = \underset{r_2}{\arg}\left\{\zeta^L(r_2) = \zeta_n^{NL}(r)\right\}$.

Having obtained the PDF of a segment, $f_{R,n}^{NL}(r)$, we can evaluate the probability, $\Pr\left[\frac{P\zeta_n^{NL}(r)h}{I_{agg}+P^N} > \gamma_0\right]$, in equation (6.25) as

$$\Pr\left[\frac{P\zeta_n^{\mathrm{NL}}(r)h}{I_{\mathrm{agg}} + P^{\mathrm{N}}} > \gamma_0\right] = \mathbb{E}_{[I_{\mathrm{agg}}]}\left\{\Pr\left[h > \frac{\gamma_0\left(I_{\mathrm{agg}} + P^{\mathrm{N}}\right)}{P\zeta_n^{\mathrm{NL}}(r)}\right]\right\}$$

$$= \mathbb{E}_{[I_{\mathrm{agg}}]}\left\{\bar{F}_H\left(\frac{\gamma_0\left(I_{\mathrm{agg}} + P^{\mathrm{N}}\right)}{P\zeta_n^{\mathrm{NL}}(r)}\right)\right\}. \quad (6.44)$$

Since the CCDF, $\bar{F}_H(h)$, of the multi-path fast fading channel gain, h, follows an exponential distribution with unitary mean given by

$$\bar{F}_H(h) = \exp(-h),$$

equation (6.44) can be further derived as

$$\Pr\left[\frac{P\zeta_n^{\mathrm{NL}}(r)h}{I_{\mathrm{agg}} + P^{\mathrm{N}}} > \gamma_0\right] = \mathbb{E}_{[I_{\mathrm{agg}}]}\left\{\exp\left(-\frac{\gamma_0\left(I_{\mathrm{agg}} + P^{\mathrm{N}}\right)}{P\zeta_n^{\mathrm{NL}}(r)}\right)\right\}$$

$$= \exp\left(-\frac{\gamma_0 P^{\mathrm{N}}}{P\zeta_n^{\mathrm{NL}}(r)}\right)\mathbb{E}_{[I_{\mathrm{agg}}]}\left\{\exp\left(-\frac{\gamma_0}{P\zeta_n^{\mathrm{NL}}(r)}I_{\mathrm{agg}}\right)\right\}$$

$$= \exp\left(-\frac{\gamma_0 P^{\mathrm{N}}}{P\zeta_n^{\mathrm{NL}}(r)}\right)\mathscr{L}_{I_{\mathrm{agg}}}\left(\frac{\gamma_0}{P\zeta_n^{\mathrm{NL}}(r)}\right). \quad (6.45)$$

Based on the condition of NLoS signal transmission, the Laplace transform, $\mathscr{L}_{I_{\mathrm{agg}}}^{\mathrm{NL}}(s)$, can be derived as

$$\mathscr{L}_{I_{\mathrm{agg}}}^{\mathrm{NL}}(s) = \mathbb{E}_{[I_{\mathrm{agg}}]}\left\{\exp(-sI_{\mathrm{agg}})\right\}$$

$$= \mathbb{E}_{[\Phi,\{\beta_i\},\{g_i\}]}\left\{\exp\left(-s\sum_{i\in\Phi/b_o} P\beta_i g_i\right)\right\}$$

$$\stackrel{(a)}{=} \exp\left(-2\pi\tilde{\lambda}\int\left(1 - \mathbb{E}_{[g]}\left\{\exp\left(-sP\beta(u)g\right)\right\}\right)u\,du\right), \quad (6.46)$$

where

- step (a) of equation (6.46) has been explained in detail in equation (3.18), and further derived in equation (3.19) of Chapter 3.

It is important to note that, unlike equation (3.18), the detailed calculation of the Laplace transform, $\mathscr{L}_{I_{\mathrm{agg}}}^{\mathrm{NL}}(s)$, given by equation (6.46) only involves active small cell BSs with a density, $\tilde{\lambda}$, because only the active small cell BS generate inter-cell inter-ference.

Compared with equation (3.19), which considers the inter-cell interference from a single-slope path loss model, the expression, $\mathbb{E}_{[g]}\left\{\exp\left(-sP\beta(u)g\right)\right\}$, in equation (6.46) should consider the inter-cell interference from both the LoS and the NLoS paths. As a result, similar to equation (6.47), the Laplace transform, $\mathscr{L}_{I_{\mathrm{agg}}}^{\mathrm{NL}}(s)$, can be further developed as

$$\mathscr{L}_{I_{\text{agg}}}^{\text{NL}}(s) = \exp\left(-2\pi\tilde{\lambda}\int\left(1 - \mathbb{E}_{[g]}\left\{\exp\left(-sP\beta(u)g\right)\right\}\right)u\,du\right)$$

$$\overset{(a)}{=} \exp\left(-2\pi\tilde{\lambda}\int_{r_2}^{+\infty}\Pr^{\text{L}}(u)\left(1 - \mathbb{E}_{[g]}\left\{\exp\left(-sP\zeta^{\text{L}}(u)g\right)\right\}\right)u\,du\right)$$

$$\times\exp\left(-2\pi\tilde{\lambda}\int_{r}^{+\infty}\left(1 - \Pr^{\text{L}}(u)\right)\left(1 - \mathbb{E}_{[g]}\left\{\exp\left(-sP\zeta^{\text{NL}}(u)g\right)\right\}\right)u\,du\right)$$

$$\overset{(b)}{=} \exp\left(-2\pi\tilde{\lambda}\int_{r_2}^{+\infty}\frac{\Pr^{\text{L}}(u)u}{1 + \left(sP\zeta^{\text{L}}(u)\right)^{-1}}\,du\right)$$

$$\times\exp\left(-2\pi\tilde{\lambda}\int_{r}^{+\infty}\frac{\left[1 - \Pr^{\text{L}}(u)\right]u}{1 + \left(sP\zeta^{\text{NL}}(u)\right)^{-1}}\,du\right), \tag{6.47}$$

where

- in step (a), the LoS and the NLoS inter-cell interference come from a distance larger than distances, r_2 and r, respectively, and
- in step (b), the PDF, $f_G(g) = 1 - \exp(-g)$, of the Rayleigh random variable, g, has been used to calculate the expectations, $\mathbb{E}_{[g]}\left\{\exp\left(-sP\zeta^{\text{L}}(u)g\right)\right\}$ and $\mathbb{E}_{[g]}\left\{\exp\left(-sP\zeta^{\text{NL}}(u)g\right)\right\}$.

Our proof of Theorem 6.3.1 is completed by plugging equations (6.34), (6.36), (6.43) and (6.45) into equation (6.25).

Appendix B Proof of Lemma (6.3.2)

Before proceeding with the proof of Lemma (6.3.2), some of the key insights of Theorem 6.3.1 should be first emphasized. In equation (6.1), the components, T_n^{L} and T_n^{NL}, are the components of the coverage probability, p^{cov}, for the cases where the signal comes from the nth piece LoS or the nth piece NLoS path, respectively.

The calculation of the component, T_n^{L}, is based on equations (6.4) and (6.8), and they can be interpreted as follows:

- In equation (6.4), the PDF, $f_{R,n}^{\text{L}}(r)$, characterizes the geometrical density function of the typical UE with no other LoS or NLoS small cell BS providing a better link to the typical UE than its serving small cell BS – the small cell BS at such an nth piece LoS path loss.
- In equation (6.8), the term, $\exp\left(-\frac{\gamma_0 P^{\text{N}}}{P\zeta_n^{\text{L}}(r)}\right)$, measures the probability of the signal power exceeding the noise power by a factor of at least, γ_0, and the term, $\mathscr{L}_{I_{\text{agg}}}^{\text{L}}\left(\frac{\gamma_0}{P\zeta_n^{\text{L}}(r)}\right)$, further computed as equation (6.9), measures the probability of the signal power exceeding the aggregated inter-cell interference power by a factor of at least, γ_0. Since the multi-path fast fading channel gain, h, follows an exponential distribution, the product of the above two probabilities yields the

probability of the signal power exceeding the sum of the noise power and the aggregated inter-cell interference power by a factor of at least, γ_0.

The calculation of the component, T_n^{NL}, is based on equations (6.5) and (6.10), and its interpretation follows a similar approach than that of the component, T_n^{L}, presented above. Thus, we omit it for the sake of brevity.

See Appendix 4A in Chapter 4 for more details on the derivations of these components and the intuition behind them.

With this in mind, Lemma (6.3.2) is proved as follows:

- When deriving the coverage probability, p^{cov}, the signal power increases as the small cell BS density, λ, increases for both

  ☐ the case where all small cell BSs are active and
  ☐ the case where all the small cell BSs have an idle mode capability, and thus not all small cell BSs are active.

  This is because as the small cell BS density, λ, increases, the two-dimensional distance, r, has to decrease to achieve the same value for either the PDF, $f_{R,n}^{\text{L}}(r)$, in equation (6.4), or the PDF, $f_{R,n}^{\text{NL}}(r)$, in equation (6.5), thus indicating that the typical UE has to connect with a nearer small cell BS providing a stronger signal power.
- When deriving the coverage probability, p^{cov}, the small cell BS density, λ, was used in equations (6.9) and (6.11) for the case where all small cell BSs were active. Instead, the active small cell BS density, $\tilde{\lambda}$, was substituted into equations (6.9) and (6.11) for the case where all the small cell BSs had an idle mode capability, and thus not all small cell BSs were active. As a result, the coverage probability, p^{cov}, is larger in the latter case than in the former, since $\tilde{\lambda} \leq \lambda$, and because the function, $\exp(-x)$, is a decreasing function with respect to the variable, x, in equations (6.9) and (6.11). The intuition behind this is that the aggregated inter-cell interference power of the latter case, where all the small cell BSs have an idle mode capability, is always no larger than that of the former case without it. Switching off a small cell BS can only reduce the inter-cell interference.

Appendix C Proof of Theorem 6.3.3

To prove Theorem 6.3.3, we basically need to show that the active small cell BS density, $\tilde{\lambda}$, with a UE association strategy based on the strongest received signal strength with probabilistic LoS and NLoS transmissions is no smaller than the active small cell BS density, $\tilde{\lambda}^{\text{minDis}}$, with a UE association strategy based on the nearest distance with a single-slope path loss model, i.e. $\tilde{\lambda} \geq \tilde{\lambda}^{\text{minDis}}$.

In the following, we further elaborate on this proof.

Let us first consider a baseline scenario where all UEs only have NLoS paths to their serving small cell BSs. In such a scenario, the UE association strategy based on the nearest distance could be a reasonable one, as it is equivalent to that based on the strongest received signal strength.

Let us now consider a new scenario with probabilistic LoS and NLoS transmissions, and a typical UE, k, and an arbitrary small cell BS, b, located at a distance, r, from each other. Due to the probabilistic LoS and NLoS transmissions, such a small cell BS, b, can be virtually split into two probabilistic small cell BSs,

- an LoS small cell BS, b^L, with a probability, $\mathrm{Pr}^L(r)$, and
- an NLoS small cell BS, b^{NL}, with a probability, $\left(1 - \mathrm{Pr}^L(r)\right)$.

To achieve the same received signal strength performance in both the baseline and the new scenarios, the equivalent distance, r_1, between the typical UE, k, and the arbitrary LoS small cell BS, b^L, in the new scenario should satisfy the condition, $r_1 = \arg_{r_1}\left\{\zeta^{NL}(r_1) = \zeta^L(r)\right\}$, as indicated in equation (6.6). In other words, this equivalent distance, r_1, is the distance at which the LoS small cell BS, b^L, in the new scenario results in the same received signal strength at the typical UE, k, than that provided by the NLoS small cell BS, b^{NL}, in the baseline scenario. Since an LoS transmission attenuates at a slower pace than the NLoS one, we have that the LoS small cell BS, b^L, should thus be located further away than the NLoS one, b^{NL}, i.e. $r_1 < r$. As a result, due to such a larger equivalent distance, r_1, one can claim that the number of equivalent small cell BSs located inside a disc area

- centred at the typical UE, k,
- with a radius, r_1,

representing the reach of the typical UE, k, increases by at least a factor, $\mathrm{Pr}^L(r)$, with respect to the baseline case. Since this factor, $\mathrm{Pr}^L(r)$, is a non-negative value, from the viewpoint of the typical UE, k, we can equally claim that the active small cell BS density, $\tilde{\lambda}$, with a UE association strategy based on the strongest received signal strength with probabilistic LoS and NLoS transmissions must be no smaller than the active small cell BS density, $\tilde{\lambda}^{minDis}$, with a UE association strategy based on the nearest distance with a single-slope path loss model, i.e. $\tilde{\lambda} \geq \tilde{\lambda}^{minDis} \approx \lambda_0(q)$.

Intuitively speaking, the existence of LoS small cell BSs "expands" the reach of the typical UE and provides more candidate small cell BSs for a typical UE to connect with, and thus the equivalent small cell BS density increases.

Appendix D Proof of Theorem 6.3.4

The main idea behind the proof of Theorem 6.3.4 is follows:

- Given a typical UE, u, and an arbitrary small cell BS, b, located at a distance, r, from each other, first, the conditional probability, $\mathrm{Pr}\left[w \nsim b \mid r\right]$, of such a typical UE, u, *not* being associated with the arbitrary small cell BS, b, conditioned on the distance, r, should be calculated.

- Then, the unconditional probability, $\Pr[w \nsim b]$, of such a typical UE, u, *not* being associated with the arbitrary small cell BS, b, should be computed. This is performed by integrating over the conditioned distance, r.
- Subsequently, a lower bound for the probability of switching off an arbitrary small cell BS, b, i.e. of having no associated UE, is derived based on the previous results.
- Such a lower bound is finally translated into an upper bound for the active small cell BS density, $\tilde{\lambda}$.

In the following, we further elaborate on the above points.

For convenience, the PDFs, $\left\{ f_{R,n}^{L}(r) \right\}$ and $\left\{ f_{R,n}^{NL}(r) \right\}$, of the distance, r, between the typical UE, u, and the arbitrary small cell BS, b, for the LoS and the NLoS cases, respectively, are stacked into a piecewise function, $f_{R}^{Path}(r)$, written as

$$f_{R}^{Path}(r) = \begin{cases} f_{R,1}^{Path}(r), & \text{when } 0 \le r \le d_1 \\ f_{R,2}^{Path}(r), & \text{when } d_1 < r \le d_2 \\ \vdots & \vdots \\ f_{R,N}^{Path}(r), & \text{when } r > d_{N-1} \end{cases}, \qquad (6.48)$$

where

- the string variable, *Path*, takes the value of "L" and "NL" for the LoS and the NLoS cases, respectively.

Based on the stacked PDF, $f_{R}^{Path}(r)$, the CDF, $F_{R}^{Path}(r)$, of the distance, r, between the typical UE, u, and the arbitrary small cell BS, b, for the LoS and the NLoS cases can be derived as

$$F_{R}^{Path}(r) = \int_{0}^{r} f_{R}^{Path}(v)dv. \qquad (6.49)$$

We can now further define the CDF,

$$F_R(r),$$

of the UE association distance, based on the strongest received signal strength, as the sum of two CDFs, the CDF, $F_{R}^{L}(r)$, and the CDF, $F_{R}^{NL}(r)$, i.e. $F_R(r) = F_{R}^{L}(r) + F_{R}^{NL}(r)$.

From the above, it follows that $F_R(+\infty) = 1$, and thus, that the conditional probability, $\Pr[w \nsim b \mid r]$, can be calculated using equation (6.17). This is because the conditional probability, $\Pr[w \nsim b \mid r]$, equates to the sum of the probabilities of the following two events, which lead to the event, $[w \nsim b \mid r]$:

- The first event or term in equation (6.17): The link between the typical UE, u, and the arbitrary small cell BS, b, is an LoS one with a probability, $\Pr^{L}(r)$, while the typical UE, u, is associated with another LoS/NLoS small cell BS that is stronger than the small cell BS, b, with a probability, $\left[F_{R}^{L}(r) + F_{R}^{NL}(r_1) \right]$, with $F_{R}^{L}(r)$ and $F_{R}^{NL}(r_1)$ corresponding to the cases of a stronger LoS small cell BS and a stronger NLoS small cell BS, respectively.

- The second event or term in equation (6.17): The link between the typical UE, u, and the arbitrary small cell BS, b, is an NLoS one with a probability, $\left[1 - \Pr^L(r)\right]$, while the typical UE, u, is associated with another LoS/NLoS small cell BS that is stronger than the small cell BS, b, with a probability, $\left[F_R^L(r_2) + F_R^{NL}(r)\right]$, with $F_R^L(r_2)$ and $F_R^{NL}(r)$ corresponding to the cases of a stronger LoS small cell BS and a stronger NLoS small cell BS, respectively.

Continuing with the proof, for an arbitrary small cell BS, b, we assume that all its candidate UEs are randomly distributed in the disc, Ω,

- centred at such a small cell BS, b,
- with a radius, $r_{max} > 0$.

Then, for the typical UE, u, inside disc, Ω, the unconditional probability, $\Pr[u \approx b]$, can be calculated by equation (6.16), where $\frac{2r}{r_{max}^2}$ is the distribution density function of the distance, r, as suggested in equation (3.11) in Chapter 3.

To conclude the proof, it should be stated that if the number, K, of candidate UEs inside the disc, Ω, follows a Poisson distribution with a density, $\lambda_\Omega = \rho \pi r_{max}^2$, which is the case in this system model, the probability mass function (PMF), $f_K(k)$, of the number, K, of candidate UEs can be written as [221]

$$f_K(k) = \frac{\lambda_\Omega^k e^{-\lambda_\Omega}}{k!}, \quad k \in \{0, 1, 2, \ldots,\}, \tag{6.50}$$

and thus the probability of an arbitrary small cell BS, b, being idle, i.e., with no UE associated with it, can be computed as in equation (6.15).

Note that the above derivation – and thus equation (6.15) – ignores the spatial correlation between nearby UEs inside the disc, Ω, i.e. if the typical UE, k, is not associated with a small cell BS, b, this may imply that another UE, k', that is nearby the typical UE, k, should also have a large probability of *not* being associated with the small cell BS, b, due to their close proximity. As a result, equation (6.15) underestimates the probability of an small cell BS, b, being idle, and the active small cell BS density, $\tilde{\lambda}$, can thus be upper-bounded by the density, $\lambda \left(1 - Q^{off}\right)$, which concludes our proof.

7 The Impact of Ultra-Dense Wireless Networks on Multi-User Diversity

7.1 Introduction

Scarce wireless resources are operated by the base stations (BSs) of – and shared among the user equipment (UE) in – a network, as explained in the introductory chapter of this book (Chapter 1). It is thus desirable that such BSs schedule the usage of such a valuable wireless resources to such UEs as efficiently as possible to maximize the overall network performance, where the performance metric is defined by the network operator.

Cellular networks, including small cell ones, accommodate such desire, and are usually scheduled systems in both the downlink and the uplink. In fact, every small cell BS has a scheduler at the medium access control (MAC) layer, which decides when and to which UE to allocate the time, the frequency and the spatial resources of the network, and what transmission parameters to use, including, for example, modulation and coding schemes (MCSs) and data rates.

Generally speaking, scheduling can be either dynamic or semi-static:

- Dynamic scheduling is the basic mode of operation and allows the mentioned scheduler at every small cell BS to take scheduling decisions for each UE at each scheduling interval – the maximum granularity. Importantly, since scheduling intervals can be as regular as 1 ms – or less – dynamic scheduling permits the scheduler to best follow and respond to the rapid variations of both the traffic demands and the radio-channel qualities of UEs, thereby efficiently exploiting the available radio resources, and enabling higher data rates [12, 13].
- Semi-static scheduling, in contrast, offers the possibility to allocate resources in a periodic manner with one control message, during a relatively large time window, comprised of many scheduling intervals. This is particularly useful to reduce the control overhead required for resource allocation. Services such as voice over internet protocol voice over IP (VoIP), where the packet size is small and the inter-arrival time of packets is predictable, benefit from this type of scheduling. In this case, semi-statically allocating resources, all at once, every so often, is more efficient than taking decisions and communicating them at each scheduling interval [12, 13].

To follow the small cell BS commands, and realize its scheduling, whether dynamic or semi-static, each UE monitors one or several control channels, searching for a scheduling grant. Typically, a small cell BS transmits a scheduling grant once per scheduling interval in the case of dynamic scheduling. However, to support services requiring very low latencies, it is also possible to configure even more frequent monitoring. Upon detection of a valid scheduling grant provided by the small cell BS, each UE obeys the scheduling decision, and receives data in the downlink or transmits data in the uplink, according to the provided information. The scheduling grant includes, among others, data on the set of time-frequency resources upon which the UE data should be transmitted as well as MCS and hybrid automatic repeat request (HARQ)-related information.

HARQ retransmissions with incremental redundancy are widely used nowadays, where the receiver – the UE in the downlink or the small cell BS in the uplink – reports the outcome of the decoding operation to the transmitter – the small cell BS in the downlink or the UE in the uplink. If the data transmission was received successfully, a positive acknowledgement is reported in the opposite link direction. However, if the packet was decoded erroneously, the transmitter can retransmit such a packet – or part of it – to the receiver in a later scheduling interval. Subsequently, the receiver can then combine the soft information from these multiple transmission attempts. This increases the likelihood of a successful reception with every retransmission [164, 222].

When it comes to take scheduling decisions, channel-dependent scheduling is probably the most used approach in current macrocell networks, and by inheritance in small cells BS too. Fluctuations in the received signal quality, stemming from

- frequency-selective fading,
- distance-dependent path loss and
- random inter-cell interference variations due to the transmissions in other cells and by other devices,

are an inherent part in any wireless communication system. Historically, such radio-channel variations, which take place at small timescales, were seen as a problem, as they could not be tracked. However, the development of channel-dependent scheduling, which was first introduced in the later versions of high speed packet access (HSPA) [223], and is integral to long-term evolution (LTE) [224], changed this viewpoint. Channel-dependent scheduling, assisted by regular radio-channel measurements, permits the small cell BS to transmit to – or receive from – a particular UE only when its radio-channel conditions are favourable, thus opportunistically exploiting the radio-channel variations. Given a large population of UEs having data to transmit – or receive – per active small cell BS, there is a high likelihood of at least some of these UEs having favourable radio-channel conditions, and thus being able to benefit from the resulting high data rates at the corresponding scheduling interval. The performance gain obtained by transmitting to – or receiving from – UEs with favourable radio-channel conditions in each scheduling interval is commonly known as multi-user diversity. The larger the radio-channel fluctuations and the larger the population of UEs per active small cell BS, the larger the multi-user diversity gain.

As mentioned before, to be able to opportunistically benefit from radio-channel variations, and take the best scheduling decisions, the channel-dependent scheduler at the small cell BS needs relevant information about the traffic and the radio-channel conditions at each UE. The specific data depends on the scheduling strategy implemented, which is not standardized, but developed by each equipment vendor, as a differentiation feature. However, most channel-dependent schedulers use – at least – knowledge related to

- the radio-channel conditions at the UEs, including spatial-domain properties,
- the priorities and the buffer statuses of the different data flows, including the amount of data pending retransmission, and
- the inter-cell interference situation in neighbouring cells, which can be useful, if some form of inter-cell interference coordination is implemented [166].

Information about the radio-channel conditions at the UE, which is probably the main driver for taking channel-dependent scheduling decisions, can be obtained through several mechanisms. The channel state information (CSI) reports fed back from the UE to the small cell BSs, however, are typically the most used to assess the downlink channel qualities. The small cell BS can configure a wide range of CSI reports for each UE, where such a UE will take the corresponding radio-channel measurements over downlink pilot signals, and then report the radio-channel quality in the time, the frequency and/or the spatial domains to the small cell BS through such CSI reports. More specifically, CSI reports may include different combinations of channel quality indicator (CQI), rank indicator (RI) and precoding matrix indicator (PMI). Instead of a CSI report, uplink sounding signals can be used not only to access uplink channel qualities, but also to gather the downlink ones in time division duplexing (TDD) systems, making some assumptions on the downlink-uplink channel reciprocity [12, 13].

Once the relevant information about the traffic and the radio-channel conditions at each UE is available at the channel-dependent scheduler, this should devise a scheduling decision for the specific scheduling interval. The proportional fair (PF) scheduling metric is a widely used channel-dependent scheduling strategy in the industry, offering a good trade-off between maximizing the overall cell throughput and improving throughput fairness among UEs with diverse radio-channel conditions. Broadly speaking, the PF scheduler gives priority to those UEs, whose currently achievable throughputs, due to favourable radio-channel conditions, are the largest compared with respect to their own average throughputs [225]. However, the gains of the PF scheduler may be limited in ultra-dense networks due to the limited multi-user diversity gains resulting from

- the small radio-channel variations on a given time-frequency resource due to the high probability of line-of-sight (LoS) transmissions, considering the close proximity between the UEs and their serving small cell BSs and
- the small number of UEs per active small cell BS.

As a result, simpler schedulers, which are not channel-dependent, such as the round robin (RR) scheduler could be more appealing in the ultra-dense regime. For completeness, let us indicate that an RR scheduler allows UEs having data to transmit –

or receive – to take turns in accessing the radio-channel in a periodically and repeated order, regardless of what their radio-channel conditions are. This RR scheduler is simple and easy to implement. Importantly, it also provides the same opportunity to access the radio-channel to all UEs, and is work-conserving, meaning that if one UE is out of packets, the next one will take its place. This prevents wireless resources from going unused [226].

In light of this overall discussion, a fundamental question a rises:

Is the PF scheduler the right choice for an ultra-dense network, or should it be substituted by another strategy of lower complexity with no channel-dependent scheduling?

In this chapter, we answer this fundamental question by means of in-depth theoretical analyses.

The rest of this chapter is organized as follows:

- Section 7.2 introduces the system model and the assumptions taken in the theoretical performance analysis framework presented in this chapter, considering a channel-dependent scheduler with a PF scheduling metric, while embracing the channel model presented in Chapter 4, with both LoS and non-line-of-sight (NLoS) transmissions.
- Section 7.3 presents the theoretical expressions for the coverage probability and the area spectral efficiency (ASE) under the new assumptions.
- Section 7.4 provides results for a number of small cell BS deployments with different densities and characteristics. It also presents the conclusions drawn by this theoretical performance analysis, shedding new light on the impact of channel-dependent schedulers and the potential benefits of lower complexity methods in a dense network. Importantly, and for completeness, this section also studies via system-level simulations the impact of a Rician multi-path fast fading – instead of a Rayleigh one – on the derived results and conclusions.
- Section 7.5, finally, summarizes the key takeaways of this chapter.

7.2 Updated System Model

To assess the impact of the channel-dependent scheduler described in Section 7.1 on an ultra-dense network, in this section, we upgrade the system model described in Section 6.2, which has accounted for

- LoS and NLoS transmissions,
- a user association strategy, based on the strongest received signal strength,
- a finite UE density, $\rho < +\infty$, and
- a simple idle mode capability at the small cell BSs,

by adding the following feature,

- a PF scheduler at the small cell BSs.

As in Chapters 3–6, Table 7.1 provides a concise and updated summary of the system model used in this chapter.

In this system model section, we touch again on the finite UE density, $\rho < +\infty$, and the idle mode capability at the small cell BSs, developing new expressions, which are key to this chapter. Moreover, we also present the new upgrades incorporated to the system model to realize the PF scheduler. Before that, however, it is important to note that the height difference between the antenna of the typical UE and that of an arbitrary small cell BS, considered in Chapter 5, is not taken into account in this chapter. This is to better understand – in isolation – the impact of the PF scheduler at the small cell BSs on an ultra-dense network. As a result, all distances are two-dimensional in this chapter.

7.2.1   Finite User Equipment Density and Idle Mode Capability at the Small Cell Base Stations

First of all, let us review in this section some of the key concepts required to understand the framework analyzed in this chapter.

Building on Section 6.2, the downlink of a cellular network is considered, where

- the small cell BSs are deployed on a plane according to a homogeneous Poisson point process (HPPP), Φ, of intensity, λ, in BSs/km^2, and
- the UEs are distributed on such a network according to another HPPP, Φ^{UE}, of intensity, $\rho < +\infty$, in UEs/km^2.[1]

In practice, a small cell BS will enter into idle mode if there is no UE associated to it. This idle mode capability at the small cell BSs reduces the inter-cell interference to neighbouring UEs and the energy consumption of the network. As a consequence, the set of active small cell BSs should be determined considering this idle mode capability and the user association strategy in use.

As mentioned earlier, a practical user association strategy is assumed in this book, where each UE is connected to the small cell BS providing the maximum average received signal strength. With this in mind, and since UEs are randomly and uniformly distributed in the network, it is safe to assume, as shown in Chapter 6, that

- the active small cell BSs also follow another HPPP, $\tilde{\Phi}$, of intensity, $\tilde{\lambda}$, in BSs/km^2, where the two following inequalities are correct:
    - ☐ the active small cell BSs density, $\tilde{\lambda}$, is no larger than the small cell BSs density, λ, i.e. $\tilde{\lambda} \leq \lambda$; and
    - ☐ the active small cell BSs density, $\tilde{\lambda}$, is no larger than the finite UE density, $\rho < +\infty$, i.e. $\tilde{\lambda} \leq \rho$.

Let us remark at this point that one UE is served by at most one BS in the system model considered in this book – no inter-small cell BS coordination techniques are

---

[1] As a clarification, and similarly to Chapter 6, note that all UEs are considered to be active UEs, i.e. they all have packets to receive.

Table 7.1. System model

Model	Description	Reference
Transmission link		
Link direction	Downlink only	Transmissions from the small cell BSs to the UEs
Deployment		
Small cell BS deployment	HPPP with a finite density, $\lambda < +\infty$	Section 3.2.2
UE deployment	HPPP with full load, resulting in at least one UE per cell	Section 3.2.3
UE to small cell BS association		
Strongest small cell BS	UEs connect to the small cell BS providing the strongest received signal strength	Section 3.2.8 and references therein, equation (2.1.4)
Path loss		
3GPP UMi [153]	Multi-slope path loss with probabilistic LoS and NLoS transmissions	Section 4.2.1, equation (4.1)
	☐ LoS component	equation (4.2)
	☐ NLoS component	equation (4.3)
	☐ Exponential probability of LoS	equation (4.19)
Multi-path fast fading		
Rayleigh*	Highly scattered scenario	Section 3.2.7 and references therein equation (2.15)
Shadow fading		
Not modelled	For tractability reasons, shadowing is not modelled since it has no qualitative impact on the results, see Section 3.4.5	Section 3.2.6 references therein
Antenna		
Small cell BS antenna	Isotropic single-antenna element with 0 dBi gain	Section 3.2.4
UE antenna	Isotropic single-antenna element with 0 dBi gain	Section 3.2.4
Small cell BS antenna height	Not considered	–
UE antenna height	Not considered	–
Idle mode capability at small cell BSs		
Connection-aware	Small cell BSs with no UE in their coverage areas are switched off	Sections 7.2.1 and 7.2.2
Scheduler at small cell BSs		
Proportional fair	The scheduler gives priority to the UEs with the best channel conditions	Section 7.2.3

* Section 7.4.4 presents simulated results with Rician multi-path fast fading to demonstrate its impact on the obtained results.

studied. As a result, it follows that the more UEs there are, the same or more active small cell BSs are needed to serve those UEs, never less. In other words, a larger finite UE density, $\rho < +\infty$, switches on the same or a larger small cell BS density, $\tilde{\lambda}$.

It is also important to remind that, according to the analysis in Section 6.4.2, the active small cell BS density, $\tilde{\lambda}$, for this system model can be accurately approximated by

$$\tilde{\lambda} = \lambda \left[1 - \frac{1}{\left(1 + \frac{\rho}{q\lambda}\right)^q} \right], \tag{7.1}$$

where

- q is a fitting parameter and depends on the path loss model.

7.2.2   The Number of User Equipments per Active Small Cell Base Station

In the following, let us introduce how the number of UEs per active small cell BS can be modelled, a key feature for the rest of this study.

According to [217], the size, X, of the coverage area of a small cell BS can be approximately characterized by a Gamma distribution, where the probability density function (PDF) of it can be derived as

$$f_X(x) = (q\lambda)^q x^{q-1} \frac{\exp(-q\lambda x)}{\Gamma(q)}, \tag{7.2}$$

where

- $\Gamma(\cdot)$ is the Gamma function [156].

Based on such a PDF, $f_X(x)$, and denoting the number of UEs per small cell BS by the random variable, K, its probability mass function (PMF), $f_K(k)$, can be calculated as

$$f_K(k) = \Pr[K = k]$$

$$\overset{(a)}{=} \int_0^{+\infty} \frac{(\rho x)^k}{k!} \exp(-\rho x) f_X(x) dx$$

$$\overset{(b)}{=} \frac{\Gamma(k+q)}{\Gamma(k+1)\Gamma(q)} \left(\frac{\rho}{\rho+q\lambda}\right)^k \left(\frac{q\lambda}{\rho+q\lambda}\right)^q, \tag{7.3}$$

where

- $k = K(\omega) \in \{0, 1, 2, \ldots, +\infty\}$ is a realization of the random variable, K,
- step (a) follows from the HPPP distribution of the UEs and
- step (b) follows from equation (7.2).

Note that this PMF, $f_K(k)$, satisfies the normalization condition, i.e. $\sum_{k=0}^{+\infty} f_K(k) = 1$. More importantly, from equation (7.3), it can be seen that the random variable, K, representing the number of UEs per small cell BS, follows a negative binomial distribution [156],

$$K \sim \text{NB}\left(q, \frac{\rho}{\rho + q\lambda}\right). \tag{7.4}$$

As discussed in Chapter 6, we assume that a small cell BS with no UE associated to it, $k = 0$, is in idle mode and not active, and that, as a result, it can be excluded from our analyses. Thus, let us focus on the active small cell BSs now, and more specifically, on the number of UEs per *active* small cell BS, which is denoted hereafter by the positive random variable, $\tilde{K}$.

Considering equation (7.3), and the fact that the only difference between the random variables, K and $\tilde{K}$, representing the number of UEs per small cell BS and the number of UEs per active small cell BS, respectively, is that by definition the value 0 cannot occur in the latter, i.e. $\tilde{K} \neq 0$, it can be concluded that the random variable, $\tilde{K}$, representing the number of UEs per active small cell BS, follows a truncated negative binomial distribution,

$$\tilde{K} \sim \text{truncNB}\left(q, \frac{\rho}{\rho + q\lambda}\right), \tag{7.5}$$

whose PMF, $f_{\tilde{K}}(\tilde{k})$, can be written as

$$f_{\tilde{K}}(\tilde{k}) = \Pr\left[\tilde{K} = \tilde{k}\right] = \frac{f_K(\tilde{k})}{1 - f_K(0)}, \tag{7.6}$$

where

- $\tilde{k} = \tilde{K}(\omega) \in \{1, 2, \dots, +\infty\}$ is a realization of the random variable, K, and
- the denominator, $(1 - f_K(0))$, represents the probability of a small cell BS being active.

Note that this PMF, $f_{\tilde{K}}(\tilde{k})$, also satisfies the normalization condition, i.e. $\sum_{k=1}^{+\infty} f_{\tilde{K}}(\tilde{k}) = 1$.

Based on the derivation of the active small cell BS density, $\tilde{\lambda}$, in equation (7.1), and the above findings with regards to the number of per active small cell BS, one can further develop equation (7.1) as follows

$$\tilde{\lambda} = \lambda\left(1 - f_K(0)\right). \tag{7.7}$$

Example To illustrate the meaning behind the above derivations, Figure 7.1 illustrates the results of equation (7.6) with the following parameters:

- a small cell BS density, $\lambda \in \{50, 200, 1000\}$ BSs/km^2,
- a finite UE density, $\rho = 300$ UEs/km^2, and
- a fitting parameter, $q = 4.18$.[2]

---

[2] In Chapter 6, it was shown that a fitting parameter, $q = 4.18$, is appropriate for the considered system model with a finite UE density, $\rho = 300$ UEs/km^2, in the investigated 3rd Generation Partnership Project (3GPP) case study.

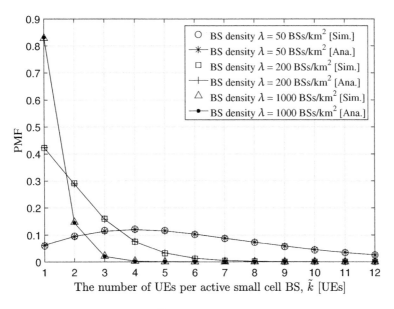

Figure 7.1 PMF of the number, $\tilde{K}$, of UEs per active small cell BS.

From this figure, we can draw the following observations:

- The analytical results based on the truncated negative binomial distribution match well with the simulation ones. More specifically, the maximum difference between the simulated PMF and the analytical one is shown to be less than 0.5%.
- The PMF, $f_{\tilde{K}}(\tilde{k})$, of the random variable, $\tilde{K}$, representing the number of UEs per active small cell BS, shows a more dominant peak at the realization, $\tilde{k} = 1$, as the small cell BS density, λ, and in turn, the active small cell BS density, $\tilde{\lambda}$, increase. This is because the ratio of the finite UE density, $\rho < +\infty$, to the active small cell BS density, $\tilde{\lambda}$, gradually decreases towards 1 as the small cell BS density, λ, increases, approaching the limit of one UE per active small cell BS in an ultra-dense network. In this particular example, for a small cell BS density, $\lambda = 1000 \, \text{BSs/km}^2$, more than 80% of the active small cell BSs will serve only one UE.

Finally, and for completeness, let us conclude this subsection by indicating that the cumulative mass function (CMF), $F_{\tilde{K}}(\tilde{k})$, of the random variable, $\tilde{K}$, representing the number of UEs per active small cell BS, can be written as

$$F_{\tilde{K}}(\tilde{k}) = \sum_{t=1}^{\tilde{k}} f_{\tilde{K}}(t). \tag{7.8}$$

7.2.3   The Proportional Fair Scheduler

According to [225], the operation of a PF scheduler can be summarized as follows:

- First, the small cell BS tracks the average throughput of each of its associated UEs using an exponential moving average.
- Second, each connected UE frequently feeds back its CSI to its serving small cell BS, and as result, such a small cell BS can calculate the ratio of the instantaneous achievable rate to the average throughput for each UE, which is defined as the PF metric for the UE selection at the small cell BS scheduler.
- Finally, for each time-frequency resource, the small cell BS selects the UE with the largest PF metric for transmission. This problem can be formulated as

$$u^* = \arg\max_{u\in\{1,2,...,\tilde{k}\}} \left\{ \frac{\tilde{R}_u}{\bar{R}_u} \right\},\qquad(7.9)$$

where
- u is the index of a UE,
- u^* is the index of the UE selected for transmission,
- $\tilde{R}_u$ is the instantaneous achievable rate of the uth UE and
- $\bar{R}_u$ is the average throughput of the uth UE.

Note that the modelling of the random variable, $\tilde{K}$, representing the number of UEs per active small cell BS, has been discussed in Section 7.2, equation (7.5).

From a network performance analysis point of view, however, it is important to note that it is very difficult – if not impossible – to derive the performance of the above introduced PF scheduler, abstracted by equation (7.9), due to the time domain correlation. This is because the objective of a performance analysis is usually to derive the average UE throughput, $\bar{R}_u$, or a related metric, but in this case, the average UE throughput, $\bar{R}_u$, is not only the final objective, but also part of the PF metric, i.e. the ratio, $\frac{\tilde{R}_u}{\bar{R}_u}$, should be known, and substituted into equation (7.9) to carry out the performance analysis of the average UE throughput, $\bar{R}_u$ – a chicken and egg problem, if the time domain is not considered.

Fortunately, there is a range of alternative metrics, based solely on the signal qualities of the UEs, which can also be used to realize the PF metric, and drive the scheduling [225, 227–229]. These metrics are much more tractable as they do not fall into the entanglement problem presented before.

In the following, the framework developed by Choi et al. in [225] is embraced, where the ratio of the instantaneous to the average signal-to-noise ratio (SNR) is used as the PF metric, instead of the original one based on the average UE throughput, $\bar{R}_u$. Importantly, this metric does not use the inter-cell interference either, which could have led to another entanglement problem. More specifically, the PF metric for the UE selection at scheduler proposed by Choi et al. is given by

$$u^* = \arg\max_{u\in\{1,2,...,\tilde{k}\}} \left\{ \frac{\tilde{Z}_u}{\bar{Z}_u} \right\},\qquad(7.10)$$

where
- $\tilde{Z}_u$ is the instantaneous SNR of the uth UE and

- $\bar{Z}_u$ is the average SNR of the uth UE.

Although the PF metric for UE selection at the scheduler in equation (7.10) is not exactly the same as that in equation (7.9), it should be noted that it captures – in principle – the most important characteristics of the PF scheduler:

- It gives preference to UEs with relatively good instantaneous channel qualities with respect to their average ones, since the instantaneous rate, $\tilde{R}_u$, of the uth UE is a strictly monotonically increasing function of the instantaneous SNR, $\tilde{Z}_u$, of such a UE.
- It allocates the same portion of time-frequency resources to each UE in the long term to enforce fairness, since the probability of the instantaneous SNR, $\tilde{Z}_u$, of the uth UE being larger than the average SNR $\bar{Z}_u$, of such a UE, i.e. $\tilde{Z}_u \geq \bar{Z}_u$, is approximately the same for all UEs, and entirely depends on the fading of the channel, which in our system model is independent and identical distributed (i.i.d.).

Because the accuracy and the practicality of the PF metric in equation (7.10) for the UE selection at scheduler have been well established in [225], this PF metric is adopted in the following sections for our analysis of the impact of a channel-dependent scheduler and the potential benefits of lower complexity methods on an ultra-dense network.

7.3   Theoretical Performance Analysis and Main Results

In this section, given the definition of the coverage probability, p^{cov}, and the ASE, A^{ASE}, summarized for convenience in Table 7.2, we present the derived expressions for these two key performance indicators, while considering the PF scheduler at the small cell BSs presented in Section 7.2. (equations (7.10) and (7.12)). These new expressions will allow one to assess the performance of an ultra-dense network considering a practical and widely used scheduler.

Table 7.2. Key performance indicators

Metric	Formulation	Reference
Coverage probability, p^{cov}	$p^{\text{cov}}(\lambda, \gamma_0) = \Pr[\Gamma > \gamma_0] = \bar{F}_\Gamma(\gamma_0)$	Section 2.1.12, equation (2.22)
ASE, A^{ASE}	$A^{\text{ASE}}(\lambda, \gamma_0) = \frac{\tilde{\lambda}}{\ln 2} \int_{\gamma_0}^{+\infty} \frac{p^{\text{cov}}(\lambda,\gamma)}{1+\gamma} d\gamma$ $+ \tilde{\lambda} \log_2 (1 + \gamma_0) p^{\text{cov}}(\lambda, \gamma_0)$	Section 6.3.3, equation (6.24)

7.3.1   Coverage Probability

Due to the existence of the PF scheduler at each small cell BS to select the UE – in this case, our typical UE – for data delivery on the time-frequency resource of study, the definition of the signal-to-interference-plus-noise ratio (SINR), Γ, of the typical UE needs to be changed as follows:

$$\Gamma = \frac{P\zeta(r)y(\tilde{k})}{I_{\text{agg}} + P^{\text{N}}}, \tag{7.11}$$

where, with respect to the definition in equation (3.3), the previous multi-path fast fading gain, h, between the typical UE and its serving small cell BS, modelled as Rayleigh fading, has been substituted by the multi-path fast fading gain, $y(\tilde{k})$.

Importantly, this multi-path fast fading gain, $y(\tilde{k})$, embraces the logic behind the PF presented in equation (7.10), and thus is defined as the maximum multi-path fast fading gain of those between the set of connected UEs and the corresponding serving small cell BS, still modelled as Rayleigh fading, but now, on condition of the random variable number, $\tilde{K}$, of UEs per active small cell BS.

To clarify this new formulation, and explain its logic, let us reformulate the PF metric for the UE selection at the scheduler in equation (7.10), based on the new definition of the multi-path fast fading gain, $y(\tilde{k})$, as

$$u^* = \underset{u \in \left\{1, 2, \dots, \tilde{k}\right\}}{\arg\max} \left\{ \frac{\frac{P\zeta(r)h_u}{P^{\text{N}}}}{\frac{P\zeta(r) \times 1}{P^{\text{N}}}} \right\} = \underset{u \in \left\{1, 2, \dots, \tilde{k}\right\}}{\arg\max} \{h_u\}, \tag{7.12}$$

where

- h_u in the numerator is an i.i.d. random variable, representing the multi-path fast fading gain between the uth UE of the mentioned set of UEs and its corresponding serving small cell BS, which follows an exponential distribution with unitary mean due to the assumption of Rayleigh multi-path fast fading in the system model (see Table 7.2),
- $\hat{h}_u = 1$ in the denominator is the average multi-path fast fading gain between the uth UE of the mentioned set of UEs and its corresponding serving small cell BS, and takes the value of 1 due to the unitary mean, and
- u^* is the UE with the maximum multi-path fast fading gain, h_u, and thus h_{u^*} is the maximum multi-path fast fading gain, which is also a random variable.

From this reformulation, we can infer that both random variables, $y(\tilde{k})$ and h_{u^*}, are equal, i.e. $y(\tilde{k}) = h_{u^*}$, indicating that the multi-path fast fading gain, $y(\tilde{k})$, of the selected typical UE can be modelled as the maximum random variable among $\tilde{k}$ i.i.d. exponential random variables.

Taking a step further, according to [230], the complementary cumulative distribution function (CCDF), $\bar{F}_{Y(\tilde{k})}(y)$, of the multi-path fast fading gain, $y(\tilde{k})$, of the typical

UE according to the previous definition, based on a Rayleigh fading formulation, can be derived as

$$\bar{F}_{Y(\tilde{k})}(y) = \Pr\left[Y(\tilde{k}) > y\right] = 1 - (1 - \exp(-y))^{\tilde{k}}. \tag{7.13}$$

It is easy to see from this expression that the probability, $\Pr[Y(\tilde{k}) > y]$, increases with the increase of the number, $\tilde{K}$, of UEs per active small cell BS, and as a result, the multi-path fast fading gain, $y(\tilde{k})$, of the typical UE also increases with it.

Importantly, it should be noted that this modelling applies to both the PF and the RR schedulers. When considering the RR scheduler, the typical UE is blindly selected by the small cell BS in terms of multi-path fast fading gain, i.e. there is no channel-dependent scheduling, as explained in the introduction of this chapter. Consequently, using the above formulation in equation (7.13), we can say that the RR scheduler can be modelled assuming that the number, $\tilde{k}$, of UE per active small cell BS equals to 1, i.e. $\tilde{k} = 1$. This leads to an extreme degeneration of the number of random variables from which the maximum random variable is extracted. As a result, the multi-path fast fading gain, $y(\tilde{k})$, of the typical UE degrades to a random variable, h_u, which is equivalent to the random variable, h, used in Chapters 3–6 – only one degree of freedom. Thus, the expressions and the results earlier obtained in these previous chapters apply to the RR scheduler and will be used as a benchmark hereafter in this chapter.

In the following, we present the new main result on the coverage probability, p^{cov}, through Theorem 7.3.1, considering the above introduced system model, where

- the finite UE density, $\rho < +\infty$,
- the simple idle mode capability at the small cell BSs, and more importantly,
- the PF scheduler at the small cell BSs

should be highlighted. Readers interested in the research article originally presenting these results are referred to [89].

THEOREM 7.3.1   *Considering the path loss model in equation (4.1), and the strongest small cell BS association presented in Section 4.2, the coverage probability, p^{cov}, can be derived as*

$$p^{cov}(\lambda, \gamma_0) = \sum_{n=1}^{N}\left(T_n^{L} + T_n^{NL}\right), \tag{7.14}$$

where

-

$$T_n^{L} = \int_{d_{n-1}}^{d_n} \mathbb{E}_{[\tilde{K}]}\left\{\Pr\left[\frac{P\zeta_n^{L}(r)y(\tilde{k})}{I_{agg} + P^{N}} > \gamma_0\right]\right\} f_{R,n}^{L}(r)dr, \tag{7.15}$$

•

$$T_n^{\mathrm{NL}} = \int_{d_{n-1}}^{d_n} \mathbb{E}_{[\tilde{K}]}\left\{\Pr\left[\frac{P\zeta_n^{\mathrm{NL}}(r)y(\tilde{k})}{I_{\mathrm{agg}} + P^{\mathrm{N}}} > \gamma_0\right]\right\} f_{R,n}^{\mathrm{NL}}(r)dr \qquad (7.16)$$

and

• *d_0 and d_N are defined as 0 and $+\infty$, respectively.*

Moreover, the PDF, $f_{R,n}^{\mathrm{L}}(r)$, and the PDF, $f_{R,n}^{\mathrm{NL}}(r)$, within the range, $(d_{n-1} < r \le d_n)$, are given by

$$f_{R,n}^{\mathrm{L}}(r) = \exp\left(-\int_0^{r_1}\left(1 - \Pr^{\mathrm{L}}(u)\right)2\pi u\lambda\,du\right)$$

$$\times \exp\left(-\int_0^r \Pr^{\mathrm{L}}(u)2\pi u\lambda\,du\right)\Pr_n^{\mathrm{L}}(r)2\pi r\lambda, \qquad (7.17)$$

and

$$f_{R,n}^{\mathrm{NL}}(r) = \exp\left(-\int_0^{r_2} \Pr^{\mathrm{L}}(u)2\pi u\lambda\,du\right)$$

$$\times \exp\left(-\int_0^r \left(1 - \Pr^{\mathrm{L}}(u)\right)2\pi u\lambda\,du\right)\left(1 - \Pr_n^{\mathrm{L}}(r)\right)2\pi r\lambda, \qquad (7.18)$$

where

• *r_1 and r_2 are determined by*

$$r_1 = \arg_{r_1}\left\{\zeta^{\mathrm{NL}}(r_1) = \zeta_n^{\mathrm{L}}(r)\right\} \qquad (7.19)$$

and

$$r_2 = \arg_{r_2}\left\{\zeta^{\mathrm{L}}(r_2) = \zeta_n^{\mathrm{NL}}(r)\right\}, \qquad (7.20)$$

respectively.

Proof   See Appendix A.   □

Digging into Theorem 7.3.1, and considering the truncated negative binomial distribution of the number, $\tilde{K}$, of UEs per active small cell BS, we present our results on the expected probability, $\mathbb{E}_{[\tilde{K}]}\left\{\Pr\left[\frac{P\zeta_n^{\mathrm{L}}(r)y(\tilde{k})}{I_{\mathrm{agg}}+P^{\mathrm{N}}} > \gamma_0\right]\right\}$, and the expected probability, $\mathbb{E}_{[\tilde{K}]}\left\{\Pr\left[\frac{P\zeta_n^{\mathrm{NL}}(r)y(\tilde{k})}{I_{\mathrm{agg}}+P^{\mathrm{N}}} > \gamma_0\right]\right\}$, in Theorem 7.3.2.

THEOREM 7.3.2   *Considering the truncated negative binomial distribution of the number, $\tilde{K}$, of UEs per active small cell BS characterized in equation (7.6), the* expected probability, $\mathbb{E}_{[\tilde{K}]}\left\{\Pr\left[\frac{P\zeta_n^{\mathrm{L}}(r)y(\tilde{k})}{I_{\mathrm{agg}}+P^{\mathrm{N}}}>\gamma_0\right]\right\}$, which is used in Theorem 7.3.1, can be calculated as

$$
\mathbb{E}_{[\tilde{K}]}\left\{\Pr\left[\frac{P\zeta_n^{\mathrm{L}}(r)y(\tilde{k})}{I_{\mathrm{agg}}+P^{\mathrm{N}}}>\gamma_0\right]\right\}
$$
$$
=\sum_{\tilde{k}=1}^{+\infty}\left[1-\sum_{t=0}^{\tilde{k}}\binom{\tilde{k}}{t}\left(-\delta_n^{\mathrm{L}}(r)\right)^t\mathscr{L}_{I_{\mathrm{agg}}}^{\mathrm{L}}\left(\frac{t\gamma_0}{P\zeta_n^{\mathrm{L}}(r)}\right)\right]f_{\tilde{K}}(\tilde{k}),\qquad(7.21)
$$

where

- $f_{\tilde{K}}(\tilde{k})$ is the PMF of the number, $\tilde{k}$, of UEs per active small cell BS, which can be calculated using equation (7.6),
- $\delta_n^{\mathrm{L}}(r)$ is a parameter defined as

$$
\delta_n^{\mathrm{L}}(r)=\exp\left(-\frac{\gamma_0 P^{\mathrm{N}}}{P\zeta_n^{\mathrm{L}}(r)}\right)\qquad(7.22)
$$

and
- $\mathscr{L}_{I_{\mathrm{agg}}}^{\mathrm{L}}(s)$ is the Laplace transform of the aggregated inter-cell interference random variable, I_{agg}, for the LoS signal transmission evaluated at the variable value, $s=\frac{\gamma_0}{P\zeta_n^{\mathrm{L}}(r)}$, which can be expressed as

$$
\mathscr{L}_{I_{\mathrm{agg}}}^{\mathrm{L}}(s)=\exp\left(-2\pi\tilde{\lambda}\int_r^{+\infty}\frac{\Pr^{\mathrm{L}}(u)u}{1+\left(sP\zeta^{\mathrm{L}}(u)\right)^{-1}}du\right)
$$
$$
\times\exp\left(-2\pi\tilde{\lambda}\int_{r_1}^{+\infty}\frac{\left[1-\Pr^{\mathrm{L}}(u)\right]u}{1+\left(sP\zeta^{\mathrm{NL}}(u)\right)^{-1}}du\right).\qquad(7.23)
$$

Similarly, the expected probability, $\mathbb{E}_{[\tilde{K}]}\left\{\Pr\left[\frac{P\zeta_n^{\mathrm{NL}}(r)y(\tilde{k})}{I_{\mathrm{agg}}+P^{\mathrm{N}}}>\gamma_0\right]\right\}$, which is used in Theorem 7.3.1, can be computed as

$$
\mathbb{E}_{[\tilde{K}]}\left\{\Pr\left[\frac{P\zeta_n^{\mathrm{NL}}(r)y(\tilde{k})}{I_{\mathrm{agg}}+P^{\mathrm{N}}}>\gamma_0\right]\right\}
$$
$$
=\sum_{\tilde{k}=1}^{+\infty}\left[1-\sum_{t=0}^{\tilde{k}}\binom{\tilde{k}}{t}\left(-\delta_n^{\mathrm{NL}}(r)\right)^t\mathscr{L}_{I_{\mathrm{agg}}}^{\mathrm{NL}}\left(\frac{t\gamma_0}{P\zeta_n^{\mathrm{NL}}(r)}\right)\right]f_{\tilde{K}}(\tilde{k}),\qquad(7.24)
$$

where

- $\delta_n^{NL}(r)$ *is a parameter defined as*

$$\delta_n^{NL}(r) = \exp\left(-\frac{\gamma_0 P^N}{P\zeta_n^{NL}(r)}\right) \tag{7.25}$$

and

- $\mathscr{L}_{I_{agg}}^{NL}(s)$ *is the Laplace transform of the aggregated inter-cell interference random variable, I_{agg}, for the NLoS signal transmission evaluated at the variable value, $s = \frac{\gamma_0}{P\zeta_n^{NL}(r)}$, which can be written as*

$$\mathscr{L}_{I_{agg}}^{NL}(s) = \exp\left(-2\pi\tilde{\lambda}\int_{r_2}^{+\infty}\frac{\Pr^L(u)u}{1+\left(sP\zeta^L(u)\right)^{-1}}du\right)$$

$$\times \exp\left(-2\pi\tilde{\lambda}\int_{r}^{+\infty}\frac{\left[1-\Pr^L(u)\right]u}{1+\left(sP\zeta^{NL}(u)\right)^{-1}}du\right). \tag{7.26}$$

Proof   See Appendix B.   □

Plugging Theorem 7.3.2 into Theorem 7.3.1, we can obtain the new theoretical result on the coverage probability, p^{cov}, while considering the PF scheduler at the small cell BSs.

From Theorems 7.3.1 and 7.3.2, an important and intuitive conclusion can be drawn, which is presented in the following Lemma (7.3.3).

LEMMA 7.3.3   *The coverage probability, p^{cov}, of the PF scheduler at the small cell BSs converges to that of the RR scheduler as the small cell BS density, λ, tends to infinity, i.e. $\lambda \to +\infty$.*

Proof   See Appendix C.   □

To ease the understanding on how to practically obtain this new coverage probability, p^{cov}, Figure 7.2 illustrates a flow chart that depicts the necessary enhancements to the stochastic geometry framework presented in Chapter 6 to compute the new results of Theorem 7.3.1.

Compared with the logic illustrated in Figure 6.1, it is important to note that the signal power is enhanced by the UE selection at the small cell BS thanks to the PF scheduler.

7.3.2   Area Spectral Efficiency

Similarly as in Chapters 4–6, plugging the coverage probability, p^{cov} – obtained from Theorem 7.3.1 – into equation (2.24) to compute the PDF, $f_\Gamma(\gamma)$, of the SINR, Γ, of the typical UE, we can obtain the ASE, A^{ASE}, by solving equation (6.24). See Table 7.2 for further reference on the ASE formulation.

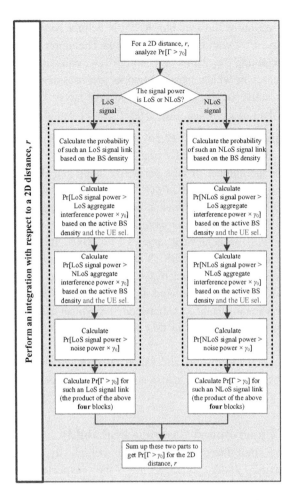

Figure 7.2 Logical steps within the standard stochastic geometry framework to obtain the results in Theorem 7.3.1, considering a PF scheduler at the small cell BSs.

7.3.3   Computational Complexity

In Theorem 7.3.2, equations (7.21) and (7.24) might be difficult to calculate as the number, $\tilde{k}$, of UEs per active small cell BS tends to infinity, i.e. $\tilde{k} \rightarrow +\infty$. In practice, it is thus useful to replace the infinite upper limit of the summations in equations (7.21) and (7.24) with the variable, $\tilde{K}^{\max}$, which is a large enough integer to ensure that

- the CMF, $F_{\tilde{K}}\left(\tilde{K}^{\max}\right)$, in equation (7.8) approximates to 1 within a gap smaller than a parameter, ϵ, and in turn, that
- the PMF, $f_{\tilde{K}}(\tilde{k})$, in equation (7.6) approximates to 0, when $\tilde{k} > \tilde{K}^{\max}$.

This allows to properly evaluate equations (7.21) and (7.24) with a finite and the smallest possible number of summands.

Although the introduction of the variable, $\tilde{K}^{\max}$, is helpful for the calculation of the coverage probability, p^{cov}, with the PF scheduler in Theorems 7.3.1 and 7.3.2, it should be noted that obtaining the results of Theorem 7.3.2 can still be computationally intensive for the cases in which sparse networks are considered, i.e. where the number, $\tilde{k}$, of UEs per active small cell BS could be very large. This leads to complex computations for the Lapace transforms, $\mathscr{L}_{I_{\mathrm{agg}}}^{\mathrm{L}}\left(\frac{t\gamma_0}{P\zeta_n^{\mathrm{L}}(r)}\right)$ and $\mathscr{L}_{I_{\mathrm{agg}}}^{\mathrm{NL}}\left(\frac{t\gamma_0}{P\zeta_n^{\mathrm{NL}}(r)}\right), t \in \{0, 1, \ldots, \tilde{K}^{\max}\}$ in equations (7.23) and (7.26), respectively. For example, when a finite UE density, $\rho = 300\,\mathrm{UEs/km^2}$, and a small cell BS density, $\lambda = 10\,\mathrm{BSs/km^2}$, are considered, the average number of UEs per active small cell BS is around $\tilde{K} = 30\,\mathrm{UEs/BS}$. However, to make the CMF, $F_{\tilde{K}}(\tilde{K}^{\max})$, in equation (7.8) approximate to 1 within a gap smaller than a parameter, $\epsilon = 0.001$, the maximum number of UEs per active small cell BS could be as large as $\tilde{K}^{\max} = 102\,\mathrm{UEs/BS}$. Note that according to equation (7.8), to achieve a target, $\epsilon = 0.001$, the maximum number, $\tilde{K}^{\max}$, of UEs per active small cell BS should be determined according to the UE density, ρ, and the small cell BS density, λ. As a result, the integrals in equations (7.23) and (7.26) need to be calculated at least 102 times for every possible value of the two-dimensional distance, r, between the typical UE and its serving small cell BS to solve equation (7.14) in this example. This is a computationally expensive task.

To address this issue, in the following, alternative and more efficient expressions to derive the coverage probability, p^{cov}, for sparse networks are presented.

A Low-Complexity Upper-Bound Coverage Probability

To reduce the above introduced computational complexity, Theorem 7.3.4 presents low-complexity upper bounds for the expected probabilities, $\mathbb{E}_{[\tilde{K}]}\left\{\Pr\left[\frac{P\zeta_n^{\mathrm{L}}(r)y(\tilde{k})}{I_{\mathrm{agg}}+P^{\mathrm{N}}} > \gamma_0\right]\right\}$ and $\mathbb{E}_{[\tilde{K}]}\left\{\Pr\left[\frac{P\zeta_n^{\mathrm{NL}}(r)y(\tilde{k})}{I_{\mathrm{agg}}+P^{\mathrm{N}}} > \gamma_0\right]\right\}$.

THEOREM 7.3.4   *The expected probability,* $\mathbb{E}_{[\tilde{K}]}\left\{\Pr\left[\frac{P\zeta_n^{\mathrm{L}}(r)y(\tilde{k})}{I_{\mathrm{agg}}+P^{\mathrm{N}}} > \gamma_0\right]\right\}$, *and the expected probability,* $\mathbb{E}_{[\tilde{K}]}\left\{\Pr\left[\frac{P\zeta_n^{\mathrm{NL}}(r)y(\tilde{k})}{I_{\mathrm{agg}}+P^{\mathrm{N}}} > \gamma_0\right]\right\}$, *can be upper-bounded by*

$$
\mathbb{E}_{[\tilde{K}]}\left\{\Pr\left[\frac{P\zeta_n^{\mathrm{L}}(r)y(\tilde{k})}{I_{\mathrm{agg}}+P^{\mathrm{N}}} > \gamma_0\right]\right\}
$$
$$
\leq \sum_{\tilde{k}=1}^{\tilde{K}^{\max}}\left\{1 - \left[1 - \delta_n^{\mathrm{L}}(r)\mathscr{L}_{I_{\mathrm{agg}}}^{\mathrm{L}}\left(\frac{\gamma_0}{P\zeta_n^{\mathrm{L}}(r)}\right)\right]^{\tilde{k}}\right\}f_{\tilde{K}}(\tilde{k}) \tag{7.27}
$$

and

$$\mathbb{E}_{[\tilde{K}]}\left\{\Pr\left[\frac{P\zeta_n^{\mathrm{NL}}(r)y(\tilde{k})}{I_{\mathrm{agg}}+P^{\mathrm{N}}}>\gamma_0\right]\right\}$$
$$\leq\sum_{\tilde{k}=1}^{\tilde{K}^{\max}}\left\{1-\left[1-\delta_n^{\mathrm{NL}}(r)\mathscr{L}_{I_{\mathrm{agg}}}^{\mathrm{NL}}\left(\frac{\gamma_0}{P\zeta_n^{\mathrm{NL}}(r)}\right)\right]^{\tilde{k}}\right\}f_{\tilde{K}}(\tilde{k}), \qquad (7.28)$$

respectively.

Proof   See Appendix D.   □

The provided upper bounds in Theorem 7.3.4 require to calculate the integrals in equations (7.23) and (7.26) *only once* for every possible value of the two-dimensional distance, r, between the typical UE and its serving small cell BS, which makes the analysis of sparse networks much more efficient. In contrast, in Theorems 7.3.1 and 7.3.2, the integrals in equations (7.23) and (7.26) need to be calculated at least $\tilde{K}^{\max}$ times for every possible value of the two-dimensional distance, r, between the typical UE and its serving small cell BS to solve equation (7.14), which is a computationally expensive task.

Consequently, plugging Theorem 7.3.4 into Theorem 7.3.1, yields yet another new theoretical result on the coverage probability, p^{cov}, in this area of work, which is particularly useful for sparse networks.

7.4   Discussion

In this section, we use numerical results from static system-level simulations to evaluate the accuracy of the above theoretical performance analysis and study the impact of a PF scheduler at the small cell BSs on an ultra-dense network in terms of the coverage probability, p^{cov}, and the ASE, A^{ASE}.

As a reminder, it should be noted that to obtain the results on the coverage probability, p^{cov}, and the ASE, A^{ASE}, one should now plug equations (4.17) and (4.19), using two-dimensional distances, into Theorem 7.3.1, and follow the subsequent derivations. In this manner, a PF scheduler at the small cell BSs can be incorporated into the analysis, while also taking into account the finite UE density, $\rho<+\infty$, and the simple idle mode capability at the small cell BSs presented in Chapter 6.

7.4.1   Case Study

To assess the impact of a PF scheduler at the small cell BSs on the performance of an ultra-dense network, the same 3GPP case study as in Chapter 4 and in the rest of the book is used to allow an apple-to-apple comparison. For readers interested in a more detailed description of this 3GPP case study, please refer to Table 7.1 and Section 4.4.1.

For the sake of presentation, please recall that the following parameters are used in our 3GPP case study:

- maximum antenna gain, $G_M = 0$ dB,
- path loss exponents, $\alpha^L = 2.09$ and $\alpha^{NL} = 3.75$ – considering a carrier frequency of 2 GHz,
- reference path losses, $A^L = 10^{-10.38}$ and $A^{NL} = 10^{-14.54}$,
- transmit power, $P = 24$ dBm, and
- noise power, $P^N = -95$ dBm – including a noise figure of 9 dB at the UE.

Antenna Configurations

In this chapter, the height difference between the antenna height of the typical UE and that of an arbitrary small cell BS is not considered, as we want to isolate the performance impact of the antenna height from that of the PF scheduler at the small cell BSs.

UE Densities

As for the finite UE density, $\rho < +\infty$, we concentrate our studies on the case, $\rho = 300$ UEs/km^2, thus using a fitting parameter, $q = 4.18$, in equations (7.1) and (7.2), as derived in Section 6.4.2.

Small Cell BS Densities

Since antenna heights are not considered in this chapter, to embrace the minimum transmitter-to-receiver distance of the selected path loss model of 10 m, while still using a simple system model based on HPPP deployments, we investigate small cell BS densities up to $\lambda = 10^4$ BSs/km^2 in our studies hereafter.

Benchmark

In this performance evaluation, we use, as the benchmark,

- the results of the analysis with a finite UE density, $\rho < +\infty$, and an idle mode capability at the small cell BSs, presented earlier in Section 6.4.

7.4.2   Coverage Probability Performance

Figure 7.3 shows the coverage probability, p^{cov}, for a SINR threshold, $\gamma_0 = 0$ dB, and the following four configurations:

- Conf. a): multi-slope path loss model with both LoS and NLoS transmissions, a finite UE density, $\rho < +\infty$, a simple idle mode capability at the small cell BSs and an RR scheduler (analytical results).

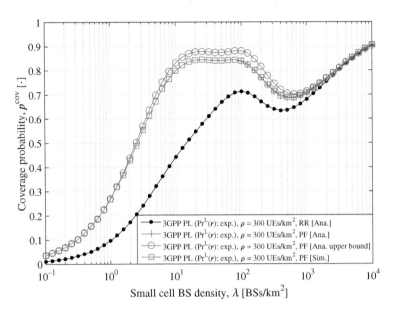

Figure 7.3 Coverage probability, p^{cov}, versus the small cell BS density, λ.

- Conf. b): multi-slope path loss model with both LoS and NLoS transmissions, a finite UE density, $\rho < +\infty$, a simple idle mode capability at the small cell BSs and a PF scheduler (analytical results).
- Conf. c): multi-slope path loss model with both LoS and NLoS transmissions, a finite UE density, $\rho < +\infty$, a simple idle mode capability at the small cell BSs and a PF scheduler (analytical upper-bound results).
- Conf. d): multi-slope path loss model with both LoS and NLoS transmissions, a finite UE density, $\rho < +\infty$, a simple idle mode capability at the small cell BSs and a PF scheduler (simulation results).

As a reminder, note that

- the theoretical results analyzing the exact performance of the coverage probability, p^{cov} – Conf. b) – are derived through Theorems 7.3.1 and 7.3.2, and for tractability reasons, only cover the higher spectrum of the small cell BS densities presented in the figure, i.e. $\lambda \geq 100\,BSs/km^2$, while
- those of the upper-bound – Conf. c) – are obtained from Theorems 7.3.1 and 7.3.4.

Also note that Conf. a) was already presented and discussed in Chapter 6, and it is incorporated for comparison purposes in Figure 7.3.

To complement Figure 7.3, Figure 7.4 also shows the ratio of the simulated coverage probability, p^{cov}, of the PF scheduler to that of the RR one for small cell BS densities larger than $\lambda = 100\,BSs/km^2$.

From these two figures, we can observe that

- the analytical results, Conf. b), match well the simulation ones, Conf. d), which validates the accuracy of Theorems 7.3.1 and 7.3.2 – and the derivations therein.

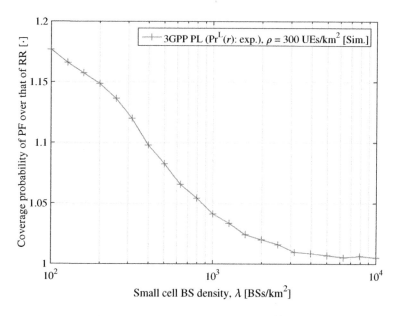

Figure 7.4 Ratio of the simulated coverage probability, p^{cov}, of the PF scheduler to that of the RR one.

However, as discussed in Section 7.3, the production of such analytical results is only efficient for dense and ultra-dense networks. Thus, in Figure 7.3, we are only able to show results in terms of the coverage probability, p^{cov}, for relative large small cell BS densities, $\lambda \geq 100\,\text{BSs/km}^2$.

- When it comes to sparser networks, i.e. $\lambda < 100\,\text{BSs/km}^2$, the alternative upper-bound formulation proposed in Theorems 7.3.1 and 7.3.4 is able to produce analytical results that successfully capture the qualitative performance trend of the PF scheduler. However, its precision is not as exact as earlier. In this case, the upper-bound analytical results, Conf. c), match the simulation results, Conf. d), with a maximum error of 0.04 in terms of coverage probability, p^{cov}, for all simulated small cell BS densities, λ, in this sparse regime.
- Importantly, as indicated in Lemma (7.3.3), although the PF scheduler shows a better performance than the RR one for all small cell BS densities, λ, such a performance gain diminishes as the network evolves into the ultra-dense regime due to the loss of multi-user diversity. As can be seen from Figure 7.4, the performance gain of the PF scheduler continuously decreases with the small cell BS density, λ. For example, it decreases from a gain around 17% to one around 0.5% when densifying from a small cell BS density, $\lambda = 10^2\,\text{BSs/km}^2$, to a small cell BS density, $\lambda = 10^4\,\text{BSs/km}^2$. This is because the PF scheduler has less and less UEs to select from. Note that the impact of the smaller radio-channel variations on a given time-frequency resource, due to the high probability of LoS transmissions in ultra-dense networks, is not considered in these results due to the

assumption of Rayleigh multi-path fast fading, but it will be considered in Section 7.4.4, when analyzing the impact of Rician fading.

A more detailed explanation of the performance behaviours in Figure 7.3 is presented in the following:

- When the small cell BS density, λ, is around, $\lambda \in [10^0, 10^1]$ BSs/km^2, the network is noise-limited, and thus the coverage probabilities, p^{cov}, of both the RR and the PF schedulers increase with the small cell BS density, λ, as the network coverage is lightened up with more and more small cell BSs.
- When the small cell BS density, λ, is around, $\lambda \in [10^1, 10^2]$ BSs/km^2, the network starts to be interference-limited, and the coverage probability, p^{cov}, of the PF scheduler shows an interesting flat trail. This is because of the following trade-off:
  - ☐ The signal power increases due to the ability of the PF scheduler to select the UE for transmission with the best channel conditions.
    - ○ While the RR scheduler chooses the UE for transmission in a time-frequency resource randomly and independently of its channel conditions, the PF scheduler selects the UE with the best instantaneous conditions with respect to its average, in this case, the one with the best multi-path fading.
  - ☐ The multi-user diversity decreases with the increase of the small cell BS density, λ.
    - ○ The above increase of the signal power is counterbalanced by the decreasing number, $\tilde{K}$, of UEs per active small cell BS with the increase of the small cell BS density, λ, which decreases the pool of UEs from which the PF scheduler can select the good UEs for transmission.
- When the small cell BS density, λ, is around, $\lambda \in [10^2, 10^3]$ BSs/km^2, the coverage probabilities, p^{cov}, of both the RR and the PF schedulers decrease with the small cell BS density, λ, as the network is pushed into the interference-limited region. As carefully analyzed in Chapter 4, the cause of this degradation is the transition of a large number of interfering links from NLoS to LoS. This phenomenon accelerates the growth of the aggregated inter-cell interference power, and in turn, decreases the SINR, Γ, of the typical UE and the coverage probability, p^{cov}.
- When the small cell BS density, λ, is considerably large, $\lambda > 10^3$ BSs/km^2, the coverage probabilities, p^{cov}, of both the RR and the PF schedulers increase with the small cell BS density, λ, as the network enters the ultra-dense regime. As studied in Chapter 6, the finite UE density, $\rho < +\infty$, and the simple idle mode capability at the small cell BSs introduce a cap on the maximum inter-cell interference power that a UE can receive in a non-coordinated ultra-dense network, while the signal power continues to grow due to the closer proximity between the UEs and their serving small cell BSs in a densifying network. This phenomenon accelerates the growth of the signal power, and in turn, increases the SINR, Γ, of the typical UE and the coverage probability, p^{cov}.

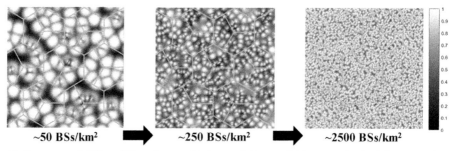

(a) Coverage probability, p^{cov}, under a multi-slope path loss model incorporating both LoS and NLoS transmissions, a finite UE density, $\rho = 300\,\text{UEs/km}^2$, a simple idle mode capability at the small cell BSs and an RR scheduler.

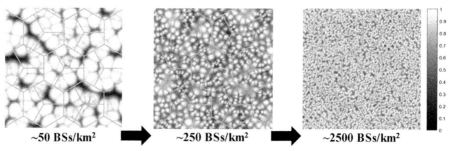

(b) Coverage probability, p^{cov}, under a multi-slope path loss model incorporating both LoS and NLoS transmissions, a finite UE density, $\rho = 300\,\text{UEs/km}^2$, a simple idle mode capability at the small cell BSs and a PF scheduler. Compared with Figure 7.5a, the SINR heat map becomes lighter as the BS density, λ, increases, showing a performance enhancement because of the multi-user diversity gain captured by the PF scheduler. However, such a performance improvement diminishes as the small cell BS density, λ, increases.

Figure 7.5 NetVisual plot of the coverage probability, p^{cov}, versus the small cell BS density, λ [Bright area: high probability, Dark area: low probability].

NetVisual Analysis

To visualize the fundamental behaviour of the coverage probability, p^{cov}, in a more intuitive manner, Figure 7.5 shows the coverage probability heat map of three different scenarios with different small cell BS densities, i.e. 50, 250 and 2500 BSs/km², while considering the multi-slope path loss model with both LoS and NLoS transmissions, the finite UE density, $\rho < +\infty$, and the simple idle mode capability at the small cell BSs with and without the PF scheduler. These heat maps are computed using NetVisual. As explained in Section 2.4, this tool is able to provide striking graphs to assess performance, which not only capture the mean, but also the standard deviation of the coverage probability, p^{cov}.

From Figure 7.5, we can see that, with respect to the case with the RR scheduler,

- the PF scheduler at the small cell BSs has a performance impact when the network is sparse and dense, i.e. when the small cell BS density, λ, is no larger than 250 BSs/km² in our example. The SINR heat map in Figure 7.5b is brighter than that in Figure 7.5a when the small cell BS density, λ, is around 250 BSs/km², and even more than that when $\lambda = 50\,\mathrm{BSs/km}^2$. This shows the performance improvement attained by the PF scheduler at the small cell BSs due to its ability to leverage the multi-user diversity. Numerically, for the case with a small cell BS density, $\lambda = 50\,\mathrm{BSs/km}^2$, the average and the standard deviation of the coverage probability, p^{cov}, go from 0.72 and 0.22 with the RR scheduler at the small cell BSs to 0.87 and 0.17 with the PF scheduler, respectively – a 20.83% improvement in mean.

- In contrast, the PF scheduler at the small cell BSs has a negligible performance impact when the network is ultra-dense, i.e. when the small cell BS density, λ, is no smaller than $2500\,\mathrm{BSs/km}^2$ in our example. The SINR maps are basically the same in both Figures 7.5a and 7.5b at a small cell BS density, $\lambda = 2500\,\mathrm{BSs/km}^2$. Since the number of small cell BSs is much larger than that of the UEs in this case, most of the active small cell BSs only serve one UE, approaching the one UE per cell limit. As a result, the multi-user diversity gain is negligible. The average and the standard deviation of the coverage probability, p^{cov}, for these two cases are both approximately 0.80 and 0.13, respectively.

Summary of Findings: Coverage Probability Remark

Remark 7.1   Considering a PF scheduler – together with a finite UE density, $\rho < +\infty$, and a simple idle mode capability at the small cell BSs – the SINR, Γ, of the typical UE – and in turn, the coverage probability, p^{cov} – improve in a sparse network with respect to that of an RR scheduler due to the multi-user diversity gain, i.e. the ability to select a better UE for transmission in each time-frequency resource. However, these multi-user diversity gains are negligible in an ultra-dense network. This is because the cell load approaches the one user per cell limit, and thus there is much less multi-user diversity gain.

7.4.3   Area Spectral Efficiency Performance

Let us now explore the behaviour of the ASE.

Figure 7.6 shows the ASE, A^{ASE}, for a SINR threshold, $\gamma_0 = 0\,\mathrm{dB}$, for the following three configurations:

- Conf. a): multi-slope path loss model with both LoS and NLoS transmissions, a finite UE density, $\rho < +\infty$, a simple idle mode capability at the small cell BSs and an RR scheduler (analytical results).
- Conf. b): multi-slope path loss model with both LoS and NLoS transmissions, a finite UE density, $\rho < +\infty$, a simple idle mode capability at the small cell BSs and a PF scheduler (analytical results).
- Conf. c): multi-slope path loss model with both LoS and NLoS transmissions, a finite UE density, $\rho < +\infty$, a simple idle mode capability at the small cell BSs and a PF scheduler (simulation results).

As the most important takeaway from this figure, we should indicate that

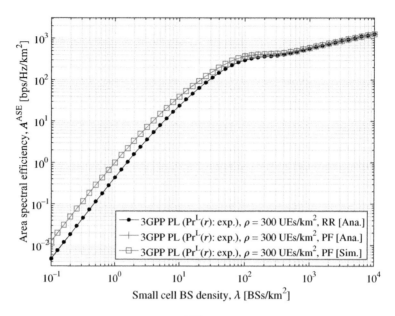

Figure 7.6 Area spectral efficiency, A^{ASE}, versus the small cell BS density, λ.

- the ASE, A^{ASE}, of the PF scheduler converges to that of the RR one in an ultra-dense network. The gap between the ASE, A^{ASE}, resulting from the PF and the RR schedulers for a small cell BS density, $\lambda = 10^2$ BSs/km², and that for a small cell BS density, $\lambda = 10^3$ BSs/km², are 70.5 bps/Hz/km² and 25.5 bps/Hz/km², respectively. This is a 23.56% and a 4.52% difference. Importantly, such a gain of the PF scheduler becomes practically 0 as the small cell BS density, λ, further increases.

These results follow from the behaviour of the coverage probability, p^{cov}, explained in Section 7.4.2.

Overall, these results show that in a practical network, where there will be a finite UE density, $\rho < +\infty$, the number of UEs, $\tilde{K}$, per active small cell BS will decrease as the small cell BS density, λ, marches into the ultra-dense regime. As a result, the multi-user diversity gains reduce, until they vanish, and the PF channel-dependent scheduler does not bring any gain in terms of the ASE, A^{ASE}. It is also important to note that the PF scheduler did not change the important conclusion of Chapter 6, i.e. in a practical network, where there is a finite UE density, $\rho < +\infty$, and a simple idle mode capability at the small cell BSs, both the coverage probability, p^{cov}, and the ASE, A^{ASE}, will increase as the small cell BS density, λ, goes ultra-dense. The fundamental reason for this behaviour – as explained before – is the cap on the maximum inter-cell interference power that a UE can receive due to the finite UE density, $\rho < +\infty$, and the idle mode capability at the small cell BSs. There can be at most as many active small cell BSs as UEs in the scenario, but never more, in this non-coordinated ultra-dense network, and the PF scheduler cannot change this fundamental fact. This inter-cell interference bound allows the SINR, Γ, of the typical UE, to grow, as the signal power continues to increase due to the closer proximity between the UEs and their serving small cell BSs in a denser network.

Summary of Findings: Area Spectral Efficiency Remark

Remark 7.2   Considering a PF scheduler – together with a finite UE density, $\rho <$ $+\infty$, and a simple idle mode capability at the small cell BSs – the ASE, A^{ASE}, improves in a sparse network with respect to that of an RR scheduler due to the multi-user diversity gain, but these gains are negligible in ultra-dense networks. This result follows from the behaviour of the coverage probability, p^{cov}.

7.4.4   Impact of the Multi-Path Fading – and Modelling – on the Main Results

In Sections 7.4.2 and 7.4.3, the impact of the number, $\tilde{K}$, of UEs per active small cell BS on the multi-user diversity was analyzed. However, as mentioned in the introduction, the channel characteristics also play a role in this type of gain. In more detail, the fluctuations in the received signal quality stemming from the multi-path fast fading gain, h, create multi-user diversity, and enable opportunistic coverage probability and ASE gains that a channel-dependent scheduler can harvest. In some cases, the multi-path fast fading components can add constructively, contributing to a better received signal strength. The degree of multi-user diversity depends on the dynamic range and magnitude of such channel fluctuations, where the objective of the channel-dependent scheduler is to leverage the high peaks. Highly scattered scenarios resulting in large angular spreads lead to larger dynamic ranges than those with a dominating LoS component, and thus facilitate larger multi-user diversity gains.

In this section, we analyze the impact of the more accurate – but also less tractable – Rician multi-path fast fading model, presented in Section 2.1.8, on the coverage probability, p^{cov}, and the ASE, A^{ASE}, when it is applied to the LoS component.

This model, in contrast to the previously used Rayleigh multi-path fast fading model, is able to capture the ratio between the power in the direct path and the power in the other scattered paths in the multi-path fast fading gain, h, as a function of the distance, r, between a UE and a small cell BS. Therefore, it can be used to model the previously mentioned dynamic range of the channel fluctuations as a function of the densification level. As a reminder, let us indicate that the dynamic range of the multi-path fast fading gain, h, is smaller under the Rician multi-path fast fading model than under the Rayleigh one – the former has larger minimum and smaller maximum values – and that importantly, the mean of the Rician multi-path fast fading gain, h, increases with the increase of the small cell BS density, λ (see Figure 2.4).

Our intention in this section is to check whether the conclusions obtained in this chapter about the PF scheduler hold when considering this more realistic – but less tractable – Rician-based multi-path fast fading model.

For the sake of mathematical complexity, we only present system-level simulation results with this alternative Rician-based multi-path fast fading model hereafter. Please

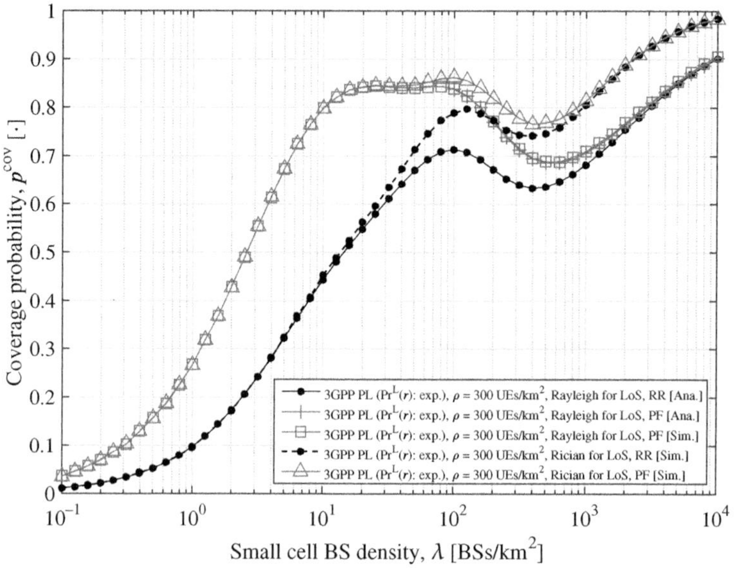
1
0.9
0.8
0.7
0.6
0.5
0.4
0.3
0.2
0.1
0
Coverage probability, p^{cov} [·]
10^{-1}
10^{0}
10^{1}
10^{2}
10^{3}
10^{4}
Small cell BS density, λ [BSs/km^2]
3GPP PL ($Pr^L(r)$: exp.), ρ = 300 UEs/km^2, Rayleigh for LoS, RR [Ana.]
3GPP PL ($Pr^L(r)$: exp.), ρ = 300 UEs/km^2, Rayleigh for LoS, PF [Ana.]
3GPP PL ($Pr^L(r)$: exp.), ρ = 300 UEs/km^2, Rayleigh for LoS, PF [Sim.]
3GPP PL ($Pr^L(r)$: exp.), ρ = 300 UEs/km^2, Rician for LoS, RR [Sim.]
3GPP PL ($Pr^L(r)$: exp.), ρ = 300 UEs/km^2, Rician for LoS, PF [Sim.]

Figure 7.7 Coverage probability, p^{cov}, versus the small cell BS density, λ, for an alternative case study with multi-path Rician fading.

refer to Section 2.1.8 for more details on both the Rayleigh and the Rician multi-path fast fading models.

Figure 7.7 illustrates the results on the coverage probability, p^{cov}, while Figure 7.8 shows those on the ASE, A^{ASE}. As a reminder, note that the assumptions and the parameters used to obtain the results in these two figures are the same as those used to get the previous results shown in this chapter, except for the Rician multi-path fast fading model for the LoS component considered here.

As one can observe from these figures, all the observations about the PF scheduler with respect to the coverage probability, p^{cov}, and the ASE, A^{ASE}, obtained under the Rayleigh-based multi-path fast fading model in Section 7.4 are qualitatively valid for those obtained under the Rician-based multi-path fast fading model. Similar trends are present.

Paying attention first to the coverage probability, p^{cov}, we can see that it first increases, then reaches a plateau, decreases and finally increases again, and only a quantitative deviation exists when considering these two different multi-path fast fading models. Similarly as in Chapter 6, the results under the Rician-based multi-path fast fading model outperform those under the Rayleigh-based one at large small cell BS densities. The reason is rooted in the impact of the idle mode capability at the small cell BSs. The idle mode capability at the small cell BSs switches off progressively more and more small cell BSs with the increase of the small cell BS density, λ, thus

- increasing the average distance between a UE and its neighbouring interfering small cell BSs, and in turn,
- delaying the NLoS to LoS inter-cell interference transition.

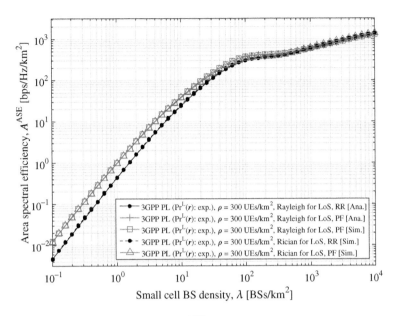

Figure 7.8 Area spectral efficiency, A^{ASE}, versus the small cell BS density, λ, for an alternative case study with multi-path Rician fading.

This latter fact exacerbates the difference in performance when considering the Rician- or the Rayleigh-based multi-path fast fading models and prolongs the difference in performance between them to larger small cell BS densities. Interested readers are encouraged to review Section 6.4.5 for further details on the better system performance under the Rician-based multi-path fast fading model.

Importantly, it should also be noted from the new results in Figure 7.7 that

- the coverage probability gap between the PF and the RR schedulers is closed at a lower small cell BS density, λ, under the Rician-based multi-path fast fading model than under the Rayleigh-based one. In more detail, under the Rician-based model, the performance gain of the PF scheduler with respect to that of the RR one is within a gap of 5% when the small cell BS density is equal or larger than $\lambda = 357.2$ BSs/km^2, while with the Rayleigh-based model, such a small cell BS density is $\lambda = 897.2$ BSs/km^2. This indicates that the advantage provided by a channel-dependent scheduler disappears earlier when considering the more realistic Rician-based multi-path fast fading model. This is due to the mentioned smaller dynamic range of the channel fluctuations under the Rician multi-path fast fading. As a result of this smaller dynamic range, the availability of the PF scheduler to select a "better" UE for transmission reduces, and in turn, the multi-user diversity.

As for the ASE, A^{ASE}, note that the behaviours under the Rayleigh- and the Rician-based multi-path fast fading models in Figure 7.8 are almost identical. This is because the relatively small change in the coverage probability, p^{cov}, results in an even smaller change in the ASE, A^{ASE}.

Overall, and despite the quantitative differences, these results on the coverage probability, p^{cov}, and the ASE, A^{ASE}, confirm our statement, i.e. the main conclusions in this chapter are general, and do not qualitatively change due to the assumption on the multi-path fast fading model.[3]

In the remainder of this book, since the modelling of the multi-path fading did not have a profound impact in terms of qualitative results, we stick to Rayleigh multi-path fast fading modelling for mathematical tractability, unless otherwise stated.

7.5   Conclusions

In this chapter, we have brought to attention the modelling of the multi-user diversity in network performance analysis. More precisely, we have described the necessary upgrades that one has to perform to the system model presented so far in this book to model the number of UEs per active small cell BS and a practical channel-dependent scheduler. We have also provided details on the derivations conducted in the theoretical performance analysis, and presented the resulting expressions for the coverage probability, p^{cov}, and the ASE, A^{ASE}. Moreover, we have shared numerical results for small cell BS deployments with different small cell BS densities and characteristics. To finalize, we have also discussed the important conclusions drawn from this work.

The most important takeaways from this study are summarized in Remarks 7.1 and 7.2, indicating to equipment manufactures and vendors that the development of a complex channel-dependent scheduler for the small cell BSs may not be necessary in ultra-dense deployments. Such complexity can be saved and used in aid of a different feature.

As in the previous chapters, the loss in multi-user diversity in denser networks has also been shown to be quantitatively – but not qualitatively – affected by the multi-path fast fading model.

Appendix A Proof of Theorem 7.3.1

To ease the understanding of the readers, it should be noted that this proof shares a lot of similarities with some parts of the proof in Appendix A. However, differences arise because now the SINR, Γ, of the typical UE depends on the number, $\tilde{K}$, of UEs per active small cell BS. The full proof is presented hereafter for completeness.

Having said that, let us first describe the main idea behind the proof of Theorem 7.3.1, and then proceed with a more detailed explanation.

In line with the guidelines provided in Section 2.3.2, to evaluate the coverage probability, p^{cov}, we need proper expressions for

---

[3] Note that this statement applies for the studied case with single-antenna UEs and small cell BSs, but may not apply to multi-antenna cases, where the fading correlation among the antennas of an array plays a key role.

- the PDF, $f_R(r)$, of the random variable, R, characterizing the distance between the typical UE and its serving small cell BS for the event that the typical UE is associated to the strongest small cell BS through either an LoS or an NLoS path and
- the expected conditional probability, $\mathbb{E}_{[\tilde{K}]}\left\{\Pr\left[\Gamma > \gamma_0 \middle| r\right]\right\}$, where
  - □ the expectation is performed over the number, $\tilde{K}$, of UEs per active small cell BS, because the SINR, Γ, depends on the random variable, $\tilde{K}$, and
  - □ $r = R(\omega)$ is a realization of the random variable, R, for both the LoS and the NLoS transmission.

It is important to note that the expected conditional probability, $\mathbb{E}_{[\tilde{K}]}\left\{\Pr\left[\Gamma > \gamma_0 \middle| r\right]\right\}$, in Theorem 7.3.1 is different from the conditional probability, $\Pr\left[\Gamma > \gamma_0 \middle| r\right]$, in Theorem 4.3.1. This is because a random UE is selected in Theorem 4.3.1, and thus the SINR, Γ, does not depend on the number, $\tilde{K}$, of UEs per active small cell BS. To highlight such a difference, the proof on the calculation of such probabilities is presented in Appendix B.

Once the above expressions are known, we can derive the coverage probability, p^{cov}, by performing the corresponding appropriate integrals, some of which will be shown in the following.

Before proceeding with the more detailed calculations, however, it is important to note that from equations (2.22) and (3.3), we can derive the coverage probability, p^{cov}, considering the number, $\tilde{K}$, of UEs per active small cell BS as

$$
p^{\text{cov}}(\lambda, \gamma_0) \overset{(a)}{=} \int_{r>0} \mathbb{E}_{[\tilde{K}]}\left\{\Pr\left[\Gamma > \gamma_0 \middle| r\right]\right\} f_R(r)\,dr
$$

$$
= \int_{r>0} \mathbb{E}_{[\tilde{K}]}\left\{\Pr\left[\frac{P\zeta(r)y(\tilde{k})}{I_{\text{agg}} + P^{\text{N}}} > \gamma_0\right]\right\} f_R(r)\,dr
$$

$$
= \int_0^{d_1} \mathbb{E}_{[\tilde{K}]}\left\{\Pr\left[\frac{P\zeta_1^{\text{L}}(r)y(\tilde{k})}{I_{\text{agg}} + P^{\text{N}}} > \gamma_0\right]\right\} f_{R,1}^{\text{L}}(r)\,dr
$$

$$
+ \int_0^{d_1} \mathbb{E}_{[\tilde{K}]}\left\{\Pr\left[\frac{P\zeta_1^{\text{NL}}(r)y(\tilde{k})}{I_{\text{agg}} + P^{\text{N}}} > \gamma_0\right]\right\} f_{R,1}^{\text{NL}}(r)\,dr
$$

$$
+ \cdots
$$

$$
+ \int_{d_{N-1}}^{\infty} \mathbb{E}_{[\tilde{K}]}\left\{\Pr\left[\frac{P\zeta_N^{\text{L}}(r)y(\tilde{k})}{I_{\text{agg}} + P^{\text{N}}} > \gamma_0\right]\right\} f_{R,N}^{\text{L}}(r)\,dr
$$

$$
+ \int_{d_{N-1}}^{\infty} \mathbb{E}_{[\tilde{K}]}\left\{\Pr\left[\frac{P\zeta_N^{\text{NL}}(r)y(\tilde{k})}{I_{\text{agg}} + P^{\text{N}}} > \gamma_0\right]\right\} f_{R,N}^{\text{NL}}(r)\,dr
$$

$$
\overset{\triangle}{=} \sum_{n=1}^{N} \left(T_n^{\text{L}} + T_n^{\text{NL}}\right), \tag{7.29}
$$

where

- R_n^L and R_n^{NL} are the piecewise distributions of the distances over which the typical UE is associated to a small cell BS through an LoS and an NLoS path, respectively:

  □ Note that these two events, i.e. the typical UE being associated with a small cell BS through an LoS or an NLoS path, are disjoint, and thus the coverage probability, p^{cov}, is the sum of the corresponding probabilities of these two events.

- $f_{R,n}^L(r)$ and $f_{R,n}^{NL}(r)$ are the piecewise PDFs of the random variables, R_n^L and R_n^{NL}, respectively:

  □ For clarity, both piecewise PDFs, $f_{R,n}^L(r)$ and $f_{R,n}^{NL}(r)$, are stacked into the PDF, $f_R(r)$, in step (a) of equation (7.29), where the stacked PDF, $f_R(r)$, takes a similar form as in equation (4.1), and is defined in equation (4.45).

  □ Moreover, since the two events, i.e. the typical UE being associated with a small cell BS through an LoS or an NLoS path, are disjoint, as mentioned before, we can rely on the following equality,

  $$\sum_{n=1}^{N} \int_{d_{n-1}}^{d_n} f_{R,n}(r)dr = \sum_{n=1}^{N} \int_{d_{n-1}}^{d_n} f_{R,n}^L(r)dr + \sum_{n=1}^{N} \int_{d_{n-1}}^{d_n} f_{R,n}^{NL}(r)dr = 1.$$

- T_n^L and T_n^{NL} are two piecewise functions, defined as

$$T_n^L = \int_{d_{n-1}}^{d_n} \mathbb{E}_{[\tilde{K}]} \left\{ \Pr \left[\frac{P\zeta_n^L(r)y(\tilde{k})}{I_{agg} + P^N} > \gamma_0 \right] \right\} f_{R,n}^L(r)dr \tag{7.30}$$

and

$$T_n^{NL} = \int_{d_{n-1}}^{d_n} \mathbb{E}_{[\tilde{K}]} \left\{ \Pr \left[\frac{P\zeta_n^{NL}(r)y(\tilde{k})}{I_{agg} + P^N} > \gamma_0 \right] \right\} f_{R,n}^{NL}(r)dr, \tag{7.31}$$

respectively, and

- d_0 and d_N are equal to 0 and $+\infty$, respectively.

$$f_R(r) = \begin{cases} f_{R,1}(r) = \begin{cases} f_{R,1}^L(r), & \text{UE associated to an LoS BS} \\ f_{R,1}^{NL}(r), & \text{UE associated to an NLoS BS} \end{cases}, & 0 \le r \le d_1 \\ f_{R,2}(r) = \begin{cases} f_{R,2}^L(r), & \text{UE associated to an LoS BS} \\ f_{R,2}^{NL}(r), & \text{UE associated to an NLoS BS} \end{cases}, & d_1 < r \le d_2 \\ \vdots & \vdots \\ f_{R,N}(r) = \begin{cases} f_{R,N}^L(r), & \text{UE associated to an LoS BS} \\ f_{R,N}^{NL}(r), & \text{UE associated to an NLoS BS} \end{cases}, & r > d_{N-1}. \end{cases} \tag{7.32}$$

Following this method, let us now dive into the more detailed derivations.

LoS-Related Calculations

Let us first look into the LoS transmissions, and show how to first calculate the PDF, $f^L_{R,n}(r)$, in equation (7.29).

To that end, we define first the two following events:

- Event B^L: The nearest small cell BS to the typical UE with an LoS path is located at a distance, $x = X^L(\omega)$, defined by the distance random variable, X^L. According to the results presented in Section 2.3.2, the PDF, $f_R(r)$, of the random variable, R, in Event B^L can be calculated by

$$f^L_X(x) = \exp\left(- \int_0^x \mathrm{Pr}^L(u)2\pi u\lambda \, du \right) \mathrm{Pr}^L(x)2\pi x\lambda. \qquad (7.33)$$

This is because, according to [52], the CCDF, $\bar{F}^L_X(x)$, of the random variable, X^L, in Event B^L is given by

$$\bar{F}^L_X(x) = \exp\left(- \int_0^x \mathrm{Pr}^L(u)2\pi u\lambda \, du \right), \qquad (7.34)$$

and taking the derivative of the cumulative distribution function (CDF), $\left(1 - \bar{F}^L_X(x)\right)$, of the random variable, X^L, with respect to the distance, x, we can get the PDF, $f^L_X(x)$, of the random variable, X^L, shown in equation (7.33). It is important to note that the derived equation (7.33) is more complex than equation (2.42) in Section 2.3.2. This is because the LoS small cell BS deployment is an inhomogeneous one due to the fact that the closer small cell BSs to the typical UE are more likely to establish LoS links than the further away ones. Compared with equation (2.42), we can see two important changes in equation (7.33).

☐ The probability of a disc of a radius, r, containing exactly 0 points, which takes the form, $\exp\left(-\pi\lambda r^2\right)$, as presented in equation (2.38) in Section 2.3.2, has changed now, and is replaced by the expression,

$$\exp\left(- \int_0^x \mathrm{Pr}^L(u)2\pi u\lambda \, du \right),$$

in equation (7.33). This is because the equivalent intensity of the LoS small cell BSs in an inhomogeneous Poisson point process (PPP) is distance-dependent, i.e. $\mathrm{Pr}^L(u)\lambda$. This adds a new integral to equation (7.33) with respect to the distance, u.

☐ The intensity, λ, of the HPPP in equation (2.42) has been replaced by the equivalent intensity, $\mathrm{Pr}^L(x)\lambda$, of the inhomogeneous PPP in equation (7.33).

- Event C^{NL} conditioned on the value, $x = X^L(\omega)$, of the random variable, X^L: The typical UE is associated to the nearest small cell BS located at a distance, $x = X^L(\omega)$, through an LoS path. If the typical UE is associated to the nearest LoS small cell BS located at a distance, $x = X^L(\omega)$, this LoS small cell BS must be at the smallest path loss, $\zeta(r)$. As a result, there must be no NLoS small cell BS inside the disc

☐ centred on the typical UE,
☐ with a radius, x_1,

where such a radius, x_1, satisfies the following condition,

$x_1 = \arg_{x_1} \{\zeta^{NL}(x_1) = \zeta^L(x)\}$. Otherwise, this NLoS small cell BS would

outperform the LoS small cell BS at the distance, $x = X^L(\omega)$.

According to [52], the conditional probability, $\Pr\left[C^{NL}\middle|X^L = x\right]$, of the Event, C^{NL}, conditioned on the realization, $x = X^L(\omega)$, of the random variable, X^L, is given by

$$\Pr\left[C^{NL}\middle|X^L = x\right] = \exp\left(-\int_0^{x_1}\left(1 - \Pr^L(u)\right)2\pi u\lambda\,du\right). \tag{7.35}$$

As a summary, note that Event B^L ensures that the path loss, $\zeta^L(x)$, associated with an arbitrary LoS small cell BS is always larger than that associated with the considered LoS small cell BS at the distance, $x = X^L(\omega)$. Besides, conditioned on such a distance, $x = X^L(\omega)$, Event C^{NL} guarantees that the path loss, $\zeta^{NL}(x)$, associated with an arbitrary NLoS small cell BS is also always larger than that associated with the considered LoS small cell BS at the distance, $x = X^L(\omega)$. With this, we can guarantee that the typical UE is associated with the strongest LoS small cell BS.

Let us thus now consider the resulting new event, in which the typical UE is associated with an LoS small cell BS, and such a small cell BS is located at the distance, $r = R^L(\omega)$. The CCDF, $\bar{F}_R^L(r)$, of such a random variable, R^L, can be derived as

$$\begin{aligned}
\bar{F}_R^L(r) &= \Pr\left[R^L > r\right] \\
&\stackrel{(a)}{=} \mathbb{E}_{[X^L]}\left\{\Pr\left[R^L > r\middle|X^L\right]\right\} \\
&= \int_0^{+\infty}\Pr\left[R^L > r\middle|X^L = x\right]f_X^L(x)dx \\
&\stackrel{(b)}{=} \int_0^r 0 \times f_X^L(x)dx + \int_r^{+\infty}\Pr\left[C^{NL}\middle|X^L = x\right]f_X^L(x)dx \\
&= \int_r^{+\infty}\Pr\left[C^{NL}\middle|X^L = x\right]f_X^L(x)dx, \tag{7.36}
\end{aligned}$$

where

- $\mathbb{E}_{[X]}\{\cdot\}$ in step (a) of equation (7.36) is the expectation operation over the random variable, X, and
- step (b) of equation (7.36) is valid because

- $\Pr\left[R^L > r \mid X^L = x\right] = 0$ for $0 < x \le r$ and
- the conditional event, $\left[R^L > r \mid X^L = x\right]$, is equivalent to the conditional event, $\left[C^{NL} \mid X^L = x\right]$, in the range, $x > r$.

Now, given the CCDF, $\bar{F}_R^L(r)$, one can find its PDF, $f_R^L(r)$, by taking the derivative, $\frac{\partial(1 - \bar{F}_R^L(r))}{\partial r}$, with respect to the distance, r, i.e.

$$f_R^L(r) = \Pr\left[C^{NL} \mid X^L = r\right] f_X^L(r). \tag{7.37}$$

Considering the distance range, $(d_{n-1} < r \le d_n)$, we can find the PDF of such a segment, $f_{R,n}^L(r)$, from such a PDF, $f_R^L(r)$, as

$$f_{R,n}^L(r) = \exp\left(-\int_0^{r_1}\left(1 - \Pr^L(u)\right)2\pi u\lambda\,du\right)$$

$$\times \exp\left(-\int_0^r \Pr^L(u)2\pi u\lambda\,du\right)\Pr_n^L(r)2\pi r\lambda,\ (d_{n-1} < r \le d_n),$$

$$\tag{7.38}$$

where

- $r_1 = \underset{r_1}{\arg}\left\{\zeta^{NL}(r_1) = \zeta_n^L(r)\right\}.$

NLoS-Related Calculations

Let us now look into the NLoS transmissions, and show how to first calculate the PDF, $f_{R,n}^{NL}(r)$, in equation (7.29).
To that end, we define the two following events:

- Event B^{NL}: The nearest small cell BS to the typical UE with an NLoS path is located at a distance, $x = X^{NL}(\omega)$, defined by the distance random variable, X^{NL}. Similarly to equation (7.33), one can find the PDF, $f_X^{NL}(x)$, as

$$f_X^{NL}(x) = \exp\left(-\int_0^x\left(1 - \Pr^L(u)\right)2\pi u\lambda\,du\right)\left(1 - \Pr^L(x)\right)2\pi x\lambda. \tag{7.39}$$

It is important to note here that the derived equation (7.39) is more complex than equation (2.42) in Section 2.3.2. This is because the NLoS small cell BS deployment is an inhomogeneous one due to the fact that the further away small cell BSs to the typical UE are more likely to establish NLoS links than the closer ones. Compared with equation (2.42), we can see two important changes in equation (7.39).

- The probability of a disc of a radius, r, containing exactly 0 points, which takes the form, $\exp\left(-\pi\lambda r^2\right)$, as presented in equation (2.38) in Section 2.3.2, has changed now, and is replaced by the expression,

$$\exp\left(-\int_0^x \left(1 - \mathrm{Pr}^L(u)\right) 2\pi u\lambda\, du\right),$$

in equation (7.39). This is because the equivalent intensity of the NLoS small cell BSs in an inhomogeneous PPP is distance-dependent, i.e. $\left(1 - \mathrm{Pr}^L(u)\right)\lambda$. This adds a new integral to equation (7.33) with respect to the distance, u.

☐ The intensity, λ, of the HPPP in equation (2.42) has been replaced by the equivalent intensity, $\left(1 - \mathrm{Pr}^L(u)\right)\lambda$, of the inhomogeneous PPP in equation (7.39).

• Event C^L conditioned on the value, $x = X^{NL}(\omega)$, of the random variable, X^{NL}: The typical UE is associated to the nearest small cell BS located at a distance, $x = X^{NL}(\omega)$, through an NLoS path. If the typical UE is associated to the nearest NLoS small cell BS located at a distance, $x = X^{NL}(\omega)$, this NLoS small cell BS must be at the smallest path loss, $\zeta(r)$. As a result, there must be no LoS small cell BS inside the disc

☐ centred on the typical UE,

☐ with a radius, x_2,

where such a radius, x_2, satisfies the following expression,

$x_2 = \arg_{x_2}\left\{\zeta^L(x_2) = \zeta^{NL}(x)\right\}$. Otherwise, this LoS small cell BS would outperform the NLoS small cell BS at the distance, $x = X^{NL}(\omega)$.

Similarly to equation (7.35), the conditional probability, $\mathrm{Pr}\left[C^L \middle| X^{NL} = x\right]$, of Event C^L conditioned on the realization, $x = X^{NL}(\omega)$, of the random variable, X^{NL}, is given by

$$\mathrm{Pr}\left[C^L \middle| X^{NL} = x\right] = \exp\left(-\int_0^{x_2} \mathrm{Pr}^L(u) 2\pi u\lambda\, du\right). \tag{7.40}$$

Let us thus now consider the resulting new event, in which the typical UE is associated with an NLoS small cell BS, and such a small cell BS is located at the distance, $r = R^{NL}(\omega)$. The CCDF, $\bar{F}_R^{NL}(r)$, of such a random variable, R^{NL}, can be derived as

$$\bar{F}_R^{NL}(r) = \mathrm{Pr}\left[R^{NL} > r\right]$$

$$= \int_r^{+\infty} \mathrm{Pr}\left[C^L \middle| X^{NL} = x\right] f_X^{NL}(x)\, dx. \tag{7.41}$$

Now, given the CCDF, $\bar{F}_R^{NL}(r)$, one can find its PDF, $f_R^{NL}(r)$, by taking the derivative, $\frac{\partial\left(1 - \bar{F}_R^{NL}(r)\right)}{\partial r}$, with respect to the distance, r, i.e.

$$f_R^{NL}(r) = \mathrm{Pr}\left[C^L \middle| X^{NL} = r\right] f_X^{NL}(r). \tag{7.42}$$

Considering the distance range, $(d_{n-1} < r \le d_n)$, we can find the PDF of such a segment, $f_{R,n}^{NL}(r)$, from such a PDF, $f_R^{NL}(r)$, as

$$f_{R,n}^{NL}(r) = \exp\left(-\int_0^{r_2} \Pr^L(u) 2\pi u \lambda \, du\right)$$

$$\times \exp\left(-\int_0^r \left(1 - \Pr^L(u)\right) 2\pi u \lambda \, du\right)\left(1 - \Pr_n^L(r)\right) 2\pi r \lambda, (d_{n-1} < r \le d_n),$$

$$(7.43)$$

where

- $r_2 = \arg_{r_2}\left\{\zeta^L(r_2) = \zeta_n^{NL}(r)\right\}.$

Our proof is thus completed.

Appendix B Proof of Theorem 7.3.2

This proof on the calculation of the expected probabilities is divided in an LoS and an NLoS subsection as follows.

LoS-Related Calculations

In Theorem 7.3.1, we have obtained the PDF of a segment, $f_{R,n}^L(r)$. Here, we evaluate the expected probability, $\mathbb{E}_{[\tilde{K}]}\left\{\Pr\left[\frac{P\zeta_n^L(r) y(\tilde{k})}{I_{agg} + P^N} > \gamma_0\right]\right\}$, in equation (7.29) as

$$\mathbb{E}_{[\tilde{K}]}\left\{\Pr\left[\frac{P\zeta_n^L(r) y(\tilde{k})}{I_{agg} + P^N} > \gamma_0\right]\right\} = \mathbb{E}_{[\tilde{K}]}\left\{\Pr\left[y(\tilde{k}) > \frac{\gamma_0\left(I_{agg} + P^N\right)}{P\zeta_n^L(r)}\right]\right\}$$

$$= \mathbb{E}_{[\tilde{K}, I_{agg}]}\left\{\bar{F}_{Y(\tilde{k})}\left(\frac{\gamma_0\left(I_{agg} + P^N\right)}{P\zeta_n^L(r)}\right)\right\},$$

$$(7.44)$$

where

- $\bar{F}_{Y(\tilde{k})}(\cdot)$ is the CCDF of the multi-path fast fading channel gain, $Y(\tilde{k})$, which is given by equation (7.13).

Thus, plugging equation (7.13) into equation (7.44) yields

$$\mathbb{E}_{[\tilde{K}]}\left\{\Pr\left[\frac{P\zeta_n^L(r) y(\tilde{k})}{I_{agg} + P^N} > \gamma_0\right]\right\}$$

$$= \mathbb{E}_{[\tilde{K}, I_{agg}]}\left\{1 - \left(1 - \exp\left(-\frac{\gamma_0\left(I_{agg} + P^N\right)}{P\zeta_n^L(r)}\right)\right)^{\tilde{k}}\right\}$$

$$= \mathbb{E}_{[\tilde{K}, I_{\text{agg}}]} \left\{ 1 - \left(1 - \exp\left(-\frac{\gamma_0 P^{\text{N}}}{P \zeta_n^{\text{L}}(r)} \right) \exp\left(-\frac{\gamma_0 I_{\text{agg}}}{P \zeta_n^{\text{L}}(r)} \right) \right)^{\tilde{k}} \right\}$$

$$\stackrel{(a)}{=} \mathbb{E}_{[I_{\text{agg}}]} \left\{ \sum_{\tilde{k}=1}^{+\infty} \left[1 - \left(1 - \exp\left(-\frac{\gamma_0 P^{\text{N}}}{P \zeta_n^{\text{L}}(r)} \right) \exp\left(-\frac{\gamma_0 I_{\text{agg}}}{P \zeta_n^{\text{L}}(r)} \right) \right)^{\tilde{k}} \right] f_{\tilde{K}}(\tilde{k}) \right\}$$

$$\stackrel{(b)}{=} \mathbb{E}_{[I_{\text{agg}}]} \left\{ \sum_{\tilde{k}=1}^{+\infty} \left[1 - \sum_{t=0}^{\tilde{k}} \binom{\tilde{k}}{t} \left(-\exp\left(-\frac{\gamma_0 P^{\text{N}}}{P \zeta_n^{\text{L}}(r)} \right) \right)^t \left(\exp\left(-\frac{\gamma_0 I_{\text{agg}}}{P \zeta_n^{\text{L}}(r)} \right) \right)^t \right] f_{\tilde{K}}(\tilde{k}) \right\}$$

$$\stackrel{(c)}{=} \mathbb{E}_{[I_{\text{agg}}]} \left\{ \sum_{\tilde{k}=1}^{+\infty} \left[1 - \sum_{t=0}^{\tilde{k}} \binom{\tilde{k}}{t} \left(-\delta_n^{\text{L}}(r) \right)^t \exp\left(-\frac{t\gamma_0}{P \zeta_n^{\text{L}}(r)} I_{\text{agg}} \right) \right] f_{\tilde{K}}(\tilde{k}) \right\}$$

$$\stackrel{(d)}{=} \sum_{\tilde{k}=1}^{+\infty} \left[1 - \sum_{t=0}^{\tilde{k}} \binom{\tilde{k}}{t} \left(-\delta_n^{\text{L}}(r) \right)^t \mathscr{L}_{I_{\text{agg}}}^{\text{L}}(s) \right] f_{\tilde{K}}(\tilde{k}), \tag{7.45}$$

where

- step (a) follows by taking the expectation over the random variable number, $\tilde{K}$, of UEs per active small cell BS,
- step (b) is a straightforward binomial expansion,
- step (c) shows a variable change, $\delta_n^{\text{L}}(r) = \exp\left(-\frac{\gamma_0 P^{\text{N}}}{P \zeta_n^{\text{L}}(r)} \right)$, and
- in step (d), $\mathscr{L}_{I_{\text{agg}}}^{\text{L}}(s)$ is the Laplace transform of the PDF of the aggregated inter-cell interference random variable, I_{agg}, conditioned on the LoS signal transmission, evaluated at the variable value, $s = \frac{t\gamma_0}{P \zeta_n^{\text{L}}(r)}$.

Based on the condition of LoS signal transmission, the Laplace transform, $\mathscr{L}_{I_{\text{agg}}}^{\text{L}}(s)$, can be derived as

$$\mathscr{L}_{I_{\text{agg}}}^{\text{L}}(s) = \mathbb{E}_{[I_{\text{agg}}]} \left\{ \exp\left(-s I_{\text{agg}} \right) \right\}$$

$$= \mathbb{E}_{[\Phi, \{\beta_i\}, \{g_i\}]} \left\{ \exp\left(-s \sum_{i \in \Phi/b_o} P \beta_i g_i \right) \right\}$$

$$\stackrel{(a)}{=} \exp\left(-2\pi\tilde{\lambda} \int \left(1 - \mathbb{E}_{[g]} \left\{ \exp\left(-s P \beta(u) g \right) \right\} \right) u \, du \right), \tag{7.46}$$

where

- Φ is the set of small cell BSs,
- b_o is the small cell BS serving the typical UE,
- b_i is the ith interfering small cell BS,
- β_i and g_i are the path loss and the multi-path fast fading gain between the typical UE and the ith interfering small cell BS and
- step (a) of equation (7.46) has been explained in detail in equation (3.18), and further derived in equation (3.19) of Chapter 3.

Importantly, note that unlike equation (3.18), the detailed calculation the Laplace transform, $\mathscr{L}^{\mathrm{L}}_{I_{\mathrm{agg}}}(s)$, given by equation (7.46) only involves active small cell BSs with a density, $\tilde{\lambda}$, because only the active small cell BS generate inter-cell interference.

Compared with equation (3.19), which considers the inter-cell interference from a single-slope path loss model, the expression, $\mathbb{E}_{[g]}\left\{\exp\left(-sP\beta(u)g\right)\right\}$, in equation (7.46) should consider the inter-cell interference from both the LoS and the NLoS paths. As a result, the Laplace transform, $\mathscr{L}^{\mathrm{L}}_{I_{\mathrm{agg}}}(s)$, can be further developed as

$$
\begin{aligned}
L^{\mathrm{L}}_{I_{\mathrm{agg}}}(s) &= \exp\left(-2\pi\tilde{\lambda}\int\left(1 - \mathbb{E}_{[g]}\left\{\exp\left(-sP\beta(u)g\right)\right\}\right)u\,du\right)\\
&\overset{(a)}{=} \exp\left(-2\pi\tilde{\lambda}\int\left[\mathrm{Pr}^{\mathrm{L}}(u)\left(1 - \mathbb{E}_{[g]}\left\{\exp\left(-sP\zeta^{\mathrm{L}}(u)g\right)\right\}\right)\right.\right.\\
&\qquad\left.\left.+ \left(1 - \mathrm{Pr}^{\mathrm{L}}(u)\right)\left(1 - \mathbb{E}_{[g]}\left\{\exp\left(-sP\zeta^{\mathrm{NL}}(u)g\right)\right\}\right)\right]u\,du\right)\\
&\overset{(b)}{=} \exp\left(-2\pi\tilde{\lambda}\int_{r}^{+\infty}\mathrm{Pr}^{\mathrm{L}}(u)\left(1 - \mathbb{E}_{[g]}\left\{\exp\left(-sP\zeta^{\mathrm{L}}(u)g\right)\right\}\right)u\,du\right)\\
&\quad\times \exp\left(-2\pi\tilde{\lambda}\int_{r_1}^{+\infty}\left(1 - \mathrm{Pr}^{\mathrm{L}}(u)\right)\left(1 - \mathbb{E}_{[g]}\left\{\exp\left(-sP\zeta^{\mathrm{NL}}(u)g\right)\right\}\right)u\,du\right)\\
&\overset{(c)}{=} \exp\left(-2\pi\tilde{\lambda}\int_{r}^{+\infty}\frac{\mathrm{Pr}^{\mathrm{L}}(u)u}{1 + \left(sP\zeta^{\mathrm{L}}(u)\right)^{-1}}\,du\right)\\
&\quad\times \exp\left(-2\pi\tilde{\lambda}\int_{r_1}^{+\infty}\frac{\left[1 - \mathrm{Pr}^{\mathrm{L}}(u)\right]u}{1 + \left(sP\zeta^{\mathrm{NL}}(u)\right)^{-1}}\,du\right),
\end{aligned}
\tag{7.47}
$$

where

- in step (a), the integration is probabilistically divided into two parts considering the LoS and the NLoS inter-cell interference,
- in step (b), the LoS and the NLoS inter-cell interference come from a distance larger than distances, r and r_1, respectively, and
- in step (c), the PDF, $f_G(g) = 1 - \exp(-g)$, of the Rayleigh random variable, g, has been used to calculate the expectations, $\mathbb{E}_{[g]}\left\{\exp\left(-sP\zeta^{\mathrm{L}}(u)g\right)\right\}$ and $\mathbb{E}_{[g]}\left\{\exp\left(-sP\zeta^{\mathrm{NL}}(u)g\right)\right\}$.

NLoS-Related Calculations

In Theorem 7.3.1, we have obtained the PDF of a segment, $f^{\mathrm{NL}}_{R,n}(r)$. Here, we evaluate the expected probability, $\mathbb{E}_{[\tilde{k}]}\left\{\mathrm{Pr}\left[\frac{P\zeta^{\mathrm{NL}}_n(r)y(\tilde{k})}{I_{\mathrm{agg}}+P_{\mathrm{N}}} > \gamma\right]\right\}$, in equation (7.29) as

$$\mathbb{E}_{[\tilde{k}]}\left\{\Pr\left[\frac{P\zeta_n^{NL}(r)y(\tilde{k})}{I_{agg}+P_N} > \gamma\right]\right\} = \mathbb{E}_{[\tilde{K}]}\left\{\Pr\left[y(\tilde{k}) > \frac{\gamma_0\left(I_{agg}+P^N\right)}{P\zeta_n^{NL}(r)}\right]\right\}$$

$$= \mathbb{E}_{[\tilde{K},I_{agg}]}\left\{\bar{F}_{Y(\tilde{k})}\left(\frac{\gamma_0\left(I_{agg}+P^N\right)}{P\zeta_n^{NL}(r)}\right)\right\},$$

$$(7.48)$$

where

- $\bar{F}_{Y(\tilde{k})}(\cdot)$ is the CCDF of the multi-path fast fading channel gain, $Y(\tilde{k})$, which is given by equation (7.13).

Similar to equation (7.49), plugging equation (7.13) into equation (7.48) yields

$$\mathbb{E}_{[\tilde{K}]}\left\{\Pr\left[\frac{P\zeta_n^{NL}(r)y(\tilde{k})}{I_{agg}+P^N} > \gamma_0\right]\right\} = \sum_{\tilde{k}=1}^{+\infty}\left[1 - \sum_{t=0}^{\tilde{k}}\binom{\tilde{k}}{t}\left(-\delta_n^{NL}(r)\right)^t \mathcal{L}_{I_{agg}}^{NL}(s)\right]f_{\tilde{K}}(\tilde{k}),$$

$$(7.49)$$

where

- the variable change, $\delta_n^{NL}(r) = \exp\left(-\frac{\gamma P_N}{P\zeta_n^{NL}(r)}\right)$, has been used and
- $\mathcal{L}_{I_{agg}}^{NL}(s)$ is the Laplace transform of the PDF of the aggregated inter-cell interference random variable, I_{agg}, conditioned on the NLoS signal transmission, evaluated at $s = \frac{t\gamma}{P\zeta_n^{NL}(r)}$.

Based on the condition of NLoS signal transmission, the Laplace transform, $\mathcal{L}_{I_{agg}}^{NL}(s)$, can be derived as

$$\mathcal{L}_{I_{agg}}^{NL}(s) = \mathbb{E}_{[I_{agg}]}\left\{\exp\left(-sI_{agg}\right)\right\}$$

$$= \mathbb{E}_{[\Phi,\{\beta_i\},\{g_i\}]}\left\{\exp\left(-s\sum_{i\in\Phi/b_o}P\beta_ig_i\right)\right\}$$

$$\overset{(a)}{=} \exp\left(-2\pi\tilde{\lambda}\int\left(1 - \mathbb{E}_{[g]}\left\{\exp\left(-sP\beta(u)g\right)\right\}\right)u\,du\right), \quad (7.50)$$

where

- step (a) of equation (7.50) has been explained in detail in equation (3.18), and further derived in equation (3.19) of Chapter 3.

Importantly, note that unlike equation (3.18), the detailed calculation of $\mathcal{L}_{I_{agg}}^{NL}(s)$ given by equation (7.50) only involves active small cell BSs with a density, $\tilde{\lambda}$, because only the active small cell BS generate inter-cell interference.

Compared with equation (3.19), which considers the inter-cell interference from a single-slope path loss model, the expression, $\mathbb{E}_{[g]}\left\{\exp\left(-sP\beta(u)g\right)\right\}$, in equation (7.50) should consider the inter-cell interference from both the LoS and the NLoS paths. As a result, similar to equation (7.51), the Laplace transform, $\mathscr{L}^{\mathrm{NL}}_{I_{\mathrm{agg}}}(s)$, can be further developed as

$$\mathscr{L}^{\mathrm{NL}}_{I_{\mathrm{agg}}}(s) = \exp\left(-2\pi\tilde{\lambda}\int\left(1 - \mathbb{E}_{[g]}\left\{\exp\left(-sP\beta(u)g\right)\right\}\right)u\,du\right)$$

$$\overset{(a)}{=} \exp\left(-2\pi\tilde{\lambda}\int_{r_2}^{+\infty}\mathrm{Pr}^{\mathrm{L}}(u)\left(1 - \mathbb{E}_{[g]}\left\{\exp\left(-sP\zeta^{\mathrm{L}}(u)g\right)\right\}\right)u\,du\right)$$

$$\times\exp\left(-2\pi\tilde{\lambda}\int_{r}^{+\infty}\left(1 - \mathrm{Pr}^{\mathrm{L}}(u)\right)\left(1 - \mathbb{E}_{[g]}\left\{\exp\left(-sP\zeta^{\mathrm{NL}}(u)g\right)\right\}\right)u\,du\right)$$

$$\overset{(b)}{=} \exp\left(-2\pi\tilde{\lambda}\int_{r_2}^{+\infty}\frac{\mathrm{Pr}^{\mathrm{L}}(u)u}{1 + \left(sP\zeta^{\mathrm{L}}(u)\right)^{-1}}\,du\right)$$

$$\times\exp\left(-2\pi\tilde{\lambda}\int_{r}^{+\infty}\frac{\left[1 - \mathrm{Pr}^{\mathrm{L}}(u)\right]u}{1 + \left(sP\zeta^{\mathrm{NL}}(u)\right)^{-1}}\,du\right), \tag{7.51}$$

where

- in step (a), the LoS and the NLoS inter-cell interference come from a distance larger than distances, r_2 and r, respectively, and
- in step (b), the PDF, $f_G(g) = 1 - \exp(-g)$, of the Rayleigh random variable, g, has been used to calculate the expectations, $\mathbb{E}_{[g]}\left\{\exp\left(-sP\zeta^{\mathrm{L}}(u)g\right)\right\}$ and $\mathbb{E}_{[g]}\left\{\exp\left(-sP\zeta^{\mathrm{NL}}(u)g\right)\right\}$.

Our proof is thus completed.

Appendix C Proof of Lemma (7.3.3)

The key of the proof for Lemma (7.3.3) lies in equations (7.21) and (7.24) of Theorem 7.3.2.

When the small cell BS density, λ, tends to infinity, $\lambda \rightarrow +\infty$, the maximum number, $\tilde{K}^{\max}$, of UEs per active small cell BS tends to 1, $\tilde{K}^{\max} \rightarrow 1$, and as a result, the probability of having one UE per active small cell BS is 1, i.e. $f_{\tilde{K}}(1) = 1$. In other words, the resulting ultra-dense network approaches the limit of one UE per active small cell BS. Thus, Theorem 7.3.2 degenerates to the results for the RR scheduler presented and analyzed in Chapter 6.

This completes the proof.

Appendix D Proof of Theorem 7.3.4

The key of the proof for Theorem 7.3.4 lies in applying the Jensen's inequality as follows [156]:

$$\mathbb{E}_{[I_{\text{agg}}]} \left\{ 1 - (1 - \exp(-x))^{\tilde{k}} \right\} \leq 1 - \left(1 - \mathbb{E}_{[I_{\text{agg}}]} \{ \exp(-x) \} \right)^{\tilde{k}}. \qquad (7.52)$$

This is because the expression, $\left[1 - (1 - \exp(-x))^{\tilde{k}} \right]$, defined in equation (7.13), is a concave function with regard to the exponential, $\exp(-x)$, when the two following conditions are met, $\tilde{k} \geq 1$ and $x \in [0, 1]$.

The proof is obtained by plugging equation (7.52) into step (a) of equation (7.45), and performing some mathematical manipulations.

Part III

Capacity Scaling Law

8   The Ultra-Dense Wireless Networks Capacity Scaling Law

8.1   Introduction

DEFINITION 8.1.1   A scaling law is a mathematical tool that describes the functional relationship between two physical quantities, which scale with each other over a significant interval.

Some of the first scaling laws pertain to the relationship between the lengths and the areas of geometrical figures. To give an example, the ratio of the areas of two squares changes in proportion with the second power of the ratio of their respective lengths. In wireless communications, a good example of a well-known scaling law is that of the capacity of a wireless channel with respect to its signal quality – the Shannon–Hartley theorem [9] – which was already introduced and described in Chapter 1. However, not all the relationships between two measurable quantities of a system are "easily" computable and can be captured through closed-form expressions. This is the case of the relationship between the performance of an ultra-dense network, for example, in terms of the coverage probability, p^{cov}, or the area spectral efficiency (ASE), A^{ASE}, and the densification level, i.e. the small cell base station (BS) density, λ. The many – random – processes involved in a large network complicate the derivations of such relationships. In general, the capacity scaling law of large wireless networks with respect to various parameters is still an open question, and a hot research topic [111].

In Chapters 3–7 of this book, we have shown how these two key performance indicators – the coverage probability, p^{cov}, and the ASE, A^{ASE} – varied with the change of the small cell BS density, λ, when considering – in isolation – different channel characteristics and network features, e.g. path loss models, antenna heights, idle mode capabilities and medium access control (MAC) layer schedulers. In some cases, and for a particular range of small cell BS densities, the coverage probability, p^{cov}, and the ASE, A^{ASE}, increased with the increase of the small cell BS density, λ, while in some others, these two key performance indicators drastically decreased with it. These are interesting contradicting behaviours, to say the least.

For convenience, let us take an overview of the following the different scaling laws observed between the ASE, A^{ASE}, and the small cell BS density, λ, when considering the different key channel characteristics and network features studied in Chapters 3–7 of this book. We focus the summary on the ASE, A^{ASE}, and not on the coverage probability, p^{cov}, for the sake of brevity, and more importantly, because the former is

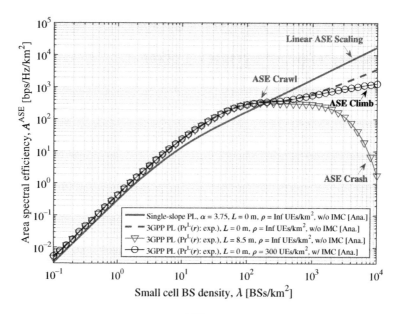

Figure 8.1 Area spectral efficiency, A^{ASE}, versus the small cell BS density, λ, with an SINR threshold, $\gamma_0 = 0$ dB. Note that all these capacity scaling laws have been presented and discussed in Chapters 3–6 of this book.

a function – a result – of the later. To assist the discussion, Figure 8.1 illustrates such scaling laws.

- A linear ASE scaling law.
  In Chapter 3, considering a single-slope path loss model, the ASE, A^{ASE}, linearly increased with the increase of the small cell BS density, λ, with each new small cell BS adding an equal contribution to it.
- The ASE Crawl.
  In Chapter 4, considering a multi-slope path loss model with both line-of-sight (LoS) and non-line-of-sight (NLoS) transmissions, the ASE, A^{ASE}, did not linearly increase with the increase of the small cell BS density, λ, as in Chapter 3. Instead, it showed a slowdown in the rate of growth – or even a decrease – as the small cell BS density, λ, increased in a particular range of such densities.

  Due to the transition of a large number of interfering links from NLoS to LoS in such a range of densities, and the resulting larger aggregated inter-cell interference, the contribution of each small cell BS to the ASE, A^{ASE}, could not be kept constant, and decreased with every new small cell BS. However, once the dominant interfering small cell BSs were in LoS, the ASE, A^{ASE}, of the ultra-dense network picked up its linear growth.
- The ASE Crash.
  In Chapter 5, considering a multi-slope path loss model with both LoS and NLoS transmissions, together with the height difference, $L > 0$, between the antennas of

the user equipment (UE) and those of the serving and interfering small cell BSs, the ASE, A^{ASE}, neither linearly increased – nor slowed down – with the increase of the small cell BS density, λ, as in Chapter 3 and Chapter 4, respectively. Instead, it crashed towards 0 as the small cell BS density, λ, increased.

Due to the cap on the maximum signal power that a UE can receive, – imposed by such an antenna height difference, $L > 0$ – and because of the increasingly overwhelming inter-cell interference from the more and closer neighbouring small cell BSs in a densifying network, the contribution of each small cell BS to the ASE, A^{ASE}, decreased towards 0 in the ultra-dense regime.

- The ASE Climb.

In Chapter 6, considering a multi-slope path loss model with both LoS and NLoS transmissions, together with a finite UE density, $\rho < +\infty$, and a simple idle mode capability at the small cell BSs,[1] the ASE, A^{ASE}, neither linearly increased – nor crashed towards 0 – with the increase of the small cell BS density, λ, as in Chapter 3 and Chapter 5, respectively. Instead, it climbed towards more positive values – at a good pace – as the small cell BS density, λ, increased.

Due to the cap on the maximum inter-cell interference power that a UE can receive – imposed by such a finite UE density, $\rho < +\infty$, and such a simple idle mode capability at the small cell BSs – and because of the increasing signal power from the shorter UE-to-serving small cell BS distances in a densifying network, the contribution of each small cell BS to the ASE, A^{ASE}, increased in the ultra-dense regime.

- The ASE Climb+.

In Chapter 7, considering a proportional fair (PF) scheduler instead of a round robin (RR) one, in addition to the assumptions in Chapter 6, the ASE, A^{ASE}, improved in the sparse network regime due to the multi-user diversity gain. However, these gains were negligible in the ultra-dense regime. This follows from the smaller number of UEs per cell and the less channel fluctuations in an ultra-dense network with smaller cell sizes.

Importantly, we should highlight that neither of these relationships between the ASE, A^{ASE}, and the small cell BS density, λ, qualitatively changed when considering more complex channel characteristics or network features. In Sections 3.4.4 and 3.4.5, the impact of a more structured small cell BS deployment as well as the consideration of a lognormal shadow fading were respectively shown to result in some quantitative – but in no qualitative – differences. Similar observations were done about the consideration of a Rician-based instead of a Rayleigh-based multi-path fast fading. Although the multi-path fast fading model in the LoS transmissions changed the quantitative results, it did not change the performance trends. Section 4.4.5, Section 5.4.5, Section 6.4.5 and Section 7.4.4 respectively provided this multi-path fast fading

---

[1] Remember that Chapter 6 did not consider the height difference, $L > 0$, between the antennas of the UE and those of the serving and interfering small cell BSs.

model analysis for each of the channel characteristics and network features studied in Chapters 4–7.

Looking closely at these capacity scaling laws, and given the contradicting behaviours in the ultra-dense regime shown by some of them, i.e.

- the negative impact on the system performance of the ASE Crawl in Chapter 4 and the ASE Crash in Chapter 5 and
- the positive one of the ASE Climb in Chapter 6,[2]

it is fair to wonder about the performance behaviour of an ultra-dense network that considers all the fundamental channel characteristics and network features analyzed so far in this book. These fundamental attributes are intrinsic to an ultra-dense network, and while they could be ignored in the performance analysis of a sparse network, as they do not play a defining role, they must be accounted for in ultra-dense deployments.

This gives rise to a fundamental question:

Will the positive effect of a simple idle mode capability on the performance of a practical ultra-dense network with a finite UE density be able to compensate for the negative impact of both the NLoS-to-LoS inter-cell interference transition and the antenna height difference?

In this chapter, we answer this fundamental question by means of in-depth theoretical analyses and present a capacity scaling law for ultra-dense networks.

The rest of this chapter is organized as follows:

- Section 8.2 briefly describes the system model and the assumptions taken in the theoretical performance analysis framework presented in this chapter, considering the most relevant features studied so far in this book.
- Section 8.3 presents the theoretical expressions for the coverage probability and the ASE under the new assumptions.
- Section 8.4 presents a new capacity scaling law for ultra-dense networks, based on the previous coverage probability and ASE formulations.
- Section 8.5 provides results for a number of small cell BS deployments with different densities and characteristics, and presents the conclusions drawn by this theoretical performance analysis, enabling a better understanding of the provided capacity scaling law.
- Section 8.6, finally, summarizes the key takeaways of this chapter.

8.2   Updated System Model

To assess the capacity scaling law of an ultra-dense network under practical conditions, this chapter embraces a system model that accounts for the most relevant channel characteristics and network features presented in Chapters 4–6 of this book, i.e.

---

[2] An RR scheduler is adopted in this chapter, since the PF one was shown to provide negligible capacity gains when the network goes ultra-dense.

- both LoS and NLoS transmissions, and a practical user association strategy, based on the strongest received signal strength, as described in Chapter 4,
- the height of the antenna of the typical UE and those of the small cell BSs, as introduced in Chapter 5, and
- a finite UE density, and a simple idle mode capability at the small cell BSs. as presented in Chapter 6.

Since exactly the same models and descriptions apply, we do not go into further details here with the system model. We refer the reader to Table 8.1 for a summary of those, and to the corresponding previous chapters for the exact details on the modelling assumptions.

It should also be noted that an RR scheduler – and not a PF one – is adopted in this chapter, as the latter was shown to provide negligible capacity gains in the ultra-dense regime in Chapter 7.

8.3 Theoretical Performance Analysis and Main Results

In this section, and similarly to Chapters 3–7, we present the derived expressions for the coverage probability, p^{cov}, and the ASE, A^{ASE}, as for the formal definitions presented in equations (2.22) and (6.24), respectively (see Table 8.2). These new expressions will be utilized to assess the capacity scaling law of an ultra-dense network under close-to-real conditions, or at least closer than any other theoretical model in the literature.

8.3.1 Coverage Probability

In the following, we present the new main result on the coverage probability, p^{cov}, through Theorem 8.3.1, considering the above introduced system model. Readers interested in the research article originally presenting these results are referred to [231].

THEOREM 8.3.1 *Considering the system model presented in Section 8.2, the coverage probability, p^{cov}, can be derived as*

$$p^{\text{cov}}(\lambda, \gamma_0) = \sum_{n=1}^{N} \left(T_n^{\text{L}} + T_n^{\text{NL}} \right), \qquad (8.1)$$

where

-

$$T_n^{\text{L}} = \int_{\sqrt{d_{n-1}^2 - L^2}}^{\sqrt{d_n^2 - L^2}} \Pr\left[\frac{P\zeta_n^{\text{L}}\left(\sqrt{r^2 + L^2}\right) h}{I_{\text{agg}} + P^{\text{N}}} > \gamma_0 \right] f_{R,n}^{\text{L}}(r) dr, \qquad (8.2)$$

Table 8.1. System model

Model	Description	Reference
Transmission link		
Link direction	Downlink only	Transmissions from the small cell BSs to the UEs
Deployment		
Small cell BS deployment	HPPP with a finite density, $\lambda < +\infty$	Section 3.2.2
UE deployment	HPPP with a finite density, $\rho < +\infty$	Section 6.2.1
UE to small cell BS association		
Strongest small cell BS	UEs connect to the small cell BS providing the strongest received signal strength	Section 3.2.8 and references therein, equation (2.1.4)
Path loss		
3GPP UMi [153]	Multi-slope path loss with probabilistic LoS and NLoS transmissions	Section 4.2.1, equation (4.1)
	☐ LoS component	equation (4.2)
	☐ NLoS component	equation (4.3)
	☐ Probability of LoS	equation (4.19)
Multi-path fast fading		
Rayleigh	Highly scattered scenario	Section 3.2.7 and references therein equation (2.15)
Shadow fading		
Not modelled	For tractability reasons, shadowing is not modelled since it has no qualitative impact on the results, see Section 3.4.5	Section 3.2.6 references therein
Antenna		
Small cell BS antenna	Isotropic single-antenna element with 0 dBi gain	Section 3.2.4
UE antenna	Isotropic single-antenna element with 0 dBi gain	Section 3.2.4
Small cell BS antenna height	Variable antenna height, $1.5 + L$ m	Section 5.2.1
UE antenna height	Fixed antenna height, 1.5 m	–
Idle mode capability at small cell BSs		
Connection-aware	Small cell BSs with no UE in their coverage areas are switched off	Section 6.2.2
Scheduler at small cell BSs		
Round robin	UEs take turns in accessing the radio channel	Section 7.1

Table 8.2. Key performance indicators

Metric	Formulation	Reference
Coverage probability, p^{cov}	$p^{\text{cov}}(\lambda, \gamma_0) = \Pr\left[\Gamma > \gamma_0\right] = \bar{F}_\Gamma\left(\gamma_0\right)$	Section 2.1.12, equation (2.22)
ASE, A^{ASE}	$A^{\text{ASE}}(\lambda, \gamma_0) = \frac{\tilde{\lambda}}{\ln 2}\int_{\gamma_0}^{+\infty}\frac{p^{\text{cov}}(\lambda,\gamma)}{1+\gamma}d\gamma$ $+\tilde{\lambda}\log_2\left(1+\gamma_0\right)p^{\text{cov}}(\lambda,\gamma_0)$	Section 6.3.3, equation (6.24)

•

$$T_n^{\text{NL}} = \int_{\sqrt{d_{n-1}^2 - L^2}}^{\sqrt{d_n^2 - L^2}} \Pr\left[\frac{P\zeta_n^{\text{NL}}\left(\sqrt{r^2+L^2}\right)h}{I_{\text{agg}} + P^{\text{N}}} > \gamma_0\right] f_{R,n}^{\text{NL}}(r)dr \quad (8.3)$$

and
• *d_0 and d_N are defined as $L \geq 0$ and $+\infty$, respectively.*

Moreover, the probability density function (PDF), $f_{R,n}^{\text{L}}(r)$, and the PDF, $f_{R,n}^{\text{NL}}(r)$, within the range, $\sqrt{d_{n-1}^2 - L^2} < r \leq \sqrt{d_n^2 - L^2}$, are given by

$$f_{R,n}^{\text{L}}(r) = \exp\left(-\int_0^{r_1}\left(1 - \Pr^{\text{L}}\left(\sqrt{u^2+L^2}\right)\right)2\pi u\lambda\, du\right)$$

$$\times \exp\left(-\int_0^r \Pr^{\text{L}}\left(\sqrt{u^2+L^2}\right)2\pi u\lambda\, du\right) \quad (8.4)$$

$$\times \Pr_n^{\text{L}}\left(\sqrt{r^2+L^2}\right)2\pi r\lambda \quad (8.5)$$

and

$$f_{R,n}^{\text{NL}}(r) = \exp\left(-\int_0^{r_2}\Pr^{\text{L}}\left(\sqrt{u^2+L^2}\right)2\pi u\lambda\, du\right)$$

$$\times \exp\left(-\int_0^r\left(1 - \Pr^{\text{L}}\left(\sqrt{u^2+L^2}\right)\right)2\pi u\lambda\, du\right) \quad (8.6)$$

$$\times \left(1 - \Pr_n^{\text{L}}\left(\sqrt{r^2+L^2}\right)\right)2\pi r\lambda, \quad (8.7)$$

where
• *r_1 and r_2 are determined by*

$$r_1 = \arg_{r_1}\left\{\zeta^{\text{NL}}\left(\sqrt{r_1^2+L^2}\right) = \zeta_n^{\text{L}}\left(\sqrt{r^2+L^2}\right)\right\} \quad (8.8)$$

and

$$r_2 = \arg_{r_2} \left\{ \zeta^{\mathrm{L}} \left(\sqrt{r_2^2 + L^2} \right) = \zeta_n^{\mathrm{NL}} \left(\sqrt{r^2 + L^2} \right) \right\}, \tag{8.9}$$

respectively.

Proof The proof of this theorem follows the same guidelines as

- that of Theorem 5.3.1 for the aspects related to the height difference, $L \geq 0$, between the antennas of the UE and those of the serving and interfering small cell BSs and
- that of Theorem 6.3.1 for the aspects related to the finite UE density, $\rho < +\infty$, and the simple idle mode capability at the small cell BSs.

Thus, we omit the proof of this theorem for brevity here. □

To aid the reader in the understanding of Theorem 8.3.1, and for the sake of clarity, the calculations of the probability, $\Pr\left[\frac{P\zeta_n^{\mathrm{L}}\left(\sqrt{r^2+L^2}\right)h}{I_{\mathrm{agg}}+P^{\mathrm{N}}} > \gamma_0 \right]$, and the probability, $\Pr\left[\frac{P\zeta_n^{\mathrm{NL}}\left(\sqrt{r^2+L^2}\right)h}{I_{\mathrm{agg}}+P^{\mathrm{N}}} > \gamma_0 \right]$, are further developed in the following lemma.

LEMMA 8.3.2 *In Theorem 8.3.1, the probability,* $\Pr\left[\frac{P\zeta_n^{\mathrm{L}}\left(\sqrt{r^2+L^2}\right)h}{I_{\mathrm{agg}}+P^{\mathrm{N}}} > \gamma_0 \right]$, *is computed as*

$$\Pr\left[\frac{P\zeta_n^{\mathrm{L}}\left(\sqrt{r^2+L^2}\right)h}{I_{\mathrm{agg}}+P^{\mathrm{N}}} > \gamma_0 \right] = \exp\left(-\frac{\gamma_0 P^{\mathrm{N}}}{P\zeta_n^{\mathrm{L}}\left(\sqrt{r^2+L^2}\right)} \right) \mathscr{L}^{\mathrm{L}}_{I_{\mathrm{agg}}}(s), \tag{8.10}$$

where

- $\mathscr{L}^{\mathrm{L}}_{I_{\mathrm{agg}}}(s)$ *is the Laplace transform of the aggregated inter-cell interference random variable,* $I^{\mathrm{L}}_{\mathrm{agg}}$, *for the LoS signal transmission evaluated at the variable value,* $s = \frac{\gamma_0}{P\zeta_n^{\mathrm{L}}(r)}$, *which can be expressed as*

$$\mathscr{L}^{\mathrm{L}}_{I_{\mathrm{agg}}}(s) = \exp\left(-2\pi\tilde{\lambda} \int_r^{+\infty} \frac{\Pr^{\mathrm{L}}\left(\sqrt{u^2+L^2}\right)u}{1+\left(sP\zeta^{\mathrm{L}}\left(\sqrt{u^2+L^2}\right)\right)^{-1}} du \right)$$

$$\times \exp\left(-2\pi\tilde{\lambda} \int_{r_1}^{+\infty} \frac{\left[1-\Pr^{\mathrm{L}}\left(\sqrt{u^2+L^2}\right)\right]u}{1+\left(sP\zeta^{\mathrm{NL}}\left(\sqrt{u^2+L^2}\right)\right)^{-1}} du \right). \tag{8.11}$$

Moreover, the probability, $\Pr\left[\dfrac{P\zeta_n^{\mathrm{NL}}\left(\sqrt{r^2+L^2}\right)h}{I_{\mathrm{agg}}+P^{\mathrm{N}}} > \gamma_0\right]$, *is computed as*

$$\Pr\left[\frac{P\zeta_n^{\mathrm{NL}}\left(\sqrt{r^2+L^2}\right)h}{I_{\mathrm{agg}}+P^{\mathrm{N}}} > \gamma_0\right] = \exp\left(-\frac{\gamma_0 P^{\mathrm{N}}}{P\zeta_n^{\mathrm{NL}}\left(\sqrt{r^2+L^2}\right)}\right)\mathscr{L}_{I_{\mathrm{agg}}}^{\mathrm{NL}}(s), \quad (8.12)$$

where

- $\mathscr{L}_{I_{\mathrm{agg}}}^{\mathrm{NL}}(s)$ *is the Laplace transform of the aggregated inter-cell interference random variable,* $I_{\mathrm{agg}}^{\mathrm{NL}}$, *for the NLoS signal transmission evaluated at the variable value,* $s = \dfrac{\gamma_0}{P\zeta_n^{\mathrm{NL}}(r)}$, *which can be written as*

$$\mathscr{L}_{I_{\mathrm{agg}}}^{\mathrm{NL}}(s) = \exp\left(-2\pi\tilde{\lambda}\int_{r_2}^{+\infty}\frac{\Pr^{\mathrm{L}}\left(\sqrt{u^2+L^2}\right)u}{1+\left(sP\zeta^{\mathrm{L}}\left(\sqrt{u^2+L^2}\right)\right)^{-1}}du\right)$$

$$\times \exp\left(-2\pi\tilde{\lambda}\int_{r}^{+\infty}\frac{\left[1-\Pr^{\mathrm{L}}\left(\sqrt{u^2+L^2}\right)\right]u}{1+\left(sP\zeta^{\mathrm{NL}}\left(\sqrt{u^2+L^2}\right)\right)^{-1}}du\right). \quad (8.13)$$

Proof The proof of this lemma follows the same guidelines as

- that of Theorem 5.3.1 for the aspects related to the height difference, $L \geq 0$, between the antennas of the UE and those of the serving and interfering small cell BSs, and
- that of Theorem 6.3.1 for the aspects related to the finite UE density, $\rho < +\infty$, and the simple idle mode capability at the small cell BSs.

Thus, we also omit the proof of this lemma for brevity here.   □

As in Chapters 3–7, to facilitate the understanding on how to practically obtain this new coverage probability, p^{cov}, Figure 8.2 illustrates the pseudo code that depicts the necessary enhancements to the traditional stochastic geometry framework presented in Chapter 3 – and updated in Chapter 4 – to compute the new results of Theorem 8.3.1.

Similarly to the logic illustrated in Figure 5.2 – and in contrast to that shown in Figure 4.2 – three-dimensional distances instead of two-dimensional ones are considered in the channel model in this new result to capture the height difference, $L \geq 0$, between the antennas of the UE and those of the serving and interfering small cell BSs. It should be noted that the outermost integration should still be performed considering two-dimensional distances, as this integration relates to the homogeneous Poisson point process (HPPP) modelling, and thus to the distances between the typical UE and the small cell BSs in the two-dimensional plane.

In addition, it is equally important to note that, in Theorem 8.3.1, similarly to the logic illustrated in Figure 6.1 – and in contrast to that shown in Figures 4.2 and 5.2 – the inter-cell interference does not come from all the small cell BSs, but only from

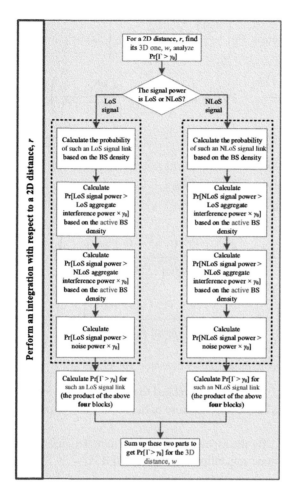

Figure 8.2 Logical steps within the standard stochastic geometry framework to obtain the results in Theorem 8.3.1, considering all the features analyzed in this book.

the active ones. Diving a little bit further into the details of this inter-cell interference modelling, let us highlight the following three aspects:

- The impact of the serving small cell BS selection on the coverage probability, p^{cov}, is measured by equations (8.5) and (8.7). These expressions are based on the small cell BS density, λ, and not on the active small cell BS density, $\tilde{\lambda}$.
- The impact of the aggregated inter-cell interference, I_{agg}, on the coverage probability, p^{cov}, is measured by equations (8.11) and (8.13). Since only the active small cell BS generate inter-cell interference, these expressions are thus based on the active small cell BS density, $\tilde{\lambda}$, and not on the small cell BS density, λ.
- The derivation of the active small cell BS density, $\tilde{\lambda}$, was presented and discussed in Section 6.3.2.

8.3.2   Area Spectral Efficiency

Similarly as in Chapters 3–7, plugging the coverage probability, p^{cov} – obtained from Theorem 8.3.1 – into equation (2.24) to compute the PDF, $f_{\Gamma}(\gamma)$, of the SINR, Γ, of the typical UE, we can obtain the ASE, A^{ASE}, by solving equation (6.24). See Table 7.2 for further reference on the ASE formulation, i.e.

$$A^{\mathrm{ASE}}(\lambda, \gamma_0) = \frac{\tilde{\lambda}}{\ln 2} \int_{\gamma_0}^{+\infty} \frac{p^{\mathrm{cov}}(\lambda, \gamma)}{1 + \gamma} d\gamma$$

$$+ \tilde{\lambda} \log_2 (1 + \gamma_0) \, p^{\mathrm{cov}}(\lambda, \gamma_0).$$

8.3.3   Computational Complexity

To calculate the coverage probability, p^{cov}, presented in Theorem 8.3.1, it should be noted that nothing has changed with respect to Chapters 4–7, i.e. three folds of integrals are still required for the calculation of $\left\{ f_{R,n}^{Path}(r) \right\}$, $\left\{ \mathscr{L}_{I_r} \left(\frac{\gamma_0}{P \zeta_n^{Path}(r)} \right) \right\}$ and $\left\{ T_n^{Path} \right\}$, respectively, where the string variable, *Path*, takes the value of "L" (for the LoS case) or "NL" (for the NLoS case). An additional fold of integral is needed for the calculation of the ASE, A^{ASE}, making it a four-fold integral computation.

8.4   Capacity Scaling Law

Given the results on the coverage probability, p^{cov}, and the ASE, A^{ASE}, obtained above with the new – and more complete – system model in this chapter, we are in a good position now to derive a new – and more accurate – capacity scaling law for ultra-dense networks.

Before turning attention to such a capacity scaling law, however, let us first elaborate on the asymptotic behaviour of the coverage probability, p^{cov}, in other words, the trend of the coverage probability, p^{cov}, when the small cell BS density, λ, tends to infinity, i.e. $\lim_{\lambda \to +\infty} p^{\mathrm{cov}}$. This will help us to better understand the new fundamental performance behaviour flourishing in the ultra-dense regime.

8.4.1   The Asymptotic Behaviour of the Coverage Probability

The following theorem presents the result on the asymptotic behaviour of the coverage probability, p^{cov}.

THEOREM 8.4.1   *Considering the system model presented in Section 8.2, the limit,* $\lim_{\lambda \to +\infty} p^{cov}$, *of the coverage probability,* p^{cov}, *when the small cell BS density,* λ, *tends to infinity can be derived as*

$$\lim_{\lambda \to +\infty} p^{\mathrm{cov}}(\lambda, \gamma_0) = \lim_{\lambda \to +\infty} \Pr \left[\frac{P \zeta_1^{\mathrm{L}}(L) h}{I_{\mathrm{agg}} + P^{\mathrm{N}}} > \gamma_0 \right]$$

$$= \exp \left(-\frac{P^{\mathrm{N}} \gamma_0}{P \zeta_1^{\mathrm{L}}(L)} \right) \lim_{\lambda \to +\infty} \mathscr{L}_{I_{\mathrm{agg}}}^{\mathrm{L}} \left(\frac{\gamma_0}{P \zeta_1^{\mathrm{L}}(L)} \right), \qquad (8.14)$$

where

- *the limit,* $\lim\limits_{\lambda \to +\infty} \mathscr{L}_{I_{\mathrm{agg}}}^{\mathrm{L}}(s)$, *of the Laplace transform,* $\mathscr{L}_{I_{\mathrm{agg}}}^{\mathrm{L}}(s)$, *of the aggregated inter-cell interference random variable,* $I_{\mathrm{agg}}^{\mathrm{L}}$, *for an LoS signal transmission evaluated at the variable value,* $s = \frac{\gamma_0}{P \zeta_1^{\mathrm{L}}(L)}$, *is given by*

$$\lim_{\lambda \to +\infty} \mathscr{L}_{I_{\mathrm{agg}}}^{\mathrm{L}}(s) = \exp \left(-2\pi\rho \int_0^{+\infty} \frac{\Pr^{\mathrm{L}} \left(\sqrt{u^2 + L^2} \right) u}{1 + \left(s P \zeta^{\mathrm{L}} \left(\sqrt{u^2 + L^2} \right) \right)^{-1}} du \right)$$

$$\times \exp \left(-2\pi\rho \int_0^{+\infty} \frac{\left[1 - \Pr^{\mathrm{L}} \left(\sqrt{u^2 + L^2} \right) \right] u}{1 + \left(s P \zeta^{\mathrm{NL}} \left(\sqrt{u^2 + L^2} \right) \right)^{-1}} du \right). \qquad (8.15)$$

Proof See Appendix A. □

From this theorem, a new coverage probability scaling law in the ultra-dense regime is shown, which is different from those presented so far in this book. Let us use the following theorem to further develop and explain this new coverage probability scaling law, referred hereafter to as the new SINR invariance law, and its implications.

THEOREM 8.4.2   The SINR invariance law:
If

- *the height difference, L, between the antennas of the UE and those of the serving and interfering small cell BSs is larger than 0, $L > 0$,*
- *the UE density, ρ, is finite, $\rho < +\infty$, and*
- *the small cell BSs are equipped with an appropriate idle mode capability,*

the asymptotic coverage probability, $\lim\limits_{\lambda \to +\infty} p^{\mathrm{cov}}$, *is independent of the small cell BS density, λ, but dependent on the finite UE density, $\rho < +\infty$.*

Proof The two terms on the right-hand side of equation (8.14) are both independent of the small cell BS density, λ. However, the second term, equation (8.15), is dependent on the finite UE density, $\rho < +\infty$. □

Theorem 8.4.2 is important, indicating that, in practical ultra-dense networks with an antenna height difference, $L > 0$, and a finite UE density, $\rho < +\infty$, the damage caused by both the ASE Crawl and the ASE Crash and the benefits brought by the ASE Climb *cancel each other out* in the presence of an appropriate idle mode capability at the small cell BSs. In other words,

- the positive impact of the ASE Climb
  - □ due to the ability of the simple idle mode capability at the small cell BSs to leverage a finite UE density, $\rho < +\infty$, to mitigate inter-cell interference counterbalances
- the negative impacts of the ASE Crawl and the ASE Crash
  - □ resulting from both the transition of a large number of inter-cell interfering links from NLoS to LoS and the antenna height difference, $L > 0$.

From Theorem 8.4.2, it is also easy to show that for a given antenna height difference, $L > 0$, and a given finite UE density, $\rho < +\infty$, the limit, $\lim_{\lambda \to +\infty} p^{cov}$, decreases with the increase of the SINR threshold, γ_0. Intuitively, as the SINR threshold, γ_0, increases, it is harder for the typical UE to achieve such a more challenging network performance entry condition. As a result, in the following two lemmas, we only elaborate – in more detail – on how the limit, $\lim_{\lambda \to +\infty} p^{cov}$, changes with the antenna height difference, $L > 0$, and the finite UE density, $\rho < +\infty$.

LEMMA 8.4.3 *For a given antenna height difference, $L > 0$, the limit, $\lim_{\lambda \to +\infty} p^{cov}$, decreases with the increase of the finite UE density, $\rho < +\infty$, according to a power law.*

More specifically,

$$\lim_{\lambda \to +\infty} p^{cov}(\lambda, \gamma_0) = c(\gamma_0) g^\rho(\gamma_0), \tag{8.16}$$

where

- *the function, $c(\gamma_0)$, equals to*

$$c(\gamma_0) = \exp\left(-\frac{P^N \gamma_0}{P \zeta_1^L(L)}\right) \tag{8.17}$$

and
- *the function, $g(\gamma_0)$, equals to*

$$g(\gamma_0) = \exp\left(-2\pi \int_0^{+\infty} \frac{\Pr^L\left(\sqrt{u^2 + L^2}\right) u}{1 + \left(sP\zeta^L\left(\sqrt{u^2 + L^2}\right)\right)^{-1}} du\right)$$

$$\times \exp\left(-2\pi \int_0^{+\infty} \frac{\left[1 - \Pr^L\left(\sqrt{u^2 + L^2}\right)\right] u}{1 + \left(sP\zeta^{NL}\left(\sqrt{u^2 + L^2}\right)\right)^{-1}} du\right), \tag{8.18}$$

where
- □ $s = \frac{\gamma_0}{P\zeta_1^L(L)}$.

Proof See Appendix B. □

The intuitions behind the results in Lemma (8.4.3) follow from what has been learnt in Chapter 6 of this book, i.e. the inter-cell interference power becomes bounded in the ultra-dense regime due to the activation of a finite small cell BS density, $\tilde{\lambda} < +\infty$, to serve a finite UE density, $\rho < +\infty$.

- The finite active small cell BS density, $\tilde{\lambda} < +\infty$, cannot be larger than the finite UE density, $\rho < +\infty$.

Importantly, a larger finite UE density, $\rho < +\infty$, results in a larger active small cell BS density, $\tilde{\lambda}$, allowing for an increased aggregated inter-cell interference power into the network, which in turn, leads to the decrease of the limit, $\lim_{\lambda \to +\infty} p^{\text{cov}}$, in Lemma (8.4.3).

8.4.2   The Asymptotic Behaviour of the ASE

Building upon Theorem 8.4.2, and the expression of the ASE in equation (6.24), a new capacity scaling law can be derived, as in Theorem 8.4.4.

THEOREM 8.4.4   The constant capacity scaling law:
If

- *the height difference, L, between the antennas of the UE and those of the serving and interfering small cell BSs is larger than 0, L > 0,*
- *the UE density, ρ, is finite, $\rho < +\infty$, and*
- *the small cell BSs are equipped with an appropriate idle mode capability,*

the asymptotic ASE, $\lim_{\lambda \to +\infty} A^{\text{ASE}}$, is independent of the small cell BS density, λ.
In more detail, the limit, $\lim_{\lambda \to +\infty} A^{\text{ASE}}$, is given by

$$\lim_{\lambda \to +\infty} A^{\text{ASE}}(\lambda, \gamma_0) = \frac{\rho}{\ln 2} \int_{\gamma_0}^{+\infty} \frac{\lim_{\lambda \to +\infty} p^{\text{cov}}(\lambda, \gamma_0)}{1 + \gamma} d\gamma$$

$$+ \rho \log_2 \left(1 + \gamma_0 \right) \lim_{\lambda \to +\infty} p^{\text{cov}}(\lambda, \gamma_0), \qquad (8.19)$$

where

- *the limit, $\lim_{\lambda \to +\infty} p^{\text{cov}}$, can be derived from Theorem 8.4.1.*

Proof   See Appendix C.   □

The implications of this new capacity scaling law in Theorem 8.4.4 are profound, as highlighted in the following remark.

Remark 8.1   This new capacity scaling law indicates that network densification is a dimension that cannot be – and should not be – continuously abused to enhance the network capacity. Instead, it should be stopped at a certain level, with such a densification level being defined by the UE density, $\rho < +\infty$. This is because both the coverage probability, p^{cov}, and the ASE, A^{ASE}, will always reach a maximum constant value, and any network densification beyond such a densification level will be translated into a waste of both invested money and energy consumption. Importantly, it should also be noted that reaching the maximum possible ASE – the value found through Theorem 8.4.4, i.e. $\lim_{\lambda \to +\infty} A^{\mathrm{ASE}}$ – will not probably be monetarily wise either. This is because the ASE gains do not linearly increase with the small cell BS density, λ, in such an ultra-dense regime.

The implications of this constant capacity scaling law will be analyzed in more detail in the remainder of this chapter, while Chapter 9 of this book will focus on several network optimization problems that can help a network operator to strike the right performance-cost trade-off.

8.5   Discussion

In this section, we use numerical results from static system-level simulations to evaluate the accuracy of the above theoretical performance analysis, and study the combined impacts of the ASE Crawl, the ASE Crash and the ASE Climb on an ultra-dense network in terms of the coverage probability, p^{cov}, and the ASE, A^{ASE}.
    As a reminder, it is important to note that to obtain the desired results on the coverage probability, p^{cov}, and the ASE, A^{ASE}, one should plug equations (4.17) and (4.19), using three-dimensional distances, into Theorem 8.3.1, and follow the subsequent derivations.

8.5.1   Case Study

To assess the network performance behaviour with all the features studied so far in this book, the same 3rd Generation Partnership Project (3GPP) case study as that presented in Chapter 4 – and used in the rest of the book – is utilized in this section to allow an apple-to-apple comparison. For readers interested in a more detailed description of this 3GPP case study, please refer to Table 8.1 and the references therein.
    For the sake of convenience, please recall that the following parameters are used in our 3GPP case study:

- maximum antenna gain, $G_{\mathrm{M}} = 0\,\mathrm{dB}$,
- path loss exponents, $\alpha^{\mathrm{L}} = 2.09$ and $\alpha^{\mathrm{NL}} = 3.75$ – considering a carrier frequency of 2 GHz,
- reference path losses, $A^{\mathrm{L}} = 10^{-10.38}$ and $A^{\mathrm{NL}} = 10^{-14.54}$,
- transmit power, $P = 24\,\mathrm{dBm}$, and
- noise power, $P^{\mathrm{N}} = -95\,\mathrm{dBm}$ – including a noise figure of 9 dB at the UE.

Antenna Configurations

As for the absolute antenna height difference, $L \geq 0$, we assume that the antenna height of the UE is 1.5 m, and that that of the small cell BSs varies between 1.5 m and 10 m. Accordingly, the antenna height difference, $L \geq 0$, takes the values, $L = \{0, 8.5\}$ m.

UE Densities

As for the finite UE density, $\rho < +\infty$, we study the following six cases in this performance evaluation, $\rho = \{100, 300, 600, 900, 2000, \infty\}$ UEs/km^2, to have a wide range of viewpoints. The optimal fitting parameters, q^*, used to calculate the active small cell BS density, $\tilde{\lambda}$, for those configurations with an idle mode capability at the small cell BSs are $q^* = \{4.73, 4.18, 3.97, 3.5, 3.5, 3.5\}$, respectively.

Small Cell BS Densities

To embrace the minimum transmitter-to-receiver distance of the selected path loss model of 10 m, and according to Chapters 3–6, the configurations with no antenna height difference, $L = 0$, extend up to a small cell BS density, $\lambda = 10^4$ BSs/km^2, while those with an antenna height difference, $L > 0$, extend up to a small cell BS density, $\lambda = 10^6$ BSs/km^2. Importantly, the larger small cell BS densities in the latter case allow one to understand and assess the capacity scaling laws in the asymptotic behaviour.

Benchmark

In this performance evaluation, we use, as the benchmark, all the partial results obtained in Chapters 3–6 of this book, as discussed in the introduction of this chapter.

8.5.2   Coverage Probability Performance

Figure 8.3 shows results for the coverage probability, p^{cov}, for a SINR threshold, $\gamma_0 = 0$ dB, and the following seven configurations:

- Conf. a): single-slope path loss model with no antenna height difference, $L = 0$ m, and an infinite number of UEs – or at least one UE per cell – thus all small cell BSs are active (analytical results).
- Conf. b): multi-slope path loss model with both LoS and NLoS transmissions, no antenna height difference, $L = 0$ m, and an infinite number of UEs – or at least one UE per cell – thus all small cell BSs are active (analytical results).
- Conf. c): multi-slope path loss model with both LoS and NLoS transmissions, an antenna height difference, $L > 0$ m, and an infinite number of UEs – or at least one UE per cell – thus all small cell BSs are active (analytical results).
- Conf. d): multi-slope path loss model with both LoS and NLoS transmissions, no antenna height difference, $L = 0$ m, a finite UE density, $\rho < +\infty$, and a simple idle mode capability at the small cell BSs, thus not all small cell BSs are active (analytical results).

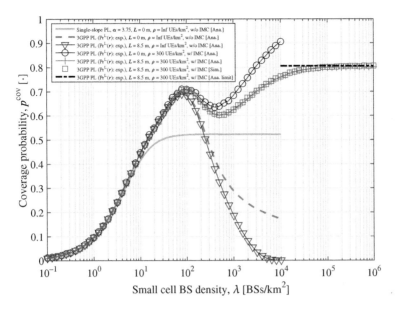

Figure 8.3 Coverage probability, p^{cov}, versus the small cell BS density, λ.

- Conf. e): multi-slope path loss model with both LoS and NLoS transmissions, an antenna height difference, $L > 0$ m, a finite UE density, $\rho < +\infty$, and a simple idle mode capability at the small cell BSs, thus not all small cell BSs are active (analytical results).
- Conf. f): multi-slope path loss model with both LoS and NLoS transmissions, an antenna height difference, $L > 0$ m, a finite UE density, $\rho < +\infty$, and a simple idle mode capability at the small cell BSs, thus not all small cell BSs are active (simulated results).
- Conf. g): the analytical limit of Conf. f) as the small cell BS density, λ, approaches infinity.

In this figure, an antenna height difference, $L = 8.5$ m, and a UE density, $\rho = 300$ UEs/km^2, are selected for the analysis. Importantly, let us indicate here that the first four configurations were already presented – and exhaustively discussed – in the corresponding Chapters 3–6. As a result, these configurations are not further discussed in the following. They are only incorporated in Figure 8.3 for comparison purposes.

As can be observed from Figure 8.3, the analytical results, Conf. e), match well the simulation results, Conf. f). This corroborates the accuracy of Theorem 8.3.1 – and the derivations therein. Due to such significant accuracy, and since the results of the ASE, A^{ASE}, are computed based on the results of the coverage probability, p^{cov}, only analytical results are considered in our discussion hereafter in this chapter.

From Figure 8.3, and with respect to configuration, Conf. e), we can also observe that,

- when the small cell BS density, λ, is sparse, and the network is noise-limited, i.e. $\lambda \leq 10^2$ BSs/km^2, the coverage probability, p^{cov}, increases with the increase of the small cell BS density, λ, as the network coverage is lightened up with the more small cell BSs, and the signal power benefits more and more from LoS transmissions.
- When the small cell BS density, λ, gets denser, and around the range, $\lambda \in (10^2, 10^3]$ BSs/km^2, the coverage probability, p^{cov}, decreases with the increase of the small cell BS density, λ. This is because of the combined effect of
  - ☐ the transition of a large number of inter-cell interfering paths from NLoS to LoS and
  - ☐ the antenna height difference, $L > 0$.

  They both slow the rate of growth of the signal power and accelerate that of the aggregated inter-cell interference. These two phenomena have been extensively discussed in Chapter 4 and Chapter 5, respectively.
- When the small cell BS density, λ, gets even denser, around the range, $\lambda \in (10^3, 10^5]$ BSs/km^2, the coverage probability, p^{cov}, increases again with the increase of the small cell BS density, λ. This is due to
  - ☐ the finite UE density, $\rho < +\infty$, and
  - ☐ the simple idle mode capability at the small cell BSs.

  They bound the aggregated inter-cell interference, since only as many small cell BSs as UEs are switched on, and the small cell BSs that are switched off – those with no active UEs – do not transmit any signalling. As a result, the SINR, Γ, of the typical UE – and in turn, the coverage probability, p^{cov} – increase due to the shorter distance between transmitter and receivers in a densifying network. This phenomenon has been extensively discussed in Chapter 6.
- Importantly, when the small cell BS density, λ, is ultra-dense, i.e. $\lambda > 10^5$ BSs/km^2, the coverage probability, p^{cov}, gradually flattens and reaches a limit (see Conf.). This verifies an important outcome of the theoretical analysis in this chapter, i.e. Theorem 8.4.2 and the discussed SINR invariance law, which stated that the asymptotic coverage probability, $\lim_{\lambda \to +\infty} p^{cov}$, is independent of the small cell BS density, λ. For this particular example, with an antenna height difference, $L = 8.5$ m, and a finite UE density, $\rho = 300$ UEs/km^2, such a limit equals to 0.806.

Let us dig more into the behaviour of this new SINR invariance law by analyzing the performance of this network using multiple UE densities in the following section.

Studying Various UE Densities

Figure 8.4 further explores the coverage probability, p^{cov}, paying special attention, this time, to the impact of various UE densities, in particular, $\rho = \{100, 300, 600, 800, 1000, 2000, 3000, 6000\}$ UEs/km^2. Since we are particularly interested in getting more insight on the asymptotic behaviour of the coverage probability, p^{cov}, in this figure, we only show results for the configuration under analysis, Conf. e), still with a multi-slope path loss model with both LoS and NLoS transmissions, an antenna

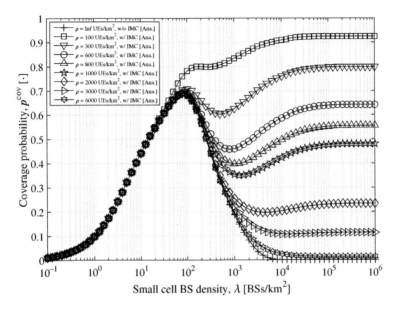

Figure 8.4 Coverage probability, p^{cov}, versus the small cell BS density, λ, for various UE densities, $\rho < +\infty$.

height difference, $L = 8.5$ m, a simple idle mode capability at the small cell BSs, and the mentioned various UE densities.

From this figure, we can observe that,

- for all UE densities, sooner or later, in terms of the small cell BS density, λ, the coverage probability, p^{cov}, gradually flattens and reaches a limit. This again verifies Theorem 8.4.2.
- As indicated by Lemma (8.4.3), the asymptotic value of the coverage probability, $\lim_{\lambda \to +\infty} p^{cov}(\lambda, \gamma_0)$, decreases with the increase of the finite UE density, $\rho < +\infty$. This is mostly due to the larger inter-cell interference generated by the more active small cell BSs switched on to serve the more UEs in the network. In this particular figure, with an antenna height difference, $L = 8.5$ m, this limit equals to 0.806 for a finite UE density, $\rho = 300$ UEs/km^2, and 0.650 for a finite UE density, $\rho = 600$ UEs/km^2. Since 0.650 is the square of 0.806, these results also corroborate the power law of the limit, $\lim_{\lambda \to +\infty} p^{cov}(\lambda, \gamma_0)$, with respect to the finite UE density, $\rho < +\infty$.

NetVisual Analysis

To visualize the fundamental behaviour of the coverage probability, p^{cov}, in a more intuitive manner, Figure 8.5 shows the coverage probability heat map of three different scenarios with different small cell BS densities, i.e. 50, 250 and 2500 BSs/km^2, while incrementally considering the key channel characteristics and network features presented in this book. As in Chapters 3–7, these heat maps are computed using

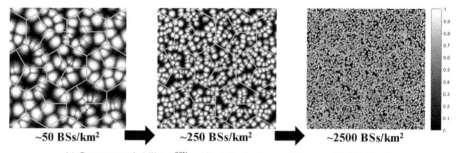

(a) Coverage probability, p^{cov}, under a single-slope path loss model, NLoS only.

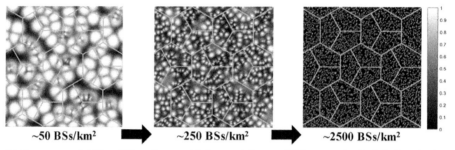

(b) Coverage probability, p^{cov}, under a multi-slope path loss model incorporating both LoS and NLoS transmissions.

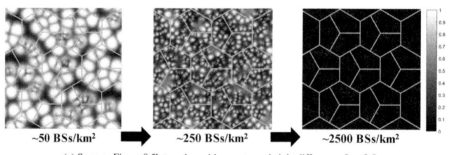

(c) Same as Figure 8.5b together with an antenna height difference, $L = 8.5$ m.

Figure 8.5 NetVisual plot of the coverage probability, p^{cov}, versus the small cell BS density, λ.

NetVisual. This tool is able to capture, not only the mean, but also the standard deviation of the coverage probability, p^{cov}, as explained in Section 2.4.

For clarity, note that Figures 8.5 (a–e) present the results of the configurations, Conf. a–e), respectively, and that the first four configurations were already presented – and exhaustively discussed – in Chapters 3–6.

From Figure 8.5, we can see that,

- compared with Figure 8.5c – the case with a multi-slope path loss function incorporating both LoS and NLoS transmissions and an antenna height difference, $L = 8.5$ m – the more complete system model considered in Figure 8.5e, which in addition accounts for a finite UE density, $\rho = 300$ UEs/km^2, and an idle mode

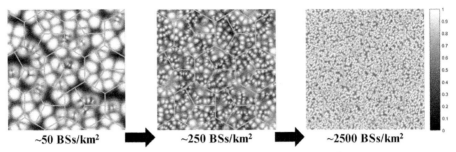

(d) Same as Figure 8.5b together with a finite UE density, $\rho = 300$ UEs/km^2, and a simple idle mode capability at the small cell BSs.

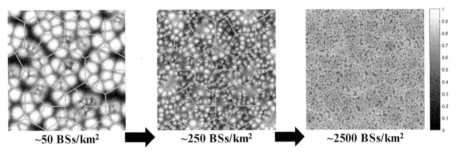

(e) Same as Figure 8.5b together with an antenna height difference, $L = 8.5$ m, a finite UE density, $\rho = 300$ UEs/km^2, and an idle mode capability at the small cell BSs.

Figure 8.5 NetVisual plot of the coverage probability, p^{cov}, versus the small cell BS density, λ. (cont.)

capability at the small cell BSs, produces a more optimistic result in the ultra-dense regime. The SINR heat map in Figure 8.5e becomes brighter when the small cell BS density, λ, is around 250 BSs/km^2, and even brighter when $\lambda = 2500$ BSs/km^2. This shows the performance improvement due to the finite UE density, $\rho < +\infty$, and the simple idle mode capability at the small cell BSs. As an example, for the case with a small cell BS density, $\lambda = 2500$ BSs/km^2, the average and the standard deviation of the coverage probability, p^{cov}, go from 0.046 and 0.054 in the former to 0.70 and 0.10 in the latter, respectively – a 14.22× improvement in mean.

- Compared with Figure 8.5d – the case with a multi-slope path loss function incorporating both LoS and NLoS transmissions, a finite UE density, $\rho = 300$ UEs/km^2, and a simple idle mode capability at the small cell BSs – the more complete system model considered in Figure 8.5e, which in addition accounts for an antenna height difference, $L = 8.5$ m, produces a more pessimistic result in the ultra-dense regime. The SINR heat map in Figure 8.5e is darker when the small cell BS density, λ, is around $\lambda = 2500$ BSs/km^2. This shows the performance hit brought about by the antenna height difference, $L > 0$. Numerically speaking, for the case with a small cell BS density, $\lambda = 2500$ BSs/km^2, the average and the standard deviation of the coverage probability, p^{cov}, go from 0.79 and 0.13 in the former to 0.70 and 0.10 in the latter, respectively – a 11.39% degradation in mean.

Summary of Findings: Coverage Probability Remark

Remark 8.2   Considering an antenna height difference, $L > 0$, a finite UE density, $\rho < +\infty$, and a simple idle mode capability at the small cell BSs, the SINR, Γ, of the typical UE becomes constant in the ultra-dense regime. The performance degradation

- generated by the antenna height difference, $L > 0$, because of the cap on the maximum signal power that a UE can receive

is counterbalanced by the performance improvement

- originated by the finite UE density, $\rho < +\infty$, and the simple idle mode capability at the small cell BSs, due to the cap on the maximum inter-cell interference power that a UE can receive.

As a result, the coverage probability, p^{cov}, becomes independent of the small cell BS density, λ, in the ultra-dense regime.

8.5.3   Area Spectral Efficiency Performance

Let us now explore the behaviour of the ASE, A^{ASE}.

Similarly as before, Figure 8.6 shows the ASE, A^{ASE}, for an SINR threshold, $\gamma = 0\,\mathrm{dB}$, for the following five configurations:

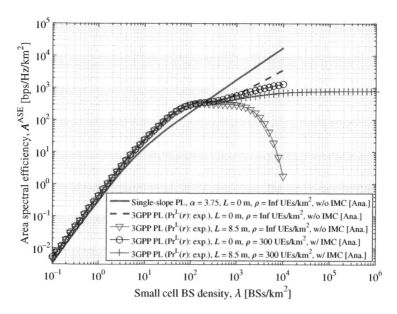

Figure 8.6 Area spectral efficiency, A^{ASE}, versus the small cell BS density, λ.

- Conf. a): single-slope path loss model with no antenna height difference, $L = 0$ m, and an infinite number of UEs – or at least one UE per cell – thus all small cell BSs are active (analytical results).
- Conf. b): multi-slope path loss model with both LoS and NLoS transmissions, no antenna height difference, $L = 0$ m, and an infinite number of UEs – or at least one UE per cell – thus all small cell BSs are active (analytical results).
- Conf. c): multi-slope path loss model with both LoS and NLoS transmissions, an antenna height difference, $L > 0$ m, and an infinite number of UEs – or at least one UE per cell – thus all small cell BSs are active (analytical results).
- Conf. d): multi-slope path loss model with both LoS and NLoS transmissions, no antenna height difference, $L = 0$ m, a finite UE density, $\rho < +\infty$, and a simple idle mode capability at the small cell BSs, thus not all small cell BSs are active (analytical results).
- Conf. e): multi-slope path loss model with both LoS and NLoS transmissions, an antenna height difference, $L > 0$ m, a finite UE density, $\rho < +\infty$, and a simple idle mode capability at the small cell BSs, thus not all small cell BSs are active (analytical results).

From Figure 8.6, and with respect to the configuration, Conf. e), we can observe that,

- when the network is sparse, i.e. $\lambda \leq 10^2$ BSs/km^2, the ASE, A^{ASE}, quickly increases with the small cell BS density, λ, because the network is generally noise-limited, and thus adding more and more small cell BSs significantly benefits the spatial reuse.
- When the small cell BS density, λ, gets denser, and around the range, $\lambda \in (10^2, 10^3]$ BSs/km^2, the ASE, A^{ASE}, shows a slowdown in the rate of growth or even a decrease as the small cell BS density, λ, increases. This is driven by the decrease of the coverage probability, p^{cov}, at these small cell BS densities, caused by both the transition of a large number of inter-cell interfering paths from NLoS to LoS, and the antenna height difference, $L > 0$, in very dense networks. These two phenomena – the ASE Crawl and the ASE Crash – have been extensively discussed in Chapter 4 and Chapter 5, respectively.
- When the small cell BS density, λ, gets even denser, around the range, $\lambda \in (10^3, 10^5]$ BSs/km^2, the ASE, A^{ASE}, picks up its growth rate with the increase of the small cell BS density, λ. This is driven by the increase of the coverage probability, p^{cov}, at these small cell BS densities, originated by the finite UE density, $\rho < +\infty$, and the simple idle mode capability at the small cell BSs. This phenomenon – the ASE Climb – has been extensively discussed in Chapter 6.
- Importantly, when the small cell BS density, λ, is ultra-dense, i.e. $\lambda > 10^5$ BSs/km^2, the ASE, A^{ASE}, gradually flattens and reaches a limit. This result verifies Theorem 8.4.4 and the discussed new constant capacity scaling law, which stated that the asymptotic ASE, $\lim_{\lambda \to +\infty} A^{\mathrm{ASE}}$, is independent of the small

cell BS density, λ. For this particular example, with an antenna height difference, $L = 8.5\,\text{m}$, and a finite UE density, $\rho = 300\,\text{UEs/km}^2$, such a limit is equal to $773.7\,\text{bps/Hz/km}^2$.

Let us explore in the following the behaviour of – and some important aspects resulting from – this new constant capacity scaling law by analyzing multiple UE densities.

Studying Various UE Densities

Figure 8.7 further explores the ASE, A^{ASE}, paying special attention, this time, to the impact of various UE densities, $\rho = \{100, 300, 600, 800, 1000, 2000, 3000, 6000\}$ UEs/km^2.

From Figure 8.7, we can see that,

- for all UE densities, sooner or later, in terms of the small cell BS density, λ, the ASE, A^{ASE}, gradually flattens and reaches a limit. This further verifies Theorem 8.4.4.
- Importantly, when the small cell BS density, λ, is dense, e.g. $\lambda > 10^3\,\text{BSs/km}^2$, and for a given small cell BS density, λ, the ASE, A^{ASE}, has a concave shape with respect to the UE density, ρ. In other words, it has a maximum. In this particular case, for a small cell BS density, $\lambda = 10^6\,\text{BSs/km}^2$, the ASE, A^{ASE}, achieves its maximum value, $928.2\,\text{bps/Hz/km}^2$, at a UE density, $\rho = 803.7\,\text{UEs/km}^2$.

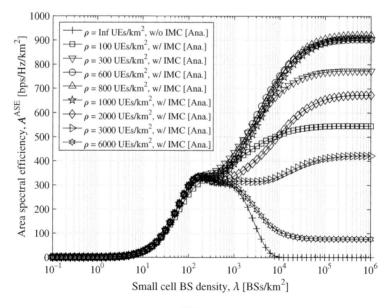

Figure 8.7 Area spectral efficiency, A^{ASE}, versus the small cell BS density, λ, for various UE densities, $\rho < +\infty$.

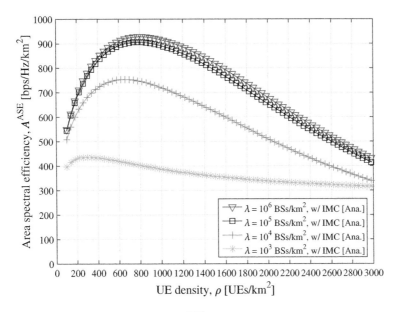

Figure 8.8 Area spectral efficiency, A^{ASE}, versus the UE density, ρ, for various small cell BS densities, λ.

To illustrate such a concave function more clearly, Figure 8.8 shows the ASE, A^{ASE}, with respect to the UE density, ρ, for various small cell BS densities, $\lambda = \{10^3, 10^4, 10^5, 10^6\}$ BSs/km^2. In this figure, we can easily see how for such various small cell BS densities, the ASE, A^{ASE}, is concave with respect to the UE density, ρ, having a maximum value, $\{435.3, 753.6, 905.6, 928.2\}$ bps/Hz/km^2, at a UE density, $\rho = \{285.1, 655.4, 783.6, 803.7\}$ UEs/km^2, respectively.

Summary of Findings: Area spectral Efficiency Remark

Remark 8.3   Considering an antenna height difference, $L > 0$, a finite UE density, $\rho < +\infty$, and a simple idle mode capability at the small cell BSs, the ASE, A^{ASE}, neither decreases nor increases with the increase of the small cell BS density, λ, but becomes constant in the ultra-dense regime. The negative impact of the ASE Crawl and the ASE Crash and the positive impact of the ASE Climb *cancel each other out*. As a result, there is no incentive to continue densifying the network once the ASE, A^{ASE}, reaches its maximum and flattens.

8.6   Conclusions

In this chapter, we have analyzed the scaling laws of both the coverage probability, p^{cov}, and the ASE, A^{ASE}, with respect to the small cell BS density, λ, while accounting for a complete system model, considering the key and most relevant channel characteristics and network features of an ultra-dense deployment according to the findings of the previous chapters, namely

- both LoS and NLoS transmissions,
- a practical user association strategy, based on the strongest received signal strength,
- the height of the antenna of the typical UE and those of the small cell BSs,
- a finite UE density and
- a simple idle mode capability at the small cell BSs.

Importantly, the scaling laws resulting from this new study have also been compared with those derived in Chapters 3–6 of this book, where only a subset of these channel characteristics and network features were considered. Compared to the state of the art, this comparison reveals a significantly different understanding of the performance of a small cell network in the ultra-dense regime, "moving" it

- from the just deploy more small cells and everything will be fine – in Chapter 3 – because the ASE, A^{ASE}, linearly increased with the increase of the small cell BS density, λ,
- to the watch out with your small cell deployments idea – in this chapter – since the ASE, A^{ASE}, reaches a maximum independent of the small cell BSs density, and thus deploying more small cell BSs after a given densification level is just a waste of both invested money and consumed energy.

Overall, a significantly different message, which has great implications for network operators and service providers and brings the spotlight to the importance of network planning and optimization. For reference, the most important conclusions from this study are summarized in Remarks 8.2 and 8.3.

Finally, and from a theoretical viewpoint, let us also use the conclusions of this chapter to bring the attention to the importance of an appropriate modelling in general, accounting for the right channel characteristics and network features, to extract the right conclusions. Otherwise, the analysis may be after all misleading. A clear example is the results of Chapter 3, which are appropriate for a sparse network, but do not apply to an ultra-dense one.

Appendix A Proof of Theorem 8.4.1

In the following, we sketch the proof of Theorem 8.4.1.

As the small cell BS density, λ, tends to infinity, i.e. $\lambda \to +\infty$, we have that

- the two-dimensional distance, r, from the typical UE to its serving BS tends to 0, i.e. $\lim_{\lambda \to +\infty} r = 0$, and thus
- the three-dimensional distance, d, from the typical UE to its serving small cell BS tends to the antenna height difference, i.e. $\lim_{\lambda \to +\infty} d = \lim_{r \to +\infty} \sqrt{r^2 + L^2} = L$.

As a consequence, the path loss $\zeta(L)$, of this link should be dominantly characterized by the first-piece LoS path loss function, i.e. $\zeta_1^L(L)$, according to the formulation in equation (5.2). For a more concrete example, the threshold of the first-piece of the LoS path loss function is $d_1 = 67.75$ m in the 3GPP case study of this book [153], which is larger than the values that the antenna height difference, L, usually takes. In [153], the antenna height difference is $L = 8.5$ m, and thus the first-piece LoS path loss function, $\zeta_1^L(L)$, should be used in such a case.

With this in mind, the limit, $\lim\limits_{\lambda \to +\infty} p^{\text{cov}}(\lambda, \gamma_0)$, can be derived as

$$
\lim_{\lambda \to +\infty} p^{\text{cov}}(\lambda, \gamma_0) = \lim_{\lambda \to +\infty} \Pr\left[\text{SINR} > \gamma_0 \,\middle|\, \zeta(w) = \zeta_1^L(L) \right]
$$

$$
\overset{(a)}{=} \lim_{\lambda \to +\infty} \Pr\left[\frac{P\zeta_1^L(L)\,h}{I_{\text{agg}} + P^N} > \gamma_0 \right]
$$

$$
= \lim_{\lambda \to +\infty} \Pr\left[h > \frac{(I_{\text{agg}} + P^N)\,\gamma_0}{P\zeta_1^L(L)} \right]
$$

$$
\overset{(b)}{=} \lim_{\lambda \to +\infty} \Pr\left[h > \frac{P^N \gamma_0}{P\zeta_1^L(L)} \right] \Pr\left[h > \frac{I_{\text{agg}} \gamma_0}{P\zeta_1^L(L)} \right]
$$

$$
= \exp\left(-\frac{P^N \gamma_0}{P\zeta_1^L(L)} \right) \lim_{\lambda \to +\infty} \mathbb{E}_{[I_{\text{agg}}]}\left\{ \exp\left(-\frac{I_{\text{agg}} \gamma_0}{P\zeta_1^L(L)} \right) \right\}
$$

$$
\overset{(c)}{=} \exp\left(-\frac{P^N \gamma_0}{P\zeta_1^L(L)} \right) \lim_{\lambda \to +\infty} \mathscr{L}_{I_{\text{agg}}}^L(s), \tag{8.20}
$$

where

- the definition of the SINR of the typical UE in equation (3.3) is plugged into step (a) of equation (8.20) and
- step (b) is derived from the complementary cumulative distribution function (CCDF) of the multi-path fast fading gain random variable, h. In more detail, we assumed that the multi-path fast fading gain random variable, h, is an exponentially distributed random variable – Rayleigh fading – and thus, we can write its CCDF as

$$
\bar{F}_H(h) = \Pr[H > h] = \exp(-h). \tag{8.21}
$$

From the distribution of this random variable, h, we can further infer that

$$
\bar{F}_H(x_1 + x_2) = \Pr[H > x_1 + x_2]
$$
$$
= \exp(-x_1)\exp(-x_2)
$$
$$
= \Pr[H > x_1]\Pr[H > x_2]. \tag{8.22}
$$

- Finally, step (c) follows from the definition of the Laplace transform, $\mathscr{L}_{I_{\text{agg}}}^L(s)$, of the aggregated inter-cell interference random variable, I_{agg}, for an LoS signal transmission evaluated at the variable value, $s = \frac{\gamma_0}{P\zeta_1^L(L)}$.

Importantly, the following intuition can be derived from the result in step (c) of equation (8.20):

- The exponential factor, $\exp\left(-\frac{P^N \gamma_0}{P\zeta_1^L(L)}\right)$, measures the probability that the signal power exceeds the noise power by a factor of at least γ_0, while
- the limit, $\lim\limits_{\lambda \to +\infty} \mathscr{L}_{I_{agg}}^L(s)$, measures the probability that the signal power exceeds the aggregated inter-cell interference power by a factor of at least γ_0.

It should be noted that, in practice, the exponential factor, $\exp\left(-\frac{P^N \gamma_0}{P\zeta_1^L(L)}\right)$, is a value very close to 1, because modern wireless communication systems are usually interference-limited and not noise-limited, i.e. $P\zeta_1^L(L) \gg P^N$.

With regard to the remaining Laplace transform, $\mathscr{L}_{I_{agg}}^L(s)$, we can further develop it from its definition in equation (2.46) as

$$
\mathscr{L}_{I_{agg}}^L(s) = \mathbb{E}_{[I_{agg}]}\left\{\exp\left(-sI_{agg}\right)\right\}
$$

$$
\overset{(a)}{=} \mathbb{E}_{\left[\tilde{\Phi}\backslash b_o, \{\beta_i\}, \{g_i\}\right]}\left\{\exp\left(-s\sum_{i\in\Phi/b_o} P\beta_i g_i\right)\right\}
$$

$$
\overset{(b)}{=} \exp\left(-2\pi\tilde{\lambda}\int_0^{+\infty}\left(1 - \mathbb{E}_{[g]}\left\{\exp\left(-sP\beta\left(\sqrt{u^2+L^2}\right)g\right)\right\}\right)u\,du\right)
$$

$$
\overset{(c)}{=} \exp\left(-2\pi\tilde{\lambda}\int_0^{+\infty}\frac{\Pr^L\left(\sqrt{u^2+L^2}\right)u}{1+\left(sP\zeta^L\left(\sqrt{u^2+L^2}\right)\right)^{-1}}du\right)
$$

$$
\times \exp\left(-2\pi\tilde{\lambda}\int_0^{+\infty}\frac{\left[1 - \Pr^L\left(\sqrt{u^2+L^2}\right)\right]u}{1+\left(sP\zeta^{NL}\left(\sqrt{u^2+L^2}\right)\right)^{-1}}du\right), \tag{8.23}
$$

where

- the definition of the aggregated inter-cell interference, I_{agg}, in equation (3.4) is plugged into step (a) of equation (8.23),
- step (b) is derived from the definition of the probability generating functional (PGFL) of a Poisson point process (PPP) in equation (2.36), which is a direct consequence of Campbell's theorem (see Section 2.3.1) and
- step (c) follows from the change,

$$
\mathbb{E}_{[g]}\left\{\exp\left(-sxg\right)\right\} = \int_0^{+\infty}\exp\left(-sxg\right)\exp\left(-g\right)dg = \frac{1}{1+sx}, \tag{8.24}
$$

where the PDF, $f_H(h)$, of the Rayleigh multi-path fast fading gain, h, has been substituted, i.e. $f_H(h) = \exp(-h)$.

Importantly, it should be noted that, in step (c) of (8.23), the inter-cell interference from both the LoS and the NLoS paths are considered, as in equation (4.13).

Finally, from equation (6.22), we can obtain the result, $\lim\limits_{\lambda \to +\infty} \tilde{\lambda} = \rho$, which leads to the following closed-from expression for the limit, $\lim\limits_{\lambda \to +\infty} \mathscr{L}^{L}_{I_{agg}}(s)$, i.e.

$$
\begin{aligned}
\lim_{\lambda \to +\infty} \mathscr{L}^{L}_{I_{agg}}(s) &= \lim_{\lambda \to +\infty} \exp\left(-2\pi\tilde{\lambda} \int_{0}^{+\infty} \frac{\Pr^{L}\left(\sqrt{u^2 + L^2}\right) u}{1 + \left(sP\zeta^{L}\left(\sqrt{u^2 + L^2}\right)\right)^{-1}} du\right) \\
&\times \exp\left(-2\pi\tilde{\lambda} \int_{0}^{+\infty} \frac{\left[1 - \Pr^{L}\left(\sqrt{u^2 + L^2}\right)\right] u}{1 + \left(sP\zeta^{NL}\left(\sqrt{u^2 + L^2}\right)\right)^{-1}} du\right) \\
&= \exp\left(-2\pi\rho \int_{0}^{+\infty} \frac{\Pr^{L}\left(\sqrt{u^2 + L^2}\right) u}{1 + \left(sP\zeta^{L}\left(\sqrt{u^2 + L^2}\right)\right)^{-1}} du\right) \\
&\times \exp\left(-2\pi\rho \int_{0}^{+\infty} \frac{\left[1 - \Pr^{L}\left(\sqrt{u^2 + L^2}\right)\right] u}{1 + \left(sP\zeta^{NL}\left(\sqrt{u^2 + L^2}\right)\right)^{-1}} du\right). \quad (8.25)
\end{aligned}
$$

Combining equations (8.20) and (8.25) yields the proof.

Appendix B Proof of Lemma (8.4.3)

In the following, we sketch the proof of Lemma (8.4.3).

In equation (8.14), two observations can be done when fixing both the antenna height difference, $L \geq 0$, and the SINR threshold, γ_0:

- The exponential term, $\exp\left(-\frac{P^{N}\gamma_0}{P\zeta^{L}_{1}(L)}\right)$, can be redefined by a function, $c(\gamma_0)$, where, in practice, such a function, $c(\gamma_0)$, approximates to 1, i.e. $\exp\left(-\frac{P^{N}\gamma_0}{P\zeta^{L}_{1}(L)}\right) \approx 1$, because the inequality, $P\zeta^{L}_{1}(L) \gg P^{N}$, holds in an interference-limited ultra-dense network.

- The limit, $\lim\limits_{\lambda \to +\infty} \mathscr{L}^{L}_{I_{agg}}\left(\frac{\gamma_0}{P\zeta^{L}_{1}(L)}\right)$, can be reformulated by a function, $g^{\rho}(\gamma)$, presented in equation (8.18), where the UE density, $\rho < +\infty$, has been pulled out from the exponential terms in equation (8.25).

Importantly, it should be noted that the function, $g(\gamma)$, ranges from 0 to 1 because the arguments inside the exponential terms in equation (8.18) are negative values, since the integrals are all performed over positive values. Thus, we can conclude that the limit, $\lim\limits_{\lambda \to +\infty} p^{\text{cov}}(\lambda, \gamma_0)$, decreases with the increase of the finite UE density, $\rho < +\infty$ increases, according to a power law.

Therefore, we can safely conclude that equation (8.14) decreases with the increase of the finite UE density, $\rho < +\infty$.

Overall, the intuition is that the bounded inter-cell interference power increases as the finite UE density, $\rho < +\infty$, increases.

Appendix C Proof of Theorem 8.4.4

In the following, we sketch the proof of Theorem 8.4.4.

As the small cell BS density, λ, tends to infinity, i.e. $\lambda \to +\infty$, we can show that the ASE, A^{ASE}, in equation (6.24) approaches a limit that is independent of the small cell BS density, λ. This is because when taking such a limit over equation (6.24), we have that

- the limit, $\lim_{\lambda \to +\infty} \tilde{\lambda}$, tends to the UE density, ρ, i.e. $\lim_{\lambda \to +\infty} \tilde{\lambda} = \rho$, as explained in Lemma (6.3.6), and thus it is independent of the small cell BS density, λ.
- The limit, $\lim_{\lambda \to +\infty} p^{\text{cov}}$, is also independent of the small cell BS density, λ, as was shown by Theorem 8.4.1.

As a result, the limit, $\lim_{\lambda \to +\infty} A^{\text{ASE}}$, in equation (8.19) is independent of the small cell BS density, λ.

This completes our proof.

9   System-Level Network Optimization

9.1   Introduction

As discussed in the introduction of this book (Chapter 1), from the year 1950 to the year 2000, the wireless network capacity increased around one million-fold. During this period, most of the gains – a $2700\times$ capacity increase – were achieved through an aggressive spatial spectrum reuse, via network densification, using smaller and smaller cells, which went from a radius of a few kilometres to hundreds of metres [11]. Generally speaking, spatial spectrum reuse is the term used to refer to those scenarios where multiple cells within an area of interest simultaneously reuse a spectrum resource. As also explained in Chapter 1, if the spatial spectrum reuse linearly increases, i.e. if the number of cells in the area of interest reusing such a given chunk of spectrum linearly increases, the wireless network capacity also has the potential to linearly increase, provided that the signal qualities of the user equipment (UE) do not decrease. In other words, every cell can make an independent and equal contribution to the network capacity, while reusing the same spectrum, if the inter-cell interference does not increase with the deployment of more cells. This is the reasoning behind the conclusions of Chapter 3, and the aforementioned $2700\times$ capacity gain stands as a testimony to the fulfilment of such potential.

As we head down the path of network densification and gradually enter the realm of ultra-dense networks, however, things start to deviate from this traditional and optimistic understanding, in which the area spectral efficiency (ASE), A^{ASE}, linearly increases with the increase of the small cell base station (BS) density, λ. As carefully analyzed in Chapter 8, Theorem 8.4.2 – and its presented constant capacity scaling law – indicate that the ASE, A^{ASE}, will not linearly increase with the small cell BS density, λ, in the ultra-dense regime. Instead, it will asymptotically reach a constant value, which is independent of it, and a function, among others, of the channel characteristics and the UE density, ρ. As a consequence, and plainly speaking, for a set of channel characteristics and network features, densifying the network beyond a certain small cell BS density, λ, will be a waste of both invested money and energy consumption.

Equipped with this understanding, a number of fundamental questions and optimization problems arise with regards to the deployment and operation on an ultra-dense network. For example:

- *Small cell BS deployment/activation:*
  For a given UE density, ρ, which is the optimal small cell BS density, λ^*, that can maximize the ASE, A^{ASE}?
- *Network-wide UE admission/scheduling:*
  For a given small cell BS density, λ, which is the optimal UE density, ρ^*, that can maximize the ASE, A^{ASE}?
- *Spatial spectrum reuse:*
  For a given UE density, ρ, and for a given small cell BS density, λ, we ask whether activating all small cell BSs on the same time/frequency resource is the best strategy, as has been practised during the last half century, with a universal frequency reuse factor of 1, or is there an optimal frequency reuse strategy that can maximize the ASE, A^{ASE}?

In this chapter, we develop these three questions further, and provide an answer to them via different theoretical analyses, using the results obtained in Chapter 8.

The rest of this chapter is organized as follows:

- Section 9.2 explores the small cell base station deployment/activation optimization problem, while
- Section 9.3 analyzes the network-wide UE admission/scheduling optimization problem, and
- Section 9.4 studies the spatial spectrum reuse optimization problem.
- Section 9.5, finally, summarizes the key takeaways of this chapter.

9.2   Small Cell Base Station Deployment/Activation

From Theorem 8.4.4 and the results in Section 8.5.3, it was concluded that, for a given UE density, ρ, the network densification should not be indefinitely abused. Instead, it should be *stopped* at a certain small cell BS density, λ. This is because both the coverage probability, p^{cov}, and the ASE, A^{ASE}, will asymptotically reach a constant value, and any network densification beyond such a level is a waste of both invested money and energy consumption.

Given this result, the following small cell BS deployment problem can be naturally formulated to find an optimal small cell BS density, λ^*, which effectively realizes most of the capacity gains.

DEFINITION 9.2.1   For a given UE density, ρ, there exists an optimal small cell BS density, λ^*, which can achieve an ASE, A^{ASE}, that is within a relative performance difference, ϵ, from the asymptotic ASE, $\lim\limits_{\lambda \to +\infty} A^{\text{ASE}}$, i.e.

$$\underset{\lambda}{\text{maximize}} 1$$

$$\text{s.t.} \frac{\left| \lim\limits_{\lambda \to +\infty} A^{\text{ASE}}(\lambda, \gamma_0) - A^{\text{ASE}}(\lambda, \gamma_0) \right|}{\lim\limits_{\lambda \to +\infty} A^{\text{ASE}}(\lambda, \gamma_0)} = \epsilon. \qquad (9.1)$$

Importantly, let us differentiate between the presented deployment optimization problem and the equivalent activation optimization problem for completeness hereafter

- By deployment problem, we consider the cases where the network operator has either a green field with no deployment or a scenario with some already deployed small cell BSs, and such a network operator is willing to expand its capacity by deploying more small cell BSs.
- By activation problem, we refer to the case where the network operator has a given deployment, and due to, for example, traffic variations across time, is willing to reconfigure the active small cell BSs density, $\tilde{\lambda}$, to meet a given network capacity demand at a given point in time.

These two problems can be seen as different sides of the same coin, and can be solved by the above formulated problem, conceptually substituting the small cell BSs density, λ, in the former by the active small cell BSs density, $\tilde{\lambda}$, in the latter.

Importantly, the solution to this small cell BS deployment/activation problem can provide good guidance to network operators when deciding the densification level of their networks. Any densification further than the optimal small cell BS density, λ^*, will not provide more than a gain, ϵ, with respect to the asymptotic ASE, $\lim\limits_{\lambda \to +\infty} A^{\mathrm{ASE}}$.

More generally, the solution to this problem answers an important fundamental question:

> For a given UE density, ρ, how dense an ultra-dense network should be in terms of the ASE, A^{ASE}?

9.2.1   Problem Solution

As can be derived from equation (9.1), this small cell BS deployment/activation problem has an intricate form due to the non-trivial formulation of both the ASE, A^{ASE}, in Section 8.3.2 and the asymptotic ASE, $\lim\limits_{\lambda \to +\infty} A^{\mathrm{ASE}}$, in equation (8.19). However, as for such a problem definition, the optimal small cell BS density, λ^*, can be found using a numerical search over the expression of the former.

Algorithm 1 illustrates a simple but yet practical algorithm, based on bisection, to calculate such an optimal small cell BS density, λ^*. It should be noted that this algorithm is provided for guidance, and that more sophisticated ones apply.

Importantly, it should also be noted that, for the large small cell BS densities in the ultra-dense regime, where the ASE, A^{ASE}, monotonically increases with the increase of the small cell BS density, λ, the found optimal small cell BS density, λ^*, is unique, and can be used to characterize the maximum network capacity with a relative performance difference, ϵ, from the ASE limit, $\lim\limits_{\lambda \to +\infty} A^{\mathrm{ASE}}$ (λ, γ_0).

Algorithm 1 Example algorithm to obtain the optimal small cell BS density, λ^*.

Step 1: Initialization

- Find a sufficiently large small cell BS density, λ^{right}, which meets the following inequality, $\dfrac{\left| \lim\limits_{\lambda \to +\infty} A^{\text{ASE}}(\lambda, \gamma_0) - A^{\text{ASE}}(\lambda^{\text{right}}, \gamma_0) \right|}{\lim\limits_{\lambda \to +\infty} A^{\text{ASE}}(\lambda, \gamma_0)} < \epsilon.$

- Initialize the following variables, $\lambda^{\text{left}} = 0$ and $\lambda^{\text{mid}} = \dfrac{\lambda^{\text{left}} + \lambda^{\text{right}}}{2}$.

Step 2: Processing

- Compute the ASE, $A^{\text{ASE}}\left(\lambda^{\text{mid}}, \gamma_0\right)$, using equation (8.14).

- If the following inequality is met, $\dfrac{\left| \lim\limits_{\lambda \to +\infty} A^{\text{ASE}}(\lambda, \gamma_0) - A^{\text{ASE}}\left(\lambda^{\text{mid}}, \gamma_0\right) \right|}{\lim\limits_{\lambda \to +\infty} A^{\text{ASE}}(\lambda, \gamma_0)} > \epsilon,$ update

  the variable, λ^{left}, as follows, $\lambda^{\text{left}} = \lambda^{\text{mid}}$.
  Otherwise, update the variable, λ^{right}, as follows, $\lambda^{\text{right}} = \lambda^{\text{mid}}$.

Step 3: Termination

- If the following inequalities are met,

  $(1 - \delta_0)\,\epsilon < \dfrac{\left| \lim\limits_{\lambda \to +\infty} A^{\text{ASE}}(\lambda, \gamma_0) - A^{\text{ASE}}\left(\lambda^{\text{mid}}, \gamma_0\right) \right|}{\lim\limits_{\lambda \to +\infty} A^{\text{ASE}}(\lambda, \gamma_0)} < (1 + \delta_0)\,\epsilon,$ where the threshold,

  δ_0, sets a precision condition to terminate the numerical search, and takes a small value, e.g. 10^{-3}, go to Step 4.
  Otherwise, go to Step 2.

Step 4: Output

- Set the optimal small cell BS density, λ^*, as $\lambda^* = \lambda^{\text{mid}}$.

9.2.2   Example and Discussion of Results

In this section, we further leverage the performance evaluation framework used in Chapter 8 – and in more detail, in Section 8.5 – to analyze the presented small cell BS deployment/activation optimization problem.

For completeness, let us recall that the following parameters are used in this, our 3rd Generation Partnership Project (3GPP) case study:

- maximum antenna gain, $G_{\text{M}} = 0\,\text{dB}$,
- path loss exponents, $\alpha^{\text{L}} = 2.09$ and $\alpha^{\text{NL}} = 3.75$ – considering a carrier frequency of $2\,\text{GHz}$,
- reference path losses, $A^{\text{L}} = 10^{-10.38}$ and $A^{\text{NL}} = 10^{-14.54}$,
- transmit power, $P = 24\,\text{dBm}$, and
- noise power, $P^{\text{N}} = -95\,\text{dBm}$ – including a noise figure of $9\,\text{dB}$ at the UE.

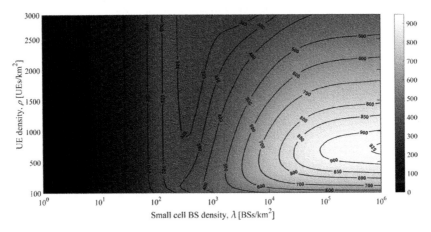

Figure 9.1 ASE, A^{ASE}, for various values of the small cell BS density, λ, and the UE density, ρ.

With this in mind, Figure 9.1 plots a two-dimensional map of the ASE, A^{ASE}, with respect to both the small cell BS, λ, and the UE density, ρ, using a signal-to-interference-plus-noise ratio (SINR) threshold, $\gamma = 0\,\text{dB}$, and an optimal fitting parameter, $q^* = 4.18$, for the most complete system model presented in the previous Chapter 8, i.e.

- Conf. e): multi-slope path loss model with both line-of-sight (LoS) and non-line-of-sight (NLoS) transmissions, an antenna height difference, $L > 0\,\text{m}$, a finite UE density, $\rho < +\infty$, and a simple idle mode capability at the small cell BSs, thus not all small cell BSs are active (analytical results).

It should also be noted that, for the particular results in Figure 9.1, an antenna height difference, $L = 8.5\,\text{m}$, is adopted, and that this plot comprehends the information already presented in Figures 8.6 and 8.7 of Chapter 8, further enhancing it with more sampling points.

From Figure 9.1, we can see that, when the small cell BS density, λ, is ultra-dense, i.e. $\lambda > 10^5\,\text{BSs/km}^2$, for a given UE density, ρ, the ASE, A^{ASE}, gradually saturates and reaches a limit with the increase of the small cell BS density, λ. This guarantees the convergence of Algorithm 1 to the optimal small cell BS density, λ^*, as discussed. This result is in line with those of Chapter 8, verifies Theorem 8.4.4 and the discussed new constant capacity scaling law, which stated that the asymptotic ASE, $\lim\limits_{\lambda \to +\infty} A^{\text{ASE}}$, is independent of the small cell BS density, λ, under the conditions of study.

As for the presented small cell BS deployment/activation optimization problem, setting a relative performance difference, $\epsilon = 0.05$, and using Algorithm 1 to numerically search the solution to Problem 9.2.1 over the presented results of the ASE, A^{ASE}, in Figure 9.1, it can be found that, for the finite UE density, $\rho = 300\,\text{UEs/km}^2$, the optimal small cell BS density is $\lambda^* = 29080\,\text{BSs/km}^2$. This solution leads to an ASE, $A^{\text{ASE}} = 740.5\,\text{bps/Hz/km}^2$, which is exactly 5.0% away from its asymptotic ASE,

Table 9.1. Small cell BS deployment/activation optimization problem

		Results	
UE density ρ	Small cell BS density λ^*	ASE A^{ASE}	Asymptotic ASE $\lim_{\lambda \to +\infty} A^{ASE}$
[UEs/km^2]	[BSs/km^2]	[bps/Hz/km^2]	[bps/Hz/km^2]
100	15,238	521.0	548.4
300	29,080	740.5	773.7
600	42,227	864.5	910.0
800	49,156	881.8	928.1
1000	55,337	870.0	915.8
2000	75,828	645.8	679.8

$\lim_{\lambda \to +\infty} A^{ASE} = 773.7$ bps/Hz/km^2. Table 9.1 presents equivalent results for various other UE densities.

9.3   Network-Wide User Equipment Admission/Scheduling

As shown in Chapter 5, in the presence of a very large – or an infinite – UE density, $\rho = \infty$, scattered across an area, the network cannot opportunistically take advantage of any inter-cell interference mitigation through a simple idle mode capability at the small cell BSs, and the ASE, A^{ASE}, crashes towards 0 when the network enters the ultra-dense regime, mostly due to the consequences unleashed by the antenna height difference, L.

It is important to note, however, that even in the presence of a very large – or an infinite – UE density, $\rho = \infty$, the ASE crash can still be avoided by taking proactive intelligent admission control and/or scheduling decisions. For example, time division multiple access (TDMA) and/or frequency division multiple access (FDMA) can be used to

- divide this very large set of UEs into smaller subsets, each with a finite and moderate UE density, ρ', and
- schedule them into the available time and/or frequency resources.

Through a proper dimensioning and scheduling of these subsets of UEs on the available time and/or frequency resources, an optimal network capacity can be achieved, as for the logic of Theorem 8.4.4.

This strategy can be seen as – or involves – a network-wide admission control or scheduling policy, depending on the time frame in which the decision-making is performed. In other words, instead of letting the network capacity crash trying to simultaneously serve a very large – or an infinite – UE density, $\rho = \infty$, the constant capacity scaling law in Theorem 8.4.4 indicates that the network could proactively choose – and serve – a subset of UEs on each time and/or frequency resource to maximize capacity. This leads to a finite UE density, $\rho < +\infty$ – and in turn, to a finite active small cell BS density, $\tilde{\lambda}$ – on each time and/or frequency resource, which provides an opportunistic but controlled inter-cell interference mitigation, helps to avoid the ASE crash and maximizes the capacity.

To find such an optimal finite UE density, $\rho^* < +\infty$, per time and/or frequency resource, we build on our previous observations in Section 8.5.3 (see Figure 8.8), indicating that the asymptotic ASE, $\lim\limits_{\lambda \to +\infty} A^{ASE}$, is a concave function with respect to the UE density, ρ, and define the following network-wide UE admission/scheduling problem.

DEFINITION 9.3.1 For a given small cell BS density, λ, there exists an optimal finite UE density, $\rho^* < +\infty$, which can maximize the asymptotic ASE, $\lim\limits_{\lambda \to +\infty} A^{ASE}$, i.e.

$$\underset{\rho}{\text{maximize}}\, A^{ASE}(\lambda, \gamma_0). \tag{9.2}$$

The solution to this network-wide UE admission/scheduling problem can be subject to some specific constraints, which we do not discuss in this chapter, but in a general sense, such a solution provides good guidance to network operators for the functioning of their dense networks, indicating the optimal finite UE density, $\rho^* < +\infty$, per time and/or frequency resource to maximize the ASE, A^{ASE}. Let us reemphasize again that the solution to this problem will lead to the acceptance in the network or scheduling of a subset of UEs, i.e. $\rho^* \leq \rho < +\infty$, and as a by-product, some small cell BSs will be opportunistically deactivated through the idle mode capability at the small cell BSs to avoid the potential threatening inter-cell interference overload.

9.3.1 Problem Solution

In the following, and for the sake of completeness, we present the algorithm used in this chapter to solve this network-wide UE admission/scheduling problem, which is illustrated in Algorithm 2. Importantly, it should be noted that other more sophisticated algorithms apply.

First, let us consider the case where the small cell BS density, λ, tends to infinity, i.e. $\lambda \to +\infty$, and recall Lemma (8.4.3) from Chapter 8. Under this consideration, we can rewrite equation (8.19) as

$$\lim_{\lambda \to +\infty} A^{ASE}(\lambda, \gamma_0) = \frac{\rho}{\ln 2} \int\limits_{\gamma_0}^{+\infty} \frac{c(\gamma)g^{\rho}(\gamma)}{1+\gamma} d\gamma$$

$$+ \log_2(1 + \gamma_0)\, \rho c\,(\gamma_0)\, g^{\rho}\,(\gamma_0). \tag{9.3}$$

Then, taking the derivative of the limit, $\lim\limits_{\lambda \to +\infty} A^{ASE}$, with respect to the UE density, ρ, and denoting such a derivative function by the variable, D^{ASE}, it can be found that

---

Algorithm 2 Example algorithm to calculate the optimal UE density, ρ^*.

Step 1: Initialization

- Initialize the following variables, $\rho^{\text{left}} = 0$, $\rho^{\text{right}} = \lambda$, $\rho^{\text{mid}} = \frac{\rho^{\text{left}} + \rho^{\text{right}}}{2}$.

Step 2: Processing

- Compute the derivative ASEs, $D^{\text{ASE}}\left(\rho^{\text{left}}, \gamma_0\right)$, $D^{\text{ASE}}\left(\rho^{\text{right}}, \gamma_0\right)$ and $D^{\text{ASE}}\left(\rho^{\text{mid}}, \gamma_0\right)$, using equation (9.4).
- If the following inequality is met, $D^{\text{ASE}}\left(\rho^{\text{mid}}, \gamma_0\right) > 0$, update the variable, ρ^{left}, as follows, $\rho^{\text{left}} = \rho^{\text{mid}}$.
  Otherwise, update the variable, ρ^{right}, as follows, update $\rho^{\text{right}} = \rho^{\text{mid}}$.

Step 3: Termination

- If the following inequality is met, $\left|D^{\text{ASE}}\left(\rho^{\text{mid}}, \gamma_0\right)\right| < \delta_0$, where the threshold, δ_0, sets a precision condition to terminate the numerical search, and takes a small value, e.g. 10^{-3}, go to Step 4.
  Otherwise, go to Step 2.

Step 4: Output

- Set the optimal UE density, ρ^*, as $\rho^* = \rho^{\text{mid}}$.

$$
\begin{aligned}
D^{\text{ASE}}\left(\rho, \gamma_0\right) &\triangleq \frac{\partial \left[\lim\limits_{\lambda \to +\infty} A^{\text{ASE}}(\lambda, \gamma_0)\right]}{\partial \rho} \\[2mm]
&= \frac{\partial \left[\frac{\rho}{\ln 2} \int_{\gamma_0}^{+\infty} \frac{c(\gamma) g^\rho(\gamma)}{1+\gamma} d\gamma + \log_2\left(1 + \gamma_0\right) \rho c\left(\gamma_0\right) g^\rho\left(\gamma_0\right)\right]}{\partial \rho} \\[2mm]
&= \frac{1}{\ln 2} \int_{\gamma_0}^{+\infty} \frac{c(\gamma) g^\rho(\gamma)\left(1 + \rho \ln g(\gamma)\right)}{1+\gamma} d\gamma \\[2mm]
&\quad + \log_2\left(1 + \gamma_0\right) c\left(\gamma_0\right) g^\rho\left(\gamma_0\right)\left(1 + \rho \ln g\left(\gamma_0\right)\right).
\end{aligned}
\tag{9.4}
$$

Finally, and according to the convex optimization theory [232], we can calculate the maximum of the limit, $\lim\limits_{\lambda \to +\infty} A^{\text{ASE}}$, by setting its derivative to 0, i.e. $D^{\text{ASE}} = 0$. Since the derivative, D^{ASE}, has a closed-form expression, shown in equation (9.4), we can obtain the optimal UE density, ρ^*, using, for example, a bisection search [209].

Unfortunately, the above calculation only holds when the small cell BS density, λ, tends to infinity, i.e. $\lambda \to +\infty$. For a general case, in which the small cell BS density, λ, adopts a finite value, the derivative, D^{ASE}, cannot be expressed by a simple closed-form expression, as in equation (9.4). This is because the coverage probability, p^{cov}, in Theorem 8.3.1 has a much more complicated expression, requiring three folds of integrals to find its solution. The ASE definition in Section 8.3.2 makes things even more complex, adding an extra fold of integrals.

Having said that, it is important to note that,

- for a given small cell BS density, λ, we can still numerically evaluate the derivative, $\frac{\partial[A^{\mathrm{ASE}}]}{\partial\rho}$, and that
- as shown earlier in Figure 9.1. the ASE, A^{ASE}, found in a semi-closed-form expression in equation (8.14), was shown to be convex with respect to the UE density, ρ, in the ultra-dense regime.

Thus, for the general case, in which the small cell BS density, λ, adopts a finite but large value within the ultra-dense regime, we can still reuse Algorithm 2 to compute the optimal UE density, ρ^*, by replacing the derivative, D^{ASE}, with the derivative, $\frac{\partial[A^{\mathrm{ASE}}]}{\partial\rho}$.

9.3.2   Example and Discussion of Results

In this section, we analyze the presented network-wide UE admission/scheduling optimization problem using the same system model and performance results – those of Figure 9.1 – as in Section 9.3.1.

From Figure 9.1, we can see that, for a given small cell BS density, λ, the ASE, A^{ASE}, has a concave shape with respect to the UE density, ρ. This guarantees the convergence of Algorithm 2 to the optimal UE density, ρ^*, as discussed. This result is in line with those of Chapter 8, further extending those presented in Figure 8.8 with more sampling points.

As for the presented network-wide UE admission/scheduling optimization problem and using Algorithm 2 to numerically search the solution to Problem 9.3.1 over the presented results of the ASE, A^{ASE}, in Figure 9.1, it can be found that, for a small cell BS density, $\lambda = 10^6$ BSs/km^2, the optimal UE density is $\rho^* = 803.7$ UEs/km^2. This solution leads to an ASE, $A^{\mathrm{ASE}} = 928.2$ bps/Hz/km^2. Importantly, it should be remarked that simultaneously accepting into the network or scheduling more UEs beyond this optimal UE density, $\rho^* = 803.7$ UEs/km^2, will not generate any gain in terms of ASE, A^{ASE}. Table 9.2 presents equivalent results for various other BS densities.

9.4   Spatial Spectrum Reuse

Compared with the aggressive spatial reuse approach with a universal frequency reuse factor of 1, in which all the active small cell BSs can potentially use at the same time all the frequency resources, let us embrace in this chapter a frequency reuse scheme, where

Table 9.2. Network-wide user equipment scheduling optimization problem

	Results	
Small cell BS density	UE density	ASE
λ	ρ^*	A^{ASE}
[UEs/km^2]	[BSs/km^2]	[bps/Hz/km^2]
10^3	285.1	435.3
10^4	655.4	753.6
10^5	783.6	905.6
10^6	803.7	928.2

- the available bandwidth, B, is divided into a number, M, of channels and
- all small cell BSs are uniformly divided into groups among such M channels,

with the resulting per-channel small cell BS density, $\frac{\lambda}{M}$. Note that the bandwidth, B, is kept constant, and does not change.

With this consideration in mind, let us first make the following observations, and then focus on the formulation of the per-channel coverage probability, $\hat{p}^{cov}$:

- Since both the small cell BS density, λ, and the UE density, ρ, remain unchanged, the frequency reuse scheme does not change the active small cell BSs density, $\tilde{\lambda}$, which can be calculated as described in Section 6.3.2.
- Since both the UE-to-small cell BS association strategy and the simple idle mode capability at the small cell BSs also remain unchanged, and thus every UE is still served by the same small cell BS – that providing the strongest signal strength – the frequency reuse scheme does not change the signal power at the typical UE either. Consequently, the probability density function (PDF), $f_{R,n}^L(r)$, in equation (8.5) and the PDF, $f_{R,n}^{NL}(r)$, in equation (8.7) take the same form.
- With the presented frequency reuse scheme, however, and the resulting per-channel small cell BS density, $\frac{\lambda}{M}$, the number of interfering small cell BSs per channel is reduced by a factor, M. Therefore, the per-channel active small cell BS density, $\frac{\tilde{\lambda}}{M}$, should replace the active small cell BS density, $\tilde{\lambda}$, in the formulations of
  □ the Laplace transform, $\mathscr{L}_{I_{agg}^L}(s)$, of the aggregated inter-cell interference random variable, I_{agg}^L, for the LoS signal transmission evaluated at the variable value, $s = \frac{\gamma_0}{P\zeta_n^L(r)}$, in equation (8.11) and
  □ the Laplace transform, $\mathscr{L}_{I_{agg}^{NL}}(s)$, of the aggregated inter-cell interference random variable, I_{agg}^{NL}, for the NLoS signal transmission evaluated at the variable value, $s = \frac{\gamma_0}{P\zeta_n^{NL}(r)}$, in equation (8.13).

As a result of this reduced inter-cell interference, when this frequency reuse scheme with a reuse factor, $\frac{1}{M}$, is adopted, the SINR, $\hat{\Gamma}$, of the typical UE – and in turn, the per-channel coverage probability, $\hat{p}^{cov}$ – will improve with respect to that with a universal reuse factor of 1.

Taking this into account, we can formulate the per-channel ASE, $\hat{A}^{ASE}$, under this M-channel frequency reuse scheme as

$$\hat{A}^{ASE}\left(\lambda,\gamma_0,M\right) = M \times \frac{\tilde{\lambda}}{M} \int_{\gamma_0}^{+\infty} \log_2\left(1+\gamma\right) f_{\hat{\Gamma}}\left(\lambda,\gamma,M\right) d\gamma$$

$$= \frac{\tilde{\lambda}}{\ln 2} \int_{\gamma_0}^{+\infty} \frac{\hat{p}^{cov}\left(\lambda,\gamma,M\right)}{1+\gamma} d\gamma + \tilde{\lambda}\log_2\left(1+\gamma_0\right)\hat{p}^{cov}\left(\lambda,\gamma_0,M\right),$$

(9.5)

where, as indicated before,

- $\hat{\Gamma}$ and $f_{\hat{\Gamma}}$ are the SINR of the typical UE and its PDF under this M-channel frequency reuse scheme, respectively.

It should be noted that the first appearance of the factor, M, at the beginning of this ASE equation, indicates that the per-channel ASE, $\hat{A}^{ASE}$, is contributed by the M groups of small cell BSs in the M different channels, while its second appearance in the factor, $\frac{1}{M}$, before the integral, indicates that each group of small cell BSs only uses a fraction, $\frac{1}{M}$, of the frequency resources.

Using this formulation for the per-channel ASE, $\hat{A}^{ASE}$, the following spectrum reuse problem can be defined to find the optimal number, M^*, of channels to maximize capacity.

DEFINITION 9.4.1 For a given small cell BS density, λ, and for a given finite UE density, $\rho < +\infty$, there exits an optimal M^*-channel frequency reuse scheme, which can maximize the per-channel ASE, $\hat{A}^{ASE}$, i.e.

$$\underset{M}{\text{maximize}} \quad \hat{A}^{ASE}\left(\lambda,\gamma_0,M\right),$$

$$\text{s.t.} \quad \hat{p}^{cov}\left(\lambda,\gamma_0,M\right) \geq p_0,$$

(9.6)

$$M \geq 1,$$

where

- p_0 is the minimum acceptable coverage probability, p^{cov}, that the network operator requires and
- there should be at least one channel, i.e. $M \geq 1$.

The solution to this spectrum reuse optimization can provide good guidance to network operators on how to partition their available resources, in this case, the frequency ones, to guarantee a certain service quality, defined by the minimum acceptable coverage probability, p_0, while maximizing the per-channel ASE, $\hat{A}^{ASE}$. The larger the minimum acceptable coverage probability, p_0, the more the number, M, of channels that are required to mitigate the inter-cell interference and meet the requirement.

However, the more the number, M, of channels, the less the per-channel capacity – the less per-channel ASE, $\hat{A}^{ASE}$.

9.4.1   Problem Solution

In the following, similarly as in Sections 9.2.1 and 9.3.1, we present the algorithm used in this chapter to find the optimal M^*-channel frequency reuse scheme. This algorithm is depicted in Algorithm 3. It should be noted that this algorithm is only provided for illustration purposes, and that other more sophisticated ones apply.

Basically speaking, Algorithm 3 performs an exhaustive search over the number of channels, $M = \{1, 2, \ldots, M_{max}\}$, to find the highest ASE value. Note that for practical reasons there is maximum number, M_{max}, of channels into which the spectrum can be divided. It is important to note that this algorithm would only work if the ASE, A^{ASE}, is convex with respect to the number of channels, $M = \{1, 2, \ldots, M_{max}\}$. The convexity of the ASE, A^{ASE}, in a dense enough regime can be justified as follows. On the one hand, a small number, M, of channels implies a larger inter-cell interference because more small cell BSs will operate and transmit in the same spectrum band, making it difficult to satisfy the coverage probability requirement, $p^{cov}\left(\frac{\lambda}{M}, \gamma_0\right) \geq p_0$, which may yield an unacceptable ASE performance. On the other hand, a large number, M, of channels would mean a significant sacrifice on the bandwidth that can be used per small cell BSs to allow a high reuse factor in the network, leading to a diminishing ASE result. Therefore, the maximum ASE, A^{ASE}, will be achieved with an appropriate value of the number, M, of channels.

Algorithm 3 Example algorithm to compute the optimal number, M^*, of channels.

Step 1: Initialization
- Initialize the following variables, $A_{max}^{ASE} = 0$, $M^* = 0$.

Step 2: Iteration over the number, M, of channels

- For each configuration, $M = \{1, 2, \ldots, M_{max}\}$, set the UE density to $\rho_M = \frac{\rho}{M}$.
- Compute the per-channel coverage probability, $\hat{p}^{cov}\left(\frac{\lambda}{M}, \gamma_0\right)$, and the per-channel ASE, $A^{ASE}\left(\frac{\lambda}{M}, \gamma_0\right)$.
- If the following inequalities are met, $A^{ASE}\left(\frac{\lambda}{M}, \gamma_0\right) > A_{max}^{ASE}$ and $p^{cov}\left(\frac{\lambda}{M}, \gamma_0\right) \geq p_0$, update the variables, $A_{max}^{ASE} = A\left(\frac{\lambda}{M}, \gamma_0\right)$ and $M^* = M$. Otherwise, continue.

Step 3: Output

- Exit with the obtained optimal number, M^*, of channels and the maximum A_{max}^{ASE}.

9.4.2   Example and Discussion of Results

In this section, we study the presented M-channel frequency reuse scheme optimization problem using the same system model as in Sections 9.2.2 and 9.3.2. In more detail, we share results on the performance of the per-channel coverage probability, $\hat{p}^{cov}$, and the per-channel ASE, $\hat{A}^{ASE}$, with respect to the number, M, of channels, using an SINR threshold, $\gamma_0 = 0\,dB$, and a minimum network coverage probability, $p_0 = 0.7$, for various values of the small cell BS density, λ, and the UE density, ρ.

In Figure 9.2, it can be seen that the performance of the per-channel coverage probability, $\hat{p}^{cov}$, increases with the increase of the number, M, of channels due to the reduced inter-cell interference and the increased SINR, $\hat{\Gamma}$, of the typical UE. Using a frequency reuse factor, $M > 2$, provides a prominent improvement of the per-channel coverage probability, $\hat{p}^{cov}$, with respect to the universal frequency reuse factor of 1, $M = 1$. However, such an improvement diminishes with the increase of the number, M, of channels, as there is less and less inter-cell interference to mitigate. As an example, for the case with a small cell BS density, $\lambda = 10^3\,BSs/km^2$, and a UE density, $\rho = 600\,UEs/km^2$, the improvement from 1 to 2 channels and from 2 to 4 channels is around 45.64% and 21.85%, respectively, while from 18 to 20 channels is only 0.46%.

In contrast to the per-channel coverage probability, $\hat{p}^{cov}$, Figure 9.3 shows that there is an optimal channelization that maximizes the per-channel ASE, $\hat{A}^{ASE}$. For the same case as before with a small cell BS density, $\lambda = 10^4\,BSs/km^2$, and a UE density, $\rho = 300\,UEs/km^2$, the optimal number of channels is $M^* = 6$. Table 9.3 presents the optimal number, M, of channels for other network configurations.

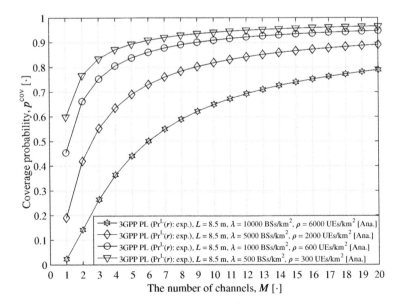

Figure 9.2 Per-channel coverage probability, $\hat{p}^{cov}$, versus the number, M, of channels.

Table 9.3. Multi-channel spectrum reuse optimization problem

	Results		
Small cell BS density λ [BSs/km^2]	UE density ρ [UEs/km^2]	Channel number M^* [·]	ASE A^{ASE}
5×10^2	300	1	363.6
10^3	600	2	422.7
5×10^3	2000	3	637.7
10^4	6000	6	718.2

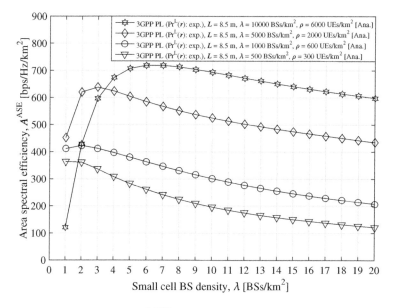

Figure 9.3 Per-channel ASE, $\hat{A}^{ASE}$, versus the number, M, of channels.

It should be noted, however, that there are cases in which deploying small cell BSs in multiple channels, $M > 1$, is not beneficial in terms of the per-channel ASE, $\hat{A}^{ASE}$. For the case with a small cell BS density, $\lambda = 500$ BSs/km^2, and a UE density, $\rho = 300$ UEs/km^2, the per-channel ASE, $\hat{A}^{ASE}$, monotonously decreases with the increase of the number, M, of channels. These results show that the trade-off between the increased SINR, $\hat{\Gamma}$, of the typical UE and the decreased available bandwidth, $\frac{B}{M}$, per channel determines the per-channel ASE performance. When the network deployment with a universal frequency reuse factor of 1, $M = 1$, already results into a large UE SINR, a co-channel deployment becomes the best policy to maximize the system capacity. In contrast, when the configuration, $M = 1$, results in a low or medium UE SINR, the multi-channel spectrum reuse can significantly boost the per-channel ASE, $\hat{A}^{ASE}$, due to its inter-cell interference reduction and the resulting enhanced UE SINR. In this case, however, note that a very large number, M, of channels may not be necessarily a good choice, despite of the potentially large UE SINR. This is because it may lead to a very limited bandwidth available for each cell, not enough to satisfy a minimum communication requirement. Thus, as previously proposed, the number, M, of channels should be carefully selected according to the channel characteristics and the network features, such as the small cell BS density, λ, and the UE density, ρ, as well as some network and/or UE performance requirements. In this case, the minimum network coverage probability, p_0, has been used as a requirement.

9.5 Conclusions

Deploying and operating a large network is expensive, and thus requires careful network dimensioning, planning and optimization to ensure a high radio resource utilization, and in turn, an optimum network performance. In this chapter, equipped with the knowledge of the constant capacity scaling law in the ultra-dense regime, derived in Theorem 8.4.2, we have shown how the theoretical performance analyses presented in Chapters 4–8 can be used to optimize both the deployment and the operation of an ultra-dense network. In more detail, we have studied the three following network-wide optimization problems:

- The small cell BS deployment/activation problem, which allows us to derive, for a given UE density, ρ, which is the optimal small cell BS density, λ^*, that can maximize the ASE, A^{ASE}.
- The network-wide UE admission/scheduling problem, which permits us to calculate, for a given small cell BS density, λ, which is the optimal UE density, ρ^*, that can maximize the ASE, A^{ASE}.
- The spatial spectrum reuse problem, which is aimed at computing, for a given UE density, ρ, and for a given small cell BS density, λ, the optimal number, M^*, of channels that can maximize the ASE, A^{ASE}, while providing a minimum coverage probability, p_0.

Importantly, while the formulations and the solutions provided in this chapter to these problems are by all means simple and non-exhaustive, they provide a good reference to the reader on how network operators can use the theoretical performance analysis tools presented in this book to derive meaningful system parameters. Refinements of these problems with a larger set of inputs and constraints can provide more accurate guidance. These solutions can also be used as initial answers to these problems, which can be further refined using more complex and tailored optimization tools, involving, for example, system-level simulations.

Part IV

Dynamic Time Division Duplexing

10 Uplink Performance Analysis of Ultra-Dense Networks

10.1 The Importance of the Uplink and Its Challenges

Although not treated in this book until now, uplink communications are key in any wireless communication system. Indeed, efficient uplink transmissions play a crucial role in the overall network performance. Recent studies have shown that, due to poor uplink performance, current cellular networks do not always consistently provide the performance required for a responsive experience, essential in the 5G era [233]. Often, the speed of the downlink determines a good time-to-content – time elapsed from the user request for online content until it is rendered on his/her smart user equipment (UE)'s display – as many popular applications receive more data in the downlink than they send in the uplink. However, once the uplink speed drops below a certain threshold, it becomes the bottleneck, limiting the speed at which content can be transferred even in the downlink. This overall network performance dependence on the uplink is further exacerbated in the newer and more demanding applications, such as teleworking, industrial internet of things (IoT), telemedicine, unmanned driving and drone services, to cite a few, which require a much more responsive experience, thus having higher requirements on the uplink data rates and network latency [233].

To satisfy the needs of such new applications and upcoming ones, the international mobile telecommunications (IMT)-2020 has set the minimum technical performance requirements of fifth generation (5G) technology high [234], with

- peak data rates of 20 Gbps in the downlink and 10 Gbps in the uplink,
- peak spectral efficiencies of 30 bit/s/Hz in the downlink and 10 bit/s/Hz in the uplink,
- user-experienced data rates of 100 Mbps in the downlink and 50 Mbps in the uplink and
- user plane latencies of 1 ms for ultra-reliable low latency communication (URLLC).

The achievement of the above key performance indicators, however, highly depends on good network coverage, particularly in the uplink. A deficient – and ultimately a lack of – uplink coverage indicates that a UE may be too far away to reach its serving base station (BS) with a sufficiently strong radio signal. In current deployments, coverage issues mainly arise because a UE can only transmit with a fraction of the transmit power available at a BS. Uplink coverage is one of the ultimate challenges both in rural areas and in densely populated urban areas, with high buildings and many other obstructions. Network densification is one of the most sensible choices for mobile operators to address the lack of uplink coverage, if not the only one. The benefits and drawbacks of network densification on uplink performance are analyzed in this chapter.

Diving in more detail into the operation of the uplink, it is important to note the following three operational challenges different from the downlink, namely

- timing,
- power control and
- inter-cell interference management.

The uplink is naturally asynchronous, i.e. the signals from different user equipments (UEs) within a cell arrive at its serving BS offset in time and frequency from each other, as they originate from different locations. These offsets can result in significant inter-symbol interference [235]. Cellular networks, however, allow for uplink inter-symbol orthogonality, implying that uplink transmissions received from the different UEs of a cell do not cause interference with one another. To realize this, the uplink slot boundaries of every UE within a cell must be time-aligned on arrival at their serving BS, and any timing misalignment between the received signals should fall within the cyclic prefix. This is practically done through a mechanism referred to as transmit-timing advance, by which every UE applies a negative time offset – the timing advance value – to the start of every uplink transmission, where the timing advance value corresponds to the amount of time a UE signal takes to reach its serving BS. The BS continuously measures the timing of the uplink signals from each UE to derive the timing advance value and communicates it to the corresponding UEs on a periodic or an aperiodic basis. UEs far from their serving BS encounter a larger propagation delay, and thus need to start their uplink transmissions proportionally earlier, compared to UEs closer to the serving BS, thus the timing advance value is larger in magnitude for the former [13].

Uplink transmit power control is the set of algorithms and tools, by which the transmit power of the uplink signals of a UE is adjusted to ensure that they are received by the serving BS at an appropriate power level, which is required for the proper decoding of the information carried by the signal at the serving BS. At the same time, the transmit power should not be unnecessarily high, as that would result in a higher power consumption than necessary, and would cause unnecessarily high inter-cell interference to the uplink transmissions in neighbouring cells. To strike the right balance between received signal power and inter-cell interference, cellular networks implement uplink transmit power control, which is a combination of:

- Open-loop transmit power control, including support for fractional path loss compensation, where the UE estimates the uplink path loss based on downlink measurements, and sets the transmit power accordingly to compensate for the estimated uplink path loss.

- Closed-loop transmit power control based on explicit power-control commands provided by the network, where the power-control commands are determined based on prior network measurements of the received signal power.

For more details on the functioning of the uplink transmit power-control schemes, which will facilitate the understanding of the rest, the reader is referred to chapter 15.1.1 of [13].

It is important to note that inter-cell interference management is also more challenging in the uplink than in the downlink. The uplink inter-cell interference originates from the UEs, whose time and space distributions are usually much harder to predict than that of the BSs due to their large volumes and mobility. Moreover, the uplink inter-cell interference can vary considerably across successive scheduling periods, as it depends on the decisions of multiple BS schedulers in a region, and its intensity is a function of the complex uplink transmit power control introduced earlier. In one scheduling period, a BS activates a subset of UEs for transmission, and in the subsequent one, it may schedule a completely different subset, thus completely changing the space-time pattern of the uplink inter-cell interference. This dependency on the BS scheduler decisions and its variability in the time domain makes the distribution of the uplink inter-cell interference much more difficult to characterize than that in the downlink, where BSs are in known locations and transmit with fixed transmit power levels.

Given these uplink unique challenges, which are not present in the downlink, the following fundamental question arises:

> Does the uplink performance of an ultra-dense network follow the same trend of the downlink performance, or do the uplink features described above change the conclusions obtained in Chapters 4–9?

In this chapter, we answer this fundamental question by means of in-depth simulation analyses. Note that although theoretical results are available in line with the framework and derivations of this book, they are omitted in this chapter for simplicity. Readers interested in such mathematical analyses are referred to [58, 236].

The rest of this chapter is organized as follows:

- Section 10.2 introduces the system model and the assumptions used in the simulation-based performance analysis presented in this chapter, paying particular attention to the uplink power control.
- Section 10.3 dives into the definition of the uplink coverage probability and the uplink area spectral efficiency (ASE).
- Section 10.4 presents and discusses simulation results showing the performance of the uplink of an ultra-dense network and verifies that the conclusions obtained in Chapters 4–9 for the downlink also hold for the uplink.
- Section 10.5, finally, summarizes the key takeaways of this chapter.

10.2   Updated System Model

To assess the performance of the uplink of an ultra-dense network, in this chapter, we upgrade the system model described in Section 8.2, which accounted for

- line-of-sight (LoS) and non-line-of-sight (NLoS) transmissions,
- a user association strategy, based on the strongest received signal strength,
- three-dimensional distances to capture the antenna heights,
- a finite UE density and
- a simple idle mode capability at the small cell BS,

with additional key features relevant to the analysis of the uplink performance, including

- an uplink transmit power-control mechanism,
- both a hexagonal and a random BS deployment and
- a new UE activation model.

In the following, we elaborate on the above three additional features of the system model for a better understanding. Table 10.1 provides a concise and updated summary of the system model used in this chapter.

10.2.1   Uplink Transmit Power Control Model

To reduce both the power consumption and the inter-cell interference, the UE transmit power, denoted by $P^{\uparrow}$ is subject to uplink transmit power control.

In this chapter, we adopt the fractional path loss compensation scheme standardized in long-term evolution (LTE) [153] and later used as the basis in new radio (NR) [13], where the UE estimates the uplink path loss based on downlink measurements, and sets the transmit power accordingly. For simplicity, we model the uplink transmit power control mechanism as

$$P^{\uparrow} = \min \left\{ P^{\uparrow\max}, 10^{\frac{P_0}{10}} \left[\zeta \left(d \right) \right]^{-\eta} N^{RB} \right\}, \tag{10.1}$$

where

- $P^{\uparrow\max}$ is the maximum total transmit power at the UE in dBm,
- P_0 is the target received power at the serving small cell BS on each time-frequency resource of study in dBm,
- ζ is the path loss in linear units,
- d is the three-dimensional distance between the UE and its serving small cell BS in kilometres,
- $\eta \in (0, 1]$ is the fractional path loss compensation factor and
- N^{RB} is the number of frequency resources in the considered bandwidth.

Note that a factor, $\eta = 0$, results in a constant uplink transmit power independent of the UE distance to its serving small cell BS, while a factor, $\eta = 1$, completely compensates for the path loss.

Table 10.1. System model

Model	Description	Reference
Transmission link		
Link direction	Uplink only	Transmissions from the UEs to the small cell BSs
Deployment		
Small cell BS deployment	HPPP with a finite density, $\lambda < +\infty$	Section 3.2.2
UE deployment	HPPP with a finite density, $\rho < +\infty$	Section 6.2.1
UE to small cell BS association		
Strongest small cell BS	UEs connect to the small cell BS providing the strongest received signal strength	Section 3.2.8 and references therein, equation (2.1.4)
Uplink power control		
Fractional power control	UEs tune their transmit powers according to path loss measurements	Section 10.2.1 and references therein, equation (10.1)
Path loss		
3GPP UMi [153]	Multi-slope path loss with probabilistic LoS and NLoS transmissions	Section 4.2.1, equation (4.1)
	☐ LoS component	equation (4.2)
	☐ NLoS component	equation (4.3)
	☐ Probability of LoS	equation (4.19)
Multi-path fast fading		
Rayleigh	Highly scattered scenario	Section 3.2.7 and references therein equation (2.15)
Shadow fading		
Not modelled	For tractability reasons, shadowing is not modelled since it has no qualitative impact on the results, see Section 3.4.5	Section 3.2.6 references therein
Antenna		
Small cell BS antenna	Isotropic single-antenna element with 0 dBi gain	Section 3.2.4
UE antenna	Isotropic single-antenna element with 0 dBi gain	Section 3.2.4
Small cell BS antenna height	Variable antenna height, $1.5 + L$ m	Section 5.2.1
UE antenna height	Fixed antenna height, 1.5 m	—
Idle mode capability at small cell BSs		
Connection-aware	Small cell BSs with no UE in their coverage areas are switched off	Section 6.2.2
Scheduler at small cell BSs		
Round robin	UEs take turns in accessing the radio channel	Section 7.1

10.2.2   Base Station Deployments

Similarly to Chapter 3, to assess the performance of small cell BS deployments, we not only consider a random BS deployment model, but also a deterministic hexagonal network layout in this chapter. The results obtained through comparing these two models with the downlink ones in Section 3.4.4 will help the understanding behind the uplink performance drivers of an ultra-dense network in our following evaluations.

Let us recall that these two small cell BS deployments, i.e. the hexagonal and the random BS ones, depicted in Figure 10.1 for completeness, have different properties. The hexagonal small cell BS deployment leads to an upper-bound performance because small cell BSs are evenly distributed in the network scenario, maximizing the minimum distance between any two small cell BSs. As a result, a very strong interference due to a very close proximity between two small cell BSs is precluded. In contrast, the random small cell BS deployment reflects a deployment that can be deemed as more realistic, capturing less symmetric networks with more challenging inter-cell interference conditions, where two small cell BSs can be arbitrarily close to each other.

10.2.3   User Equipment Activation Model

In Section 6.4.2, and in more detail, in equation (7.1) [217], the distribution of the active small cell BSs, $\tilde{\Phi}$, is characterized by a homogeneous Poisson point process (HPPP) of density, $\tilde{\lambda}$. The assumption was that each active small cell BS must have at least one active UE with data packets to transmit or receive. However, it is important

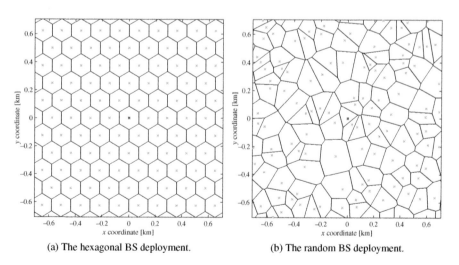

(a) The hexagonal BS deployment.   (b) The random BS deployment.

Figure 10.1 Illustration of two widely accepted types of BS deployments, the hexagonal and the random BS deployments. Here, BSs are represented by markers "*x*" and cell coverage areas for UE distribution are outlined by solid lines.

to note that the distribution of the active UEs, which is dependent of that of the small cell BSs and the user association criteria, does not necessarily follow an HPPP. More specifically, the uplink UE distribution depends on the Voronoi cell lattice resulting from the small cell BS deployment. Imagine that we have a large Voronoi cell in a certain small cell BS deployment. This assumption poses a condition, i.e. there should not be more than one active UE inside the large Voronoi cell. However, the larger the Voronoi cell, the higher the probability of having more than one UE in it. As a result, although the deployed UEs follow an HPPP, the thinning process to filter out the active UEs is not evenly applied on the plane. Due to this, the analysis of the uplink of an ultra-dense network is more complex than that of the downlink, involving some model approximations and parameter tweaking. For further reference, interested readers are encouraged to read the work in [237], and the references therein.

Equipped with the above knowledge, and in order not to lose the big picture or be tied up with complicated technical details, in this chapter, we use simulation results – instead of theoretical derivations – to

- reveal the key insights into the performance of the uplink of an ultra-dense network and
- show how the uplink of an ultra-dense network shares many of the key characteristics with its downlink counterpart, which have been discussed at length in Chapters 4–9 of this book.

Note that approximated theoretical results in line with the framework and derivations of this book are available in the indicated references.

10.3   Performance Metrics

In this section, we define the uplink coverage probability, $p^{\text{cov}\uparrow}$, of an ultra-dense network. The definition of the uplink ASE, $A^{\text{ASE}\uparrow}$, of an ultra-dense network can be obtained by substituting the corresponding uplink variables into equation (6.24), and thus is not further discussed. For completeness, the definition of these metrics is presented in Table 10.2.

Table 10.2. Key performance indicators

Metric	Formulation	Reference
Coverage probability, $p^{\text{cov}\uparrow}$	$p^{\text{cov}\uparrow}\left(\lambda, \gamma_0^\uparrow\right) = \Pr\left[\Gamma^\uparrow > \gamma_0^\uparrow\right] = \bar{F}_{\Gamma^\uparrow}\left(\gamma_0^\uparrow\right)$	Section 2.1.12, equation (2.22)
ASE, $A^{\text{ASE}\uparrow}$	$A^{\text{ASE}\uparrow}\left(\lambda, \gamma_0^\uparrow\right) = \frac{\tilde{\lambda}}{\ln 2}\int_{\gamma_0}^{+\infty}\frac{p^{\text{cov}\uparrow}(\lambda, \gamma)}{1+\gamma}d\gamma$ $+ \tilde{\lambda}\log_2\left(1 + \gamma_0^\uparrow\right)p^{\text{cov}\uparrow}\left(\lambda, \gamma_0^\uparrow\right)$	Section 6.3.3, equation (6.24)

10.3.1 Coverage Probability

Similar to the downlink, we can define the uplink coverage probability, $p^{\text{cov}\uparrow}$, of a network of density, λ, as the probability that the uplink signal-to-interference-plus-noise ratio (SINR), $\Gamma^\uparrow$, of the typical UE at the origin, o, is larger than a given SINR threshold, $\gamma_0^\uparrow$, i.e.

$$p^{\text{cov}\uparrow}\left(\lambda, \gamma_0^\uparrow\right) = \Pr\left[\Gamma^\uparrow > \gamma_0^\uparrow\right], \tag{10.2}$$

where

- $\gamma_0^\uparrow$ is the minimum working uplink SINR of the network in linear units, i.e. the uplink SINR necessary to successfully support the basic modulation and coding scheme of the particular technology in use, and
- $\gamma^\uparrow = \Gamma^\uparrow(\omega)$ is an uplink SINR realization of the typical UE, and can be calculated as

$$\gamma^\uparrow = \frac{P_{b_o}^\uparrow \zeta\left(d_{b_o}\right) h}{I_{\text{agg}}^\uparrow + P_{\text{N}}^\uparrow}, \tag{10.3}$$

where
☐ b_o is the small cell BS serving the typical UE,
☐ d_{b_o} is the distance from the typical UE to its serving small cell BS,
☐ $P_{b_o}^\uparrow$ is the transmit power of the typical UE, given by equation (10.1),
☐ h is the multi-path Rayleigh fading channel gain, modelled as an exponentially distributed random variable with a mean of 1,
☐ $P_{\text{N}}^\uparrow$ is the additive white Gaussian noise (AWGN) power at the small cell BS serving the typical UE and
☐ $I_{\text{agg}}^\uparrow$ is the aggregate uplink inter-cell interference given by

$$I_{\text{agg}}^\uparrow = \sum_{i:\, b_i \in \tilde{\Phi} \backslash b_o} P_i^\uparrow \beta_i g_i, \tag{10.4}$$

where
○ b_i is the small cell BS serving the ith interfering UE,
○ $P_i^\uparrow$ is the uplink transmit power of the ith interfering UE and
○ β_i and g_i are the path loss and the multi-path Rayleigh fading gain between the ith interfering UE and the small cell BS serving the typical UE, respectively.

Note that, in equation (10.4), only the set of active small cell BSs, except the small cell BS serving the typical UE, i.e. $\tilde{\Phi} \backslash b_o$, are actively serving UEs, and their associated UEs are generating inter-cell interference. In other words, no UEs are associated with the small cell BSs in idle mode, and thus they are not taken into account in the computation of the aggregated uplink inter-cell interference, $I_{\text{agg}}^{\text{U}}$.

As discussed in Section 10.2.3, the analysis of the performance of the uplink of an ultra-dense network is particularly difficult, and usually involves model approximations and parameter tweaking. To ease the understanding of how to practically

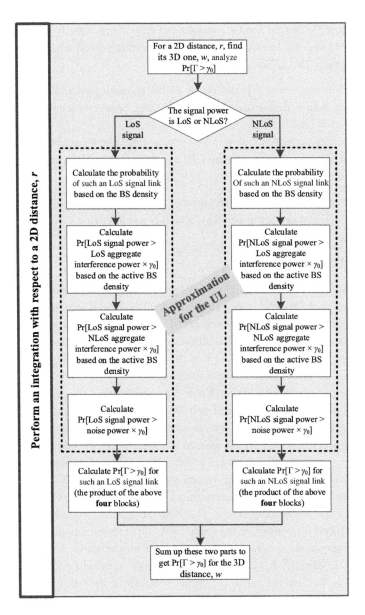

Figure 10.2 Logical steps within the standard stochastic geometry framework to obtain the results for uplink UDNs, considering the antenna height difference, the finite UE density and the idle mode capability at the small cell BS.

obtain the uplink coverage probability, $p^{\mathrm{cov}\uparrow}$, Figure 10.2 illustrates the pseudo-code that depicts the enhancements that would need to be performed over the stochastic geometry framework presented in Chapter 8 to compute the analytical results of the uplink performance of an ultra-dense network. It is important to note that, in contrast to the logic illustrated in Figure 8.2, the analytical calculation of the uplink coverage probability, $p^{\text{cov}\uparrow}$, would need to consider not only the previously mentioned approximations, but also the UE location-dependent transmit power, where such a feature is hard to capture. This is because

- the interference power depends on the uplink transmit power of the interfering UEs, which is, in turn, a function of the distance between such interfering UEs and their serving small cell BSs while
- the impact of such an uplink transmit power should be measured considering the distance between such interfering UEs and the small cell BS serving the typical UE.

Thus, an accurate analysis requires a joint characterization of the distances between an interfering UE and two small cell BSs, which is a challenging task, if not impossible.

10.3.2 Computational Complexity

Since we mostly rely on simulations to obtain the uplink coverage probability, $p^{\text{cov}\uparrow}$, and the uplink ASE, $A^{\text{ASE}\uparrow}$, due to the difficulties in conducting accurate theoretical analyses for the uplink, it would be useful to discuss the complexity of uplink simulations compared with the downlink counterparts. For the downlink simulations, once we verify the HPPP assumption of the active small cell BSs, in each simulation experiment, we can obtain the active small cell BSs by applying a uniform thinning to the deployed small cell BSs, and hence the complexity of the downlink simulations is of the order of the active small cell BS density, $\tilde{\lambda}$. However, such an efficient approach cannot be used in the uplink simulations because the active UEs are not distributed according to an HPPP. Instead, we need to model and construct all links in the simulated network, the density of which is the product of the small cell BS density, λ, and the UE density, ρ, so that we can find out which BSs/UEs are active and the uplink transmit powers of the interfering UEs. Thus, the complexity of the uplink simulations is of the order of $\lambda \times \rho$. Since the UE density, ρ is always larger than the active small cell BS density, $\tilde{\lambda}$, the complexity of the uplink simulations is larger than that of the downlink ones by at least an order of magnitude of λ. Unfortunately, the small cell BS density, λ, becomes extremely large in the ultra-dense networks, e.g. 10^3–10^6 BSs/km^2, which makes the uplink simulations 10^3–10^6 times more complex than the downlink counterparts. In our simulations for the uplink, we have used 1000 CPU cores to run 100,000 experiments up to a small cell BS density of 10^4 BSs/km^2 during a period of two months. The uplink simulations beyond 10^4 BSs/km^2 are of extremely high computational complexity.

10.4 Discussion

In this section, we use simulation results to study the performance of the uplink of an ultra-dense network and verify whether the conclusions obtained in Chapters 4–9 for the downlink hold for the uplink.

10.4.1   Case Study

To evaluate the performance of the uplink of an ultra-dense network, the same 3rd Generation Partnership Project (3GPP) case study as in Chapter 4 is used to allow an apple-to-apple comparison. For readers interested in a more detailed description of this 3GPP case study, please refer to Table 10.1 and the discussion in Section 4.4.1 and references therein.

For completeness, let us highlight that the following parameters are used in this 3GPP case study:

- maximum antenna gain, $G_M = 0$ dB,
- path loss exponents, $\alpha^L = 2.09$ and $\alpha^{NL} = 3.75$,
- reference path losses, $A^L = 10^{-10.38}$ and $A^{NL} = 10^{-14.54}$,
- target received power at the serving small cell BS, $P_0 = -76$ dBm,
- fractional path loss compensation factor, $\eta = 0.8$,
- number of frequency resources, $N^{RB} = 55$,
- maximum transmit power at the UE, $P^{\uparrow max} = 23$ dBm, and
- noise power, $P_N^\uparrow = -91$ dBm – including a noise figure of 13 dB at the small cell BS.

In addition to this 3GPP case study, and similarly to the logic and procedure in Section 3.4.5, we also define an advanced 3GPP case study in the following, considering a correlated shadow fading and a distance-dependent Rician fading for the LoS transmissions. Studying this case will help us understanding the impact of

- neglecting the shadow fading and
- the choice of the multi-path fading model in the previous 3GPP case study,

and whether their consideration is of vital importance – or can be safely ignored – in a theoretical modelling for the sake of tractability in an ultra-dense network. As a reminder, details about both the correlated shadow fading and the distance-dependent Rician fading for LoS transmissions were presented in Section 3.4.5. Note that a shadowing variance and a cross-correlation coefficient, $\sigma_s^2 = 10$ dB and $\tau = 0.5$, are respectively used.

10.4.2   Active Small Cell Base Station Density

In this subsection, the results obtained from the above introduced 3GPP case study are discussed, and the advanced 3GPP case study will be studied in the following sections.

Figure 10.3 illustrates the active small cell BS density, $\tilde{\lambda}$, as a function of the small cell BS density, λ, for the two following configurations:

- Conf. a): multi-slope path loss model with LoS and NLoS transmissions, no antenna height difference, $L = 0$ m, a finite UE density, $\rho < +\infty$, with a simple idle mode capability at the small cell BSs, and a round robin (RR) scheduler (simulation results).

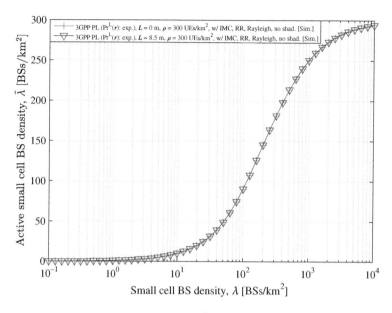

Figure 10.3 Active small cell BS density, $\tilde{\lambda}$, for the 3GPP case study under consideration with Rayleigh fading and no shadow fading.

- Conf. b): multi-slope path loss model with LoS and NLoS transmissions, an antenna height difference, $L = 8.5$ m, a finite UE density, $\rho < +\infty$, with a simple idle mode capability at the small cell BSs, and an RR scheduler (simulation results).

From Figure 10.3, we can observe that the active small cell BS density, $\tilde{\lambda}$, does not depend on the antenna height difference, L, and monotonically increases with the increase of the small cell BS density, λ. Indeed, the active small cell BS density, $\tilde{\lambda}$, is bounded by the UE density, in this case, $\rho = 300$ UEs/km^2, as there cannot be more active small cell BSs than UEs in a non-cooperative network. These results are in line with the analytical derivations presented in [217]. In more detail, the active small cell BS density, $\tilde{\lambda}$, can be analytically approximated by

$$\tilde{\lambda} = \lambda \left[1 - \frac{1}{\left(1 + \frac{\rho}{q\lambda}\right)^q} \right].$$ (10.5)

10.4.3   Coverage Probability Performance

In this subsection, Figures 10.4 and 10.5 show the uplink coverage probability, $p^{\mathrm{cov}\uparrow}$, when considering the SINR thresholds, $\gamma_0^{\uparrow} = 0$ dB and $\gamma_0^{\uparrow} = 10$ dB, respectively, for the same two configurations presented in the previous subsection.

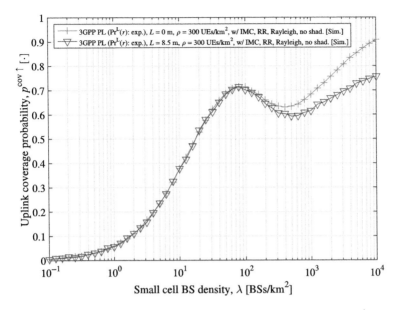

Figure 10.4 Uplink coverage probability, $p^{\text{cov}\uparrow}$, with an SINR threshold, $\gamma_0^\uparrow = 0\,\text{dB}$, for the 3GPP case study under consideration with Rayleigh fading and no shadow fading.

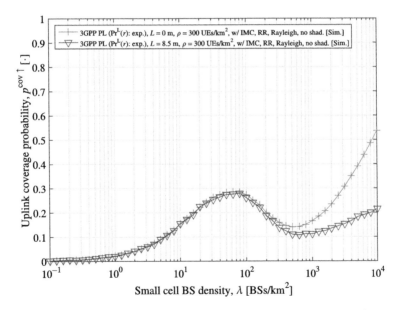

Figure 10.5 Uplink coverage probability, $p^{\text{cov}\uparrow}$, with an SINR threshold, $\gamma_0^\uparrow = 10\,\text{dB}$, for the 3GPP case study under consideration with Rayleigh fading and no shadow fading.

From Figures 10.4 and 10.5 the following observations can be made:

- When the small cell BS density, λ, is in the range of $\lambda \in \left[10^{-1}, 70\right]$ BSs/km^2, the network is noise-limited, and thus the uplink coverage probability, $p^{\text{cov}\uparrow}$, increases with the increase of the small cell BS density, λ, as the network is lightened up with coverage, and the signal power strength benefits from the LoS transmissions.
- When the small cell BS density, λ, is in the range of $\lambda \in [70, 400]$ BSs/km^2, the uplink coverage probability, $p^{\text{cov}\uparrow}$, decreases with the increase of the small cell BS density, λ. This is because of the transition of a large number of interfering paths from NLoS to LoS, which accelerates the growth of the aggregated inter-cell interference. Such a performance behaviour is in line with the downlink results presented in Chapter 4.
- When the small cell BS density, λ, is in the range of $\lambda \in \left[400, 10^4\right]$ BSs/km^2, the uplink coverage probability, $p^{\text{cov}\uparrow}$, increases with the increase of the small cell BS density, λ. This is due to the idle mode capability at the small cell BSs. In line with the downlink case, discussed in Chapter 6, the finite UE density, $\rho < +\infty$, and the idle mode capability at the small cell BS introduce a cap on the maximum interference power that a small cell BS can receive in a non-coordinated ultra-dense network, while the signal power continues to grow due to the closer proximity between transmitters and receivers in a densifying network. This phenomenon accelerates the growth of the signal power, and in turn, increases the SINR, Γ, of the typical UE.
- These two figures also show that the antenna height difference, $L > 0$, between the UEs and the small cell BSs has a significant impact on the uplink coverage probability, $p^{\text{cov}\uparrow}$. Similar to the downlink case, presented in Chapter 5, the non-zero antenna height difference, $L > 0$, slows down – and ultimately places a bound on – the continuous growth of the signal power due to the closer proximity between transmitters and receivers in a densifying network. This decreases the SINR, Γ, of the typical UE, and in turn, degrades the uplink coverage probability, $p^{\text{cov}\uparrow}$. To give an example, with a small cell BS density, $\lambda = 10^4$ BSs/km^2, and an uplink SINR threshold, $\gamma_0^{\uparrow} = 0$ dB, the uplink coverage probability, $p^{\text{cov}\uparrow}$, with an antenna height difference, $L = 8.5$ m, degrades around 13% with respect to that with no antenna height difference, $L = 0$ m. Such performance degradation further enlarges to 32% with a small cell BS density, $\lambda = 10^4$ BSs/km^2, and an uplink SINR threshold is $\gamma_0^{\uparrow} = 10$ dB. This also shows that it is harder to have good uplink coverage probability, $p^{\text{cov}\uparrow}$, when the uplink SINR threshold, $\gamma_0^{\uparrow}$, increases.

NetVisual Analysis

To visualize the fundamental behaviour of the uplink coverage probability, $p^{\text{cov}\uparrow}$, in a more intuitive manner, Figure 10.6 shows the uplink coverage probability heat map of three different scenarios with different small cell BS densities, 50, 250 and 2500 BSs/km^2. The antenna height difference is $L = 8.5$ m. The equivalent results for the downlink case are also provided for comparison purposes. As in Chapters 3–7, these heat maps are computed using NetVisual. This tool is able to capture, not only

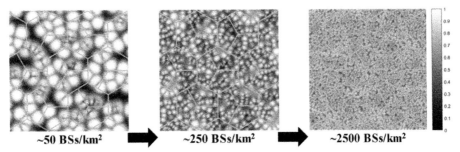

(a) Downlink performance trend with a multi-slope path loss function incorporating both LoS and NLoS transmissions, a finite UE density, $\rho = 300$ UEs/km^2, the idle mode capability at the small cell BSs and an RR scheduler.

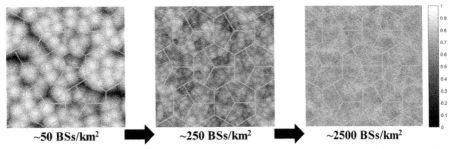

(b) Uplink performance trend with a multi-slope path loss function incorporating both LoS and NLoS transmissions, a finite UE density, $\rho = 300$ UEs/km^2, the idle mode capability at the small cell BSs and an RR scheduler.

Figure 10.6 NetVisual plot of the coverage probability, p^{cov}, with respect to the BS density, λ.

the mean, but also the standard deviation of the uplink coverage probability, $p^{\text{cov}\uparrow}$, as explained in Section 2.4.

From Figure 10.6, and comparing the downlink performance shown in part (a) – taken from Chapter 8 – to the uplink one shown in part (b), we can observe that the latter exhibits a more evenly distributed brightness across the scenario. In more detail, and taking as an example the case when the small cell BS density, λ, is around 2500 BSs/km^2, we can observe that the average coverage probability is roughly 0.7 in both the downlink and the uplink cases. However, the standard deviation decreases by 55.00% in the uplink case. It goes from 0.10 to 0.045. In short, the standard deviation of the uplink coverage probability, $p^{\text{cov}\uparrow}$, is smaller than that of the downlink counterpart, p^{cov}. This result is in line with the 3GPP system-level simulations presented Figure A.2.2–2 of [238]. The reason behind this phenomenon is the uplink transmit power-control mechanism, which tends to decrease the inter-cell interference generated by the cell-centre UEs, as they are closer to their serving cells, use less uplink transmit power and thus generate less inter-cell interference.

10.4.4 Area Spectral Efficiency Performance

Let us now explore the behaviour of the uplink ASE, $A^{\text{ASE}\uparrow}$.

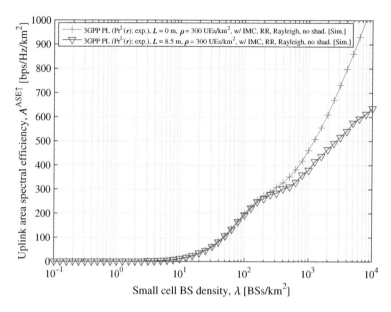

Figure 10.7 Uplink ASE, $A^{\mathrm{ASE}\uparrow}$, with an SINR threshold, $\gamma_0^{\uparrow} = 0\,\mathrm{dB}$, for the 3GPP case study under consideration with Rayleigh fading and no shadow fading.

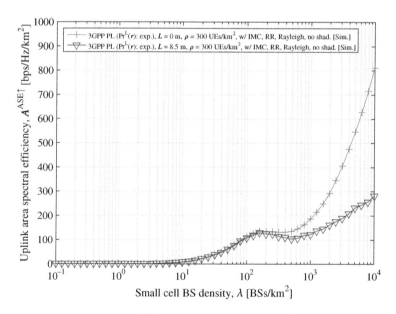

Figure 10.8 Uplink ASE, $A^{\mathrm{ASE}\uparrow}$, with an SINR threshold, $\gamma_0^{\uparrow} = 10\,\mathrm{dB}$, for the 3GPP case study under consideration with Rayleigh fading and no shadow fading.

Figures 10.7 and 10.8 show the uplink ASE, $A^{\mathrm{ASE}\uparrow}$, when considering the SINR thresholds, $\gamma_0^{\uparrow} = 0\,\mathrm{dB}$ and $\gamma_0^{\uparrow} = 10\,\mathrm{dB}$, respectively, for the same two configurations presented in Section 10.4.3.

From these two figures, we can draw the following observations:

- When the network is sparse, i.e. $\lambda \leq 10^2$ BSs/km^2, for both SINR thresholds, $\gamma_0^\uparrow = 0$ dB and $\gamma_0^\uparrow = 10$ dB, the uplink ASE, $A^{ASE\uparrow}$, quickly increases with the small cell BS density, λ. This is because the network is generally noise-limited, and thus adding more small cell BSs significantly benefits the spatial reuse.
- When the small cell BS density, λ, increases to the range, $\lambda \in (10^2, 10^3]$ BSs/km^2, the uplink ASE, $A^{ASE\uparrow}$, shows a slowdown in the rate of growth for the SINR threshold, $\gamma_0^\uparrow = 0$ dB, and even a decrease for the SINR threshold, $\gamma_0^\uparrow = 10$ dB. This is driven by the decrease of the uplink coverage probability, $p^{cov\uparrow}$, at these small cell BS densities, caused by both the transition of a large number of inter-cell interfering paths from NLoS to LoS, and the antenna height difference, $L > 0$, in dense networks. These two phenomena – the ASE Crawl and the ASE Crash – have been extensively discussed in Chapters 4 and 5, respectively. It is important to note that the condition, $\gamma_0^\uparrow = 10$ dB, indicates a higher SINR requirement compared with the condition, $\gamma_0^\uparrow = 0$ dB. The former is harder to achieve than the latter in the face of the ASE Crawl and the ASE Crash, leading to a more severe performance degradation of ASE.
- When the small cell BS density, λ, increases to the range, $\lambda \in (10^3, 10^4]$ BSs/km^2, for both threshold, $\gamma_0^\uparrow = 0$ dB and $\gamma_0^\uparrow = 10$ dB, the uplink ASE, $A^{ASE\uparrow}$, picks up its growth rate with the increase of the small cell BS density, λ. This is driven by the increase of the uplink coverage probability, $p^{cov\uparrow}$, at these small cell BS densities originated by the finite UE density, $\rho < +\infty$, and the simple idle mode capability at the small cell BSs. This phenomenon – the ASE Climb – has been extensively discussed in Chapter 6.

10.4.5   Impact of the Shadow Fading and the Multi-Path Fading – and Modelling – on the Main Results

In this subsection, the results for the advanced 3GPP case study with a hexagonal small cell BS deployment and Rician multi-path fast fading are presented.

Figure 10.9 illustrates the active small cell BS density, $\tilde{\lambda}$, as a function of the small cell BS density, λ, for the two following configurations:

- Conf. a): random small cell BS deployment, multi-slope path loss model with LoS and NLoS transmissions, an antenna height difference, $L = 8.5$ m, a finite UE density, $\rho < +\infty$, with a simple idle mode capability at the small cell BSs, and an RR scheduler (simulation results).
- Conf. b): hexagonal small cell BS deployment, multi-slope path loss model with LoS and NLoS transmissions, an antenna height difference, $L = 8.5$ m, a finite UE density, $\rho < +\infty$, with a simple idle mode capability at the small cell BSs, and an RR scheduler (simulation results).

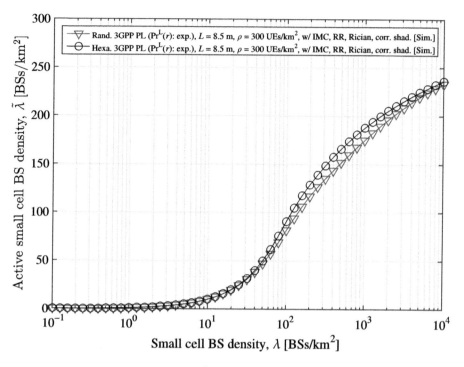

Figure 10.9 Active small cell BS density, $\tilde{\lambda}$, for the advanced 3GPP case study under consideration with Rician fading for the LoS transmissions and correlated shadow fading.

Figures 10.10 and 10.11 show the uplink coverage probability, $p^{\text{cov}\uparrow}$, when considering the SINR thresholds, $\gamma_0^\uparrow = 0\,\text{dB}$ and $\gamma_0^\uparrow = 10\,\text{dB}$, respectively, for the same two configurations, Conf. a) and Conf. b).

In addition, Figures 10.12 and 10.13 show the uplink ASE, $A^{\text{ASE}\uparrow}$, when considering the SINR thresholds, $\gamma_0^\uparrow = 0\,\text{dB}$ and $\gamma_0^\uparrow = 10\,\text{dB}$, respectively, for the same two configurations discussed earlier.

From the results shown in all these figures, we can conclude that the results obtained under Rician multi-path fading and correlated shadow fading follow the same trends as the previous results obtained under Rayleigh multi-path fading but without shadow fading. Quantitative differences exist as discussed in Section 3.4.5 when considering these different models, but qualitatively, the same fundamental behaviour prevails. This indicates that the more complex theoretical analysis in the latter case is not crucial to further understand the performance of this type of ultra-dense network.

10.5   Conclusions

In this chapter, we have analyzed the uplink performance of an ultra-dense network. We have highlighted the important differences between the modelling of the uplink and the downlink, paying attention to the uplink transmit power control and the cal-

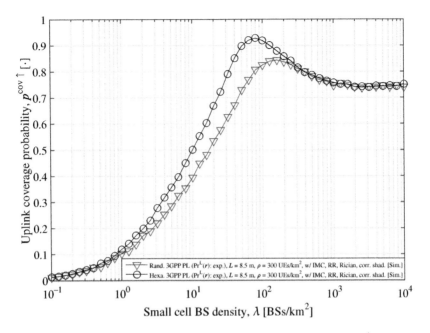

Figure 10.10 Uplink coverage probability, $p^{cov\uparrow}$, with an SINR threshold, $\gamma_0^{\uparrow} = 0\,\mathrm{dB}$, for the advanced 3GPP case study under consideration with Rician fading for the LoS transmissions and correlated shadow fading.

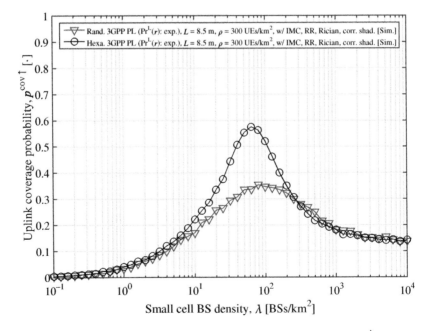

Figure 10.11 Uplink coverage probability, $p^{cov\uparrow}$, with an SINR threshold, $\gamma_0^{\uparrow} = 10\,\mathrm{dB}$, for the advanced 3GPP case study under consideration with Rician fading for the LoS transmissions and correlated shadow fading.

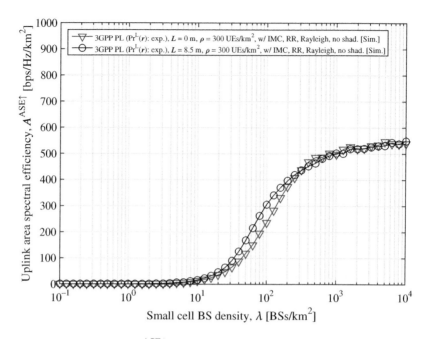

Figure 10.12 Uplink ASE, $A^{\text{ASE}\uparrow}$, with an SINR threshold, $\gamma = 0\,\text{dB}$, for the advanced 3GPP case study under consideration with Rician fading for LoS transmissions and correlated shadow fading).

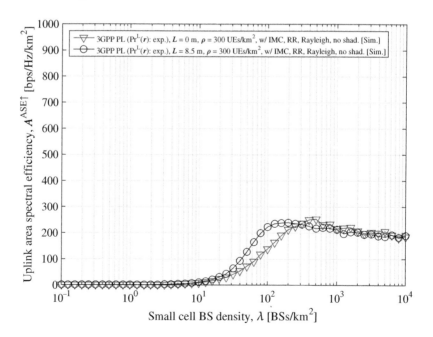

Figure 10.13 Uplink ASE, $A^{\text{ASE}\uparrow}$, with an SINR threshold, $\gamma = 10\,\text{dB}$, for the advanced 3GPP case study under consideration with Rician fading for LoS transmissions and correlated shadow fading).

culation of the key performance metrics, i.e. the uplink coverage probability and the uplink ASE. Moreover, we have shared and discussed a number of system-level simulation results for small cell BS deployments with different small cell BS densities and characteristics. Importantly, these results indicate that the performance trend of the uplink of the analyzed uncoordinated ultra-dense network follows the same behaviour as that of the downlink studied in Chapters 4–9. The ASE Crawl, Crash and Climb respectively presented in Chapter 4, Chapter 5 and Chapter 6 also appear in the uplink, and play an important role that network providers should consider when deploying an ultra-dense network.

11 Dynamic Time Division Duplexing Ultra-Dense Wireless Communication Networks

11.1 The True Value of Dynamic Time Division Duplex

It is envisaged that next-generation networks in general, and dense small cell networks in particular, will embrace time division duplexing (TDD),[1] as it does not require a pair of frequency carriers, and offers the possibility of adopting the amount of time radio resources to the downlink and the uplink traffic conditions. In this line, seven TDD configurations, each associated with a specific number of downlink and uplink subframes in a 10 ms TDD frame are available for static or semi-static selection at the network side in 3rd Generation Partnership Project (3GPP) long-term evolution (LTE) networks [153].[2] However, such a static TDD operation cannot tailor the downlink and the uplink subframe resources to the fast fluctuations in the downlink and the uplink traffic demands. These fluctuations are exacerbated in dense small cell networks due to the low number of connected user equipment (UE) per cell, and the resulting burst in their downlink and uplink traffic loads.

To overcome the drawbacks of static TDD, and allow a more *dynamic* and independent adaptation of the downlink and the uplink subframe usage to the quick variation of the downlink and uplink traffic demands in every small cell, a new technology, referred to as dynamic TDD, has emerged [91, 141, 239–241]. In dynamic TDD, the configuration of the number of downlink and uplink subframes in each cell – or a cluster of cells – can be dynamically changed on a per-frame basis, i.e. once every 10 ms [153]. Thus, dynamic TDD can provide a tailored configuration of the downlink and the uplink subframe usage for each cell – or a cluster of cells – at the expense of allowing inter-cell interlink interference, e.g. the downlink transmissions of a cell may interfere with the uplink ones of a neighbouring cell, and vice versa.

It should also be noted that dynamic TDD is a precursor of frequency domain [91, 242], which has also been identified as one of the candidate technologies for

---

[1] In frequency division duplexing (FDD), the downlink and the uplink are orthogonal and separated by the frequency domain using two different frequency carriers. In TDD, the time rather than the frequency domain is used to separate the transmission and reception of the signals, and thus a single frequency is assigned to a cell for both directions.

[2] In static or semi-static TDD, the downlink and uplink data transmissions, which occur at the same frequency band, are allocated an amount of time resources that does not change – or changes slowly – across time.

next-generation networks. In an frequency domain (FD) system, a base station (BS) can simultaneously transmit to and receive from different UEs, thus enhancing spectrum reuse, but creating not only inter-cell interlink interference, but also intra-cell interlink interference, also known as self-interference [242]. The main difference between an FD system and a dynamic TDD one is that intra-cell interlink interference does not exist in the latter [91].

The physical (PHY) layer signal-to-interference-plus-noise ratio (SINR) performance of dynamic TDD was analyzed in [141], assuming deterministic positions of small cell BSs and UEs, and in [239–241], considering stochastic positions of BSs and UEs. The main conclusion in all these works was that the uplink SINR of dynamic TDD suffers from a severe performance degradation due to the strong downlink-to-uplink interference, which is generated from one small cell BS to another. To address this challenge, cell clustering as well as full or partial interference cancellation (IC) at the small cell BS side were studied to mitigate downlink-to-uplink interference in [142, 143]. However, since clustering and IC may not be easy to realize in practical systems, such an inter-cell interference disadvantage of dynamic TDD in terms of uplink SINR triggers a fundamental question:

What is the added value of dynamic TDD? What gain can dynamic TDD achieve at the medium access control (MAC) layer to compensate the mentioned loss in the PHY one?

The answer to these questions can only be found through a thorough MAC layer analysis, which not only assesses the PHY layer performance of a dynamic TDD network, but also the performance gains brought by the MAC layer when dynamically adapting the downlink and uplink subframe usage to the rapid variation of traffic demands. This requires a different analysis than the ones carried out in Chapters 3–10 of this book, which were mostly PHY layer centric.

In this chapter, we conduct a theoretical study of the MAC layer performance of a dynamic TDD network. In more detail, we derive closed-form expressions for the downlink and uplink time resource utilizations (TRUs) when dynamic TDD operates in a synchronous manner – the mode of operation adopted in existing 3GPP LTE and new radio (NR) systems. Such results quantify the performance of dynamic TDD in terms of its MAC layer subframe usage, as small cell networks evolve into dense and ultra-dense ones. More specifically:

- we show that the downlink and the uplink TRUs varies across the different TDD subframes of a TDD frame, and that such a difference diminishes as the network densifies;
- we prove that the average TRU, i.e. the sum of the average downlink and the average uplink TRUs, in a dynamic TDD system is larger than that in a static TDD one, and that such a performance gain increases in the ultra-dense regime; and
- we derive the limit of the performance gain brought by dynamic TDD with respect to static TDD in an ultra-dense network in terms of the average TRU.

Regarding the PHY layer performance of dynamic TDD, as discussed in Chapter 10, the PHY layer uplink performance analysis of an ultra-dense network is particularly challenging, mainly because ultra-dense networks are fundamentally different from the current sparse and dense networks, and the distribution of the uplink interfering UEs is difficult to characterize analytically. Thus, in this chapter, for the sake of tractability, we also evaluate the joint MAC layer and PHY layer performance of a downlink dynamic TDD network in terms of the coverage probability and the area spectral efficiency (ASE) by means of simulations.

11.2   Updated System Model

To assess the performance of a dynamic TDD ultra-dense network, in this section, we upgrade the system model described in Section 10.2, which accounted for

- line-of-sight (LoS) and non-line-of-sight (NLoS) transmissions,
- a user association strategy, based on the strongest received signal strength,
- three-dimensional distances to capture the antenna heights,
- a finite UE density and
- a simple idle mode capability at the small cell BSs,

with the new network features discussed in the introduction of this chapter, i.e.

- a UE (de)activation traffic model based on downlink and uplink traffic demands, and
- a synchronous dynamic TDD model, considering inter-cell interlink interference in practical TDD configurations.

In the following, these two upgrades are discussed in more detail.

11.2.1   User Equipment Activation Model

In Section 6.2, the active small cell BSs distribution was characterized assuming a homogeneous Poisson point process (HPPP) distribution, $\tilde{\Phi}$, with density

$$\tilde{\lambda} = \lambda \left[1 - \frac{1}{\left(1 + \frac{\rho}{q\lambda} \right)^q} \right],$$ (11.1)

where

- q was a fitting parameter, which took values around 3.5–4 – see equation (6.12).

Importantly, it should be recalled that, in each active small cell BS of such a distribution, at least one active UE was served, and thus both active small cell BSs and UEs can generate inter-cell interference in the currently dynamic TDD network under study.

For each of those active UEs, in this chapter, the probabilities of a UE requesting downlink or uplink data in a subframe will be respectively denoted by p^D and p^U, with $p^D + p^U = 1$. Besides, note that it is also assumed that each traffic request is large enough to be transmitted for at least one TDD frame, consisting of T subframes.

We could further refine this model by considering that there may be subframes in which a UE may receive and transmit data, as will be formally presented in Section 11.2.1. The probability of such a UE using both downlink and uplink data in a subframe could be denoted by p^{D+U}, and in such a case, the following condition should hold, $p^D + p^U + p^{D+U} = 1$. It would thus be straightforward to derive the equivalent probabilities of a UE requesting downlink and uplink data as $\frac{p^D + p^{D+U}}{1 + p^{D+U}}$ and $\frac{p^U + p^{D+U}}{1 + p^{D+U}}$, respectively. For the sake of simplicity, however, we do not consider the probability, p^{D+U}, in the sequel, although the formulations presented in the following can be easily extended to comprehend it.

11.2.2   Dynamic Time Division Duplexing

We need to differentiate two types of dynamic TDD systems, i.e. asynchronous and synchronous. Most previous works in the literature have investigated dynamic TDD operating in an asynchronous manner [141, 239–241]. More specifically, in an asynchronous network, e.g. Wi-Fi systems, the TDD frames are not aligned among cells, and thus the inter-cell interlink interference can be assumed to be statistically uniform in the time domain, and solely characterized by the downlink and the uplink transmission probabilities. As a result, both the MAC and the PHY layer performance can be deemed to be uniform across all TDD frames and subframes. However, in synchronous networks, such as 3GPP LTE and NR ones, the TDD frames from different cells are aligned in the time domain to simplify the design of the radio protocols. They start and finish at the same time. Consequently, the inter-cell interlink interference, and in turn the performance of dynamic TDD, also becomes a function of the TDD configuration structure.

Figure 11.1 illustrates an example of such a synchronous TDD configuration structure in 3GPP LTE [153]. The TDD frame is composed of 10 subframes, and the time length of each subframe is 1 ms [153]. In each TDD frame, there are at least one

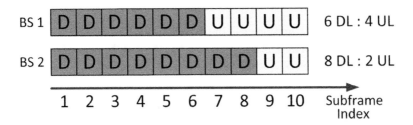

Figure 11.1 Example of some LTE TDD configurations, where one TDD frame is composed of $T = 10$ subframes. In this case, "D" and "U" denote a downlink and an uplink subframe, respectively.

downlink and one uplink subframe to accommodate for both downlink and uplink control channels. In addition, although not shown in the figure, note that there exists a downlink-to-uplink transition subframe between the downlink and the uplink subframes, in which UEs receive and transmit data. Such a transition subframe incurs an overhead [153], and because of it, there is no advantage in designing, for example, a TDD configuration structure with alternating downlink and uplink subframes. For the sake of analytical tractability, and to investigate the full potential of dynamic TDD, we do not consider either the non-zero constraint on the number of downlink and uplink subframes in a TDD frame, or the downlink-to-uplink transition subframe in this chapter.

To illustrate the difference between the synchronous dynamic TDD system in Figure 11.1 and an asynchronous one, let us assume a synchronous system where small cell BS 1 and small cell BS 2 use the TDD configurations with six and eight downlink subframes, respectively. For this synchronous dynamic TDD network with frame alignment, we can see that the first six and the last two subframes of small cell BS 1 will not see any uplink-to-downlink or downlink-to-uplink interference from small cell BS 2. However, this is not the case for the 7th and the 8th subframes of small cell BS 1, which are uplink subframes, and collide with the downlink subframes of small cell BS 2 with a 100% probability. Thus, the average probability of the downlink-to-uplink interference for the uplink in small cell BS 1 is 50% while that of the uplink-to-downlink interference for the downlink is 0. In contrast, for an asynchronous network, each downlink subframe of small cell BS 1 will experience downlink-to-uplink interference from small cell BS 2 with an 80% probability, since small cell BS 2 transmits downlink subframes 80% of the time.

As can be derived from this example, ignoring the TDD frame alignment in synchronous networks can lead to a considerable overestimation of inter-cell interlink interference. In the sequel, we will focus on synchronous dynamic TDD as these are the choice in cellular systems, such as 3GPP LTE or NR, and refer to it as dynamic TDD for simplicity. We will explicitly state asynchronous dynamic TDD when needed. For ease of understanding, the above discussion has been summarized in the following two remarks.

Remark 11.1   The first few subframes of a TDD frame are more likely to carry downlink transmissions than the last few. The opposite observation applies on the uplink. This implies a subframe-dependent MAC layer performance of dynamic TDD.

Remark 11.2   The subframes in the middle of the TDD frame are more likely to be subject to inter-cell interlink interference than those at the two ends of the TDD frame. This leads to a subframe-dependent PHY layer performance of dynamic TDD.

11.3   Performance Metrics

In this section, we define MAC layer performance metrics and redefine the coverage probability and the ASE ones to accommodate for dynamic TDD.

11.3.1   Medium Access Control Layer Time Resource Utilization

For the lth subframe ($l \in \{1, 2, \ldots, T\}$), we define the subframe-dependent down-link and uplink TRUs, q_l^D and q_l^U, as the probability of a small cell BS transmitting downlink traffic and that of its UEs transmitting uplink traffic in such a subframe, respectively. In general, we may not always achieve the equality, $q_l^D + q_l^U = 1$, because some subframe resources might be unused due to the lack of traffic or wasted due to the inability of the small cell BS to match subframe and traffic types as in static TDD, e.g. there is no downlink traffic to be served by a downlink subframe.

Moreover, we define the average downlink TRU, κ^D, and the average uplink TRU, κ^U, as the mean values of the subframe-dependent downlink and uplink TRUs, q_l^D and q_l^U, across all of the T subframes of a TDD frame, i.e.

$$\begin{cases} \kappa^D = \frac{1}{T} \sum_{l=1}^{T} q_l^D \\ \kappa^U = \frac{1}{T} \sum_{l=1}^{T} q_l^U. \end{cases} \tag{11.2}$$

Finally, we also define the average total TRU, κ, as the sum of the average downlink and the average uplink TRUs, κ^D and κ^U, i.e.

$$\kappa = \kappa^D + \kappa^U. \tag{11.3}$$

In the following sections, we will investigate the performance of the above presented variables, q_l^{Link}, κ^{Link} and κ, while considering Remark 1, where the string variable, *Link*, denotes the link direction, and takes the value of "D" and "U" for the downlink and the uplink, respectively.

Based on such definitions, the downlink and uplink ASEs in bps/Hz/km^2 can be further computed by

$$\frac{\tilde{\lambda}}{T} \sum_{l=1}^{T} q_l^{Link} r_l^{Link}, \tag{11.4}$$

where

- r_l^{Link} is the per-BS PHY layer data rate in bps/Hz/BS for the lth subframe. As explained in Remark 11.2, the derivation of the rate, r_l^{Link}, must consider the subframe-dependent inter-cell interlink interference.

  In the following, we will discuss how to calculate the rate, r_l^{Link}, in our simulations. For the ease of mathematical notation, we omit the subscript, l, since all of the following calculations are meant for the lth subframe.

11.3.2   Physical Layer Coverage Probability

To proceed with the performance analysis of a synchronous dynamic TDD system, in the following, we embrace the same system model presented in Chapters 3–10 of this book.

In more detail, we use the general path loss model presented in Section 5.2.1, where the string variable, *Path*, takes the value of "L" (for the LoS case) or "NL" (for the NLoS case), and in this dynamic TDD case, the string variable, *Dir*, may use "B2U," "B2B" or "U2U," depending on whether we are referring to the BS-to-UE path loss, the BS-to-BS path loss or the UE-to-UE path loss, respectively. As for the UE association strategy, let us recall that each UE is connected to the small cell BS at the smallest path loss. Also note that each small cell BS and UE is equipped with an isotropic antenna, and that the multi-path fading between a transmitter and a receiver is modelled as independent and identical distributed (i.i.d.) Rayleigh fading.

Based on such a system model, we can define the coverage probability, $p^{\text{cov},Link}$, of the typical UE as the probability that its downlink or its uplink SINR, γ^{Link}, is above a designated threshold, γ_0, i.e.

$$p^{\text{cov},Link}\left(\lambda,\gamma\right) = \Pr\left[\gamma^{Link} > \gamma_0\right], \tag{11.5}$$

where

- the string variable, *Link*, denotes the link direction, and takes the value of "D" and "U" for the downlink and the uplink, respectively, and
- the downlink and uplink SINR, γ^{Link}, can calculated as

$$\gamma^{Link} = \frac{P^{Link}\,\zeta_{b_o}^{\text{B2U}}\,(d)\,h}{I_{\text{agg}}^{\text{D}} + I_{\text{agg}}^{\text{U}} + P_{\text{N}}^{Link}}, \tag{11.6}$$

where

- ☐ P^{Link} is the transmit power,
- ☐ d is the distance from the typical UE to its serving small cell BS, denoted by b_o,
- ☐ h is the multi-path fast fading gain between the typical UE and its serving small cell BS, modelled as Rayleigh fading,
- ☐ P_{N}^{Link} is the received additive white Gaussian noise (AWGN) power and
- ☐ $I_{\text{agg}}^{\text{D}}$ and $I_{\text{agg}}^{\text{U}}$ are the aggregated downlink and uplink originated inter-cell interference.

In more detail, it is important to note that

- P^{Link} is the small cell BS transmit power when considering the downlink, i.e. P^{D}, or the UE transmit power when considering the uplink, i.e. P^{U}. Note that the latter is usually subject to a semi-static power control, such as that presented and investigated in Chapter 10, and for the sake of tractability, and to avoid the overwhelming downlink-to-uplink interference problem generating from the large difference in transmit power between the downlink and the uplink, we assume that the downlink transmit power, P^{U}, is a cell-specific constant comparable to the downlink transmit power, P^{D}. This assumption is in line with the uplink power boosting concept presented and discussed in [91].

- P_N^{Link} is the AWGN at the UE side when considering the downlink, i.e. P_N^D, or the AWGN at the BS side when considering the uplink, i.e. P_N^U.
- The aggregated downlink and uplink originated inter-cell interference, I_{agg}^D and I_{agg}^U, are contributed by the interfering small cell BSs and the interfering small cell UEs, respectively, where the sets of such interfering small cell BSs and UEs are denoted by $\tilde{\Phi}^D$ and $\tilde{\Phi}^U$, respectively, and it is easy to show that
  - $\tilde{\Phi}^D \cap \tilde{\Phi}^U = \emptyset$, because full duplex is not considered, and that
  - $\tilde{\Phi}^D \cup \tilde{\Phi}^U = \tilde{\Phi} \setminus b_o$, because only the active small cell BSs in the set, $\tilde{\Phi} \setminus b_o$, can schedule downlink and/or uplink transmissions that inject effective inter-cell interference into the network.
- When considering static TDD, let us remind that $I_{agg}^D = 0$ when considering the uplink and that $I_{agg}^U = 0$ when considering the downlink, due to the absence of interlink interference.

11.3.3 Joint Medium Access Control and Physical Layer Area Spectral Efficiency

Similarly as in Chapters 3–10, plugging the coverage probability, $p^{cov,Link}(\lambda, \gamma)$ – obtained from equation (11.5) – into equation (2.24) to compute the probability density function (PDF), $f_\Gamma(\gamma)$, of the SINR, Γ, of the typical UE, we can obtain the ASE, $A^{ASE}(\lambda, \gamma_0)$, by solving equation (6.24), i.e.

$$A^{ASE,Link}(\lambda, \gamma_0)$$

$$= \frac{1}{T} \sum_{l=1}^{T} q_l^{Link} \tilde{\lambda} \left[\frac{1}{\ln 2} \int_{\gamma_0}^{+\infty} \frac{p^{cov,Link}(\lambda, \gamma)}{1 + \gamma} d\gamma + \log_2 (1 + \gamma_0) p^{cov,Link}(\lambda, \gamma_0) \right],$$

where

- q_l^D and q_l^U characterize how much subframe time is used for the downlink and the uplink transmission, respectively, as discussed in Section 11.3.1, and
- the term, $\frac{1}{\ln 2} \int_{\gamma_0}^{+\infty} \frac{p^{cov,Link}(\lambda, \gamma)}{1+\gamma} d\gamma + \log_2 (1 + \gamma_0) p^{cov,Link}(\lambda, \gamma_0)$, represents the rate, r_l^{Link}, as presented in Section 11.3.1.

11.4 Main Results

The main goal of this section is to share theoretical results on the characterization of the downlink and the uplink TRUs defined in Section 11.3.1.

For a static TDD network, i.e. with a fixed TDD frame configuration across all small cell BSs, the subframe-dependent downlink and uplink TRUs, q_l^D and q_l^U, are either 1 or 0.

For a dynamic TDD system, on the other hand, computing such subframe-dependent downlink and uplink TRUs, q_l^D and q_l^U, is a non-trivial task, because it involves the following distributions:

- the distribution of the number of UEs in an active small cell BS, which will be shown to follow a truncated Negative Binomial distribution in the following Section 11.4.1;
- the distribution of the number of downlink and uplink data requests in an active small cell BS, which will be shown to follow a Binomial distribution in Section 11.4.2;
- the dynamic TDD subframe splitting strategy and the corresponding distribution of the number of downlink and uplink subframes in a TDD frame, which will be shown to follow an aggregated Binomial distribution in Section 11.4.3; and finally
- prior information about the TDD frame structure, such as the downlink-before-uplink structure adopted in LTE [153] (see Figure 11.1), which will lead to subframe-dependent results for the probabilities, q_l^D and q_l^U, as it will be presented in Section 11.4.4.

Based on such derived subframe-dependent downlink and uplink TRUs, q_l^D and q_l^U, we will then be a position to study the analytical results on the average downlink, the average uplink and the average total TRUs, κ^D, κ^U and κ, in Section 11.4.6, which will allow one to better understand the fundamental behaviour of a dynamic TDD small cell network. Numerical examples will be provided where necessary to help readers understand the logic flow.

Let us develop on the aforementioned distribution in the following.

11.4.1   Distribution of the Number of User Equipments per Active Small Cell Base Station

In this section, we derive the distribution of the number of UEs per active small cell BS, which is key to later understanding its traffic load.

According to [217], considering both the active and the inactive small cell BSs, the coverage area size, X, of a small cell BS can be approximately characterized by a Gamma distribution, and its PDF can be written as

$$f_X(x) = (q\lambda)^q \, x^{q-1} \frac{\exp(-q\lambda x)}{\Gamma(q)}, \tag{11.7}$$

where

- q is a distribution parameter and
- $\Gamma(\cdot)$ is the Gamma function [156].

The number of UEs per small cell BS can then be denoted by a random variable (RV), K, and its probability mass function (PMF) can be written as

$$f_K(k) = \Pr[K = k]$$

$$\overset{(a)}{=} \int_0^{+\infty} \frac{(\rho x)^k}{k!} \exp(-\rho x) f_X(x) dx$$

$$\overset{(b)}{=} \frac{\Gamma(k+q)}{\Gamma(k+1)\Gamma(q)} \left(\frac{\rho}{\rho + q\lambda} \right)^k \left(\frac{q\lambda}{\rho + q\lambda} \right)^q, \tag{11.8}$$

where

- step (a) is due to the HPPP distribution of UEs and
- step (b) is obtained from equation (11.7).

Note that the PMF, $f_K(k)$, satisfies the normalization condition, i.e. $\sum_{k=0}^{+\infty} f_K(k) = 1$, and that the RV, K, follows a Negative Binomial distribution, i.e. $K \sim \text{NB}\left(q, \frac{\rho}{\rho+q\lambda}\right)$ [156].

As discussed in Section 11.2.1, we assume that small cell BSs with no active UE, $K = 0$, are idle, and that they can be safely ignored in our analysis as they do not generate any type of inter-cell interference. Thus, in the following, we focus on the active small cell BSs.

The number of UEs per active small cell BS can then be denoted by another RV, $\tilde{K}$, and considering both equation (11.8), and the fact that the only difference between the RV, K, and the RV, $\tilde{K}$, lies in $\tilde{K} \neq 0$, we can conclude that the RV, $\tilde{K}$, follows a truncated Negative Binomial distribution, i.e. $\tilde{K} \sim \text{truncNB}\left(q, \frac{\rho}{\rho+q\lambda}\right)$, where its PMF can be written as

$$f_{\tilde{K}}(\tilde{k}) = \Pr\left[\tilde{K} = \tilde{k}\right] = \frac{f_K(\tilde{k})}{1 - f_K(0)}. \tag{11.9}$$

Note that based on the definition of the active small cell BS density, $\tilde{\lambda}$, in Section 11.2.1, we have that $\tilde{\lambda} = (1 - f_K(0))\lambda$, and that the PMF, $f_{\tilde{K}}(\tilde{k})$, satisfies the normalization condition, i.e. $\sum_{k=1}^{+\infty} f_{\tilde{K}}(\tilde{k}) = 1$.

Example: As an example, we plot the PMF, $f_{\tilde{K}}(\tilde{k})$, of the number, $\tilde{K}$, of UEs per active small cell BS in Figure 11.2 with the following parameter values: small cell BS density, $\lambda \in \{50, 200, 1000\}$ BSs/km², UE density, $\rho = 300$ UEs/km², and $q = 3.5$. From this figure, we can draw the following observations:

- The analytical results based on the truncated Negative Binomial distribution match well with the simulation results. More specifically, the maximum difference between the simulated and analytical PMFs, $f_{\tilde{K}}(\tilde{k})$, is shown to be less than 0.5 percentile.
- For the active small cell BSs, the PMF, $f_{\tilde{K}}(\tilde{k})$, shows a more dominant peak at $\tilde{k} = 1$ as the small cell BS density, λ, – and in turn, the active small cell BS density, $\tilde{\lambda}$ – increase. This is because the ratio of the UE density, ρ, to the active small cell BS density, $\tilde{\lambda}$, gradually decreases towards 1, as the small cell BS density, λ, approaches the limit of one UE per active small cell BS in an ultra-dense network. In particular, when the small cell BS density equals, $\lambda = 1000$ BSs/km², more than 80% of the active BSs will serve only one UE. Intuitively speaking, each of those small cell BSs should dynamically engage all subframes for downlink or uplink transmissions based on the specific traffic demand of its served UE. Dynamic TDD makes this possible, avoiding any waste of the subframe resources and maximizing their utilization.

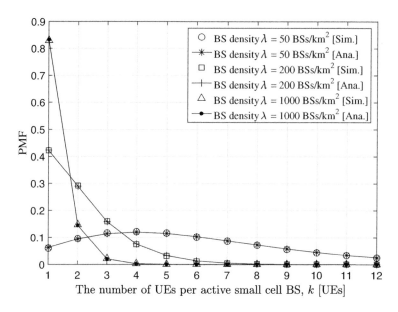

Figure 11.2 PMF of the number, $\tilde{K}$, of UEs per active small cell BS ($\rho = 300$ UEs/km^2, $q = 3.5$, various λ).

11.4.2   Distribution of the Number of Downlink and Uplink Data Requests in an Active Small Cell Base Station

After analyzing the distribution, $f_{\tilde{K}}(\tilde{k})$, of the number, $\tilde{K}$, of UEs per active small cell BS, we further study the distribution of the numbers of downlink and uplink data requests in such an active small cell BS. Such numbers reflect the downlink and uplink traffic loads in the active small cell BS under study, and their understanding is necessary to derive a tailored dynamic TDD configuration.

For clarity, the number of downlink and uplink data requests in an active small cell BS are denoted by the RVs, M^{D} and M^{U}, respectively. Since we assume that each UE generates either one downlink or one uplink data request (see Section 11.2.1), it is easy to show that

$$M^{\mathrm{D}} + M^{\mathrm{U}} = \tilde{K}. \tag{11.10}$$

As discussed in Section 11.2.1, for each UE in an active small cell BS, the probabilities of it requesting downlink or uplink data in a subframe are respectively denoted by p^{D} and p^{U}. Thus, for a given number, $\tilde{k}$, of UEs per active small cell BS, the RVs, M^{D} and M^{U}, follow Binomial distributions, i.e. $M^{\mathrm{D}} \sim \mathrm{Bi}(\tilde{k}, p^{\mathrm{D}})$ and $M^{\mathrm{U}} \sim \mathrm{Bi}(\tilde{k}, p^{\mathrm{U}})$ [156], where their PMFs can be written as

$$f_{M^{\mathrm{D}}}\left(m^{\mathrm{D}}\right) = \binom{\tilde{k}}{m^{\mathrm{D}}} \left(p^{\mathrm{D}}\right)^{m^{\mathrm{D}}} \left(1 - p^{\mathrm{D}}\right)^{\tilde{k} - m^{\mathrm{D}}} \tag{11.11}$$

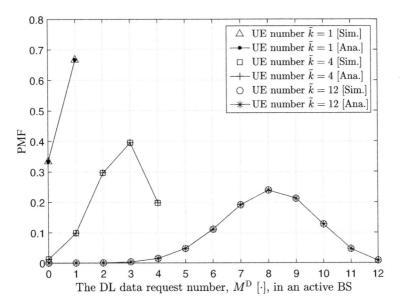

Figure 11.3 PMF of the number, M^D, of downlink data requests in an active small cell BS ($p^D = \frac{2}{3}$, various $\tilde{k}$).

and

$$f_{M^U}\left(m^U\right) = \binom{\tilde{k}}{m^U}\left(p^U\right)^{m^U}\left(1-p^U\right)^{\tilde{k}-m^U}, \tag{11.12}$$

respectively.

Example: As an example, we plot the PMF, $f_{M^D}\left(m^D\right)$, of the number, M^D, of downlink data requests in an active small cell BS in Figure 11.3 with the following parameter values: $\tilde{k} \in \{1, 4, 12\}$ and $p^D = \frac{2}{3}$. For brevity, we omit displaying the PMF, $f_{M^U}\left(m^U\right)$, of the number, M^U, of uplink data requests in an active small cell BS due to its duality with that of the downlink, as shown in equation (11.10).

From this figure, we can draw the following observations:

- The Binomial distribution accurately characterizes the distribution of the number, M^D, of downlink data requests in an active small cell BS, as the analytical PMFs, $f_{M^D}\left(m^D\right)$, match well the simulated ones.
- The RV, M^D, has an average value around, $p^D\tilde{k}$, which is in line with intuition.

11.4.3   Distribution of the Number of Downlink/Uplink Subframes in a Dynamic Time Division Duplexing Frame

After analyzing the distribution, $f_{M^D}(m^D)$, of the number, M^D, of downlink data requests in an active small cell BS, we are in a good position to understand how such data requests are translated into downlink and uplink subframe usage. Therefore, in the following, we further study the distribution of the number of downlink and uplink subframes scheduled in an active small cell BS.

For a given number, $\tilde{k}$, of UEs per active small cell BS, the number of downlink subframes in such an active small cell BS is denoted by the RV, N^D. In the following, we assume a particular dynamic TDD algorithm to choose the number of downlink subframes, which matches the downlink subframe ratio with the downlink data request ratio [91]. In more detail, for a given number, $\tilde{k}$, of UEs per active small cell BS, and for a given number, m^D, of downlink data requests, the number, $n(m^D, \tilde{k})$, of downlink subframes in an active small cell BS can be determined by

$$n(m^D, \tilde{k}) = \text{round}\left(\frac{m^D}{\tilde{k}}T\right), \tag{11.13}$$

where

- round(x) is an operator that rounds a real value, x, to its nearest integer.

In equation (11.13), the ratio, $\frac{m^D}{\tilde{k}}$, can be deemed as the downlink data request ratio, as the number, $\tilde{k}$, of UEs per active small cell BS is equal to the total number, $m^D + m^U$, of downlink and uplink data requests. As a result, the variable, $\frac{m^D}{\tilde{k}}T$, yields the number of downlink subframes that matches the downlink subframe ratio with the downlink data request ratio, where such a variable uses the round operator in equation (11.13) due to the integer nature of the subframe allocation.

Based on equation (11.13), the PMF, $f_{N^D}(n^D), n^D \in \{0, 1, \ldots, T\}$, of the number, N^D, of downlink subframes in an active small cell BS can be derived as

$$f_{N^D}\left(n^D\right) = \Pr\left[N^D = n^D\right]$$

$$\stackrel{(a)}{=} \sum_{m^D=0}^{\tilde{k}} I\left\{\text{round}\left(\frac{m^D}{\tilde{k}}T\right) = n^D\right\} f_{M^D}\left(m^D\right), \tag{11.14}$$

where

- equation (11.13) is used in step (a) and
- $I\{X\}$ is an indicator function, which outputs 1 when the variable, X, is true, 0, otherwise.

Note that the PMF, $f_{M^D}\left(m^D\right)$, is calculated using equation (11.11), and that due to the existence of the sum and the indicator function in the PMF, $f_{N^D}\left(n^D\right)$, of equation (11.14), this last one can be seen as an aggregated PMF of Binomial PMFs. This is because the RV, N^D, is computed from the RV, M^D, according to the many-to-one mapping specified in equation (11.13). Also note that the PMF, $f_{N^D}\left(n^D\right)$, satisfies the normalization condition, i.e. $\sum_{n^D=0}^{T} f_{N^D}\left(n^D\right) = 1$.

Because the total number of subframes in a TDD frame is T, and each subframe should be either a downlink or an uplink one, it is apparent that $N^D + N^U = T$, and thus we can conclude that

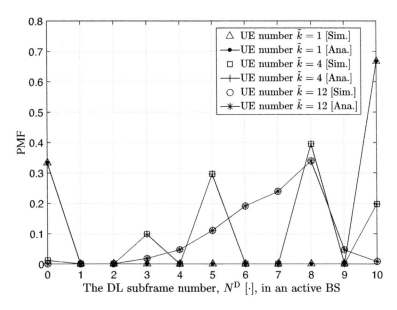

Figure 11.4 PMF of the downlink subframe number, N^{D}, in an active small cell BS ($p^{\mathrm{D}} = \frac{2}{3}$, $T = 10$, various $\tilde{k}$).

$$f_{N^{\mathrm{U}}}\left(n^{\mathrm{U}}\right) = f_{N^{\mathrm{D}}}\left(T - n^{\mathrm{U}}\right). \tag{11.15}$$

Example: As an example, we plot the PMF, $f_{N^{\mathrm{D}}}\left(n^{\mathrm{D}}\right)$, of the number, N^{D}, of downlink subframes in an active small cell BS in Figure 11.4 with the following parameter values: $\tilde{k} \in \{1, 4, 12\}$, $p^{\mathrm{D}} = \frac{2}{3}$ and $T = 10$.

From this figure, we can draw the following observations:

- When the number, $\tilde{k}$, of UEs per active small cell BS is equal to 1, i.e. $\tilde{k} = 1$, the number, N^{D}, of downlink subframes is set to either $T = 10$ or 0, meaning that the active small cell BS dynamically invest all subframes in downlink transmissions or uplink ones based on the instantaneous data request being a downlink (with a probability of p^{D}) or an uplink one (with a probability of p^{U}). Such a strategy fully uses the subframe resources in a dynamic TDD network.

- When the number, $\tilde{k}$, of UEs per active small cell BS is larger than 1, i.e. $\tilde{k} > 1$, the PMF, $f_{N^{\mathrm{D}}}\left(n^{\mathrm{D}}\right)$, of the number, N^{D}, of downlink subframes turns out to be rather complex, because multiple different numbers, M^{D}, of downlink data requests may be mapped to the same number, N^{D}, of downlink subframes, as a result of the dynamic TDD algorithm depicted by equation (11.13). For example, when the number, $\tilde{k}$, of UEs equals to $\tilde{k} = 12$, we have that $f_{N^{\mathrm{D}}}$ (8) $= 0.339$, as exhibited in Figure 11.4. This is because:

  □ according to equation (11.13), both $M^{\mathrm{D}} = 9$ and $M^{\mathrm{D}} = 10$ result in $N^{\mathrm{D}} = 8$, since round $\left(\frac{9}{12} \times 10\right) = 8$ and round $\left(\frac{10}{12} \times 10\right) = 8$. In other words, with dynamic TDD, 8 subframes will be allocated for the downlink, if either 9 or 10 data requests out of $\tilde{k} = 12$ data requests are downlink.

☐ As shown in Figure 11.3, $f_{M^\mathrm{D}}(9) = 0.212$ and $f_{M^\mathrm{D}}(10) = 0.127$ and thus
☐ from equation (11.14), we have that $f_{N^\mathrm{D}}(8) = f_{M^\mathrm{D}}(9) + f_{M^\mathrm{D}}(10) = 0.339$. This means that, with dynamic TDD, the probability of allocating 8 downlink subframes equals to the sum of the probabilities of observing 9 and 10 downlink data requests in $\tilde{k} = 12$ data requests.

11.4.4   Subframe-Dependent Downlink and Uplink Time Resource Utilizations

Considering the downlink-before-uplink structure of the synchronous TDD configurations, and based on the previously derived results, i.e. $f_{\tilde{K}}(\tilde{k})$, $f_{M^\mathrm{D}}(m^\mathrm{D})$, $f_{M^\mathrm{U}}(m^\mathrm{U})$, $f_{N^\mathrm{D}}(n^\mathrm{D})$ and $f_{N^\mathrm{U}}(n^\mathrm{U})$, in this subsection, we present our main results on the subframe-dependent downlink and uplink TRUs, q_l^D and q_l^U, for dynamic TDD, i.e. on the probability of each subframe being a downlink or an uplink one. These results are given in Theorem 11.4.1.

THEOREM 11.4.1   *In a dynamic TDD network, the subframe-dependent downlink and uplink TRUs, q_l^D and q_l^U, are given by*

$$
\begin{cases}
q_l^\mathrm{D} = \displaystyle\sum_{\tilde{k}=1}^{+\infty} \left(1 - \sum_{i=0}^{l-1} f_{N^\mathrm{D}}(i)\right) f_{\tilde{K}}(\tilde{k}) \\[4mm]
q_l^\mathrm{U} = \displaystyle\sum_{\tilde{k}=1}^{+\infty} \sum_{i=0}^{l-1} f_{N^\mathrm{D}}(i) f_{\tilde{K}}(\tilde{k})
\end{cases}
, \qquad (11.16)
$$

where

- $f_{N^\mathrm{D}}(i)$ *and* $f_{\tilde{K}}(\tilde{k})$ *are given by equations (11.14) and (11.9), respectively.*

Proof   Based on equation (11.14), and conditioned on $\tilde{K} = \tilde{k}$, the probability of performing a downlink transmission in the lth subframe ($l \in \{1, 2, \ldots, T\}$) can be calculated as

$$
\begin{aligned}
q_{l,\tilde{k}}^\mathrm{D} &= \Pr\left[Y_l = \text{``D''} | \tilde{K} = \tilde{k}\right] \\
&\overset{(a)}{=} \Pr\left[N^\mathrm{D} \geq l\right] \\
&= 1 - F_{N^\mathrm{D}}(l-1),
\end{aligned}
\qquad (11.17)
$$

where

- Y_l is the link direction of the transmission on the lth subframe, which takes the value of "D" and "U" for the downlink and the uplink,
- step (a) is due to the TDD configuration structure shown in Figure 11.1 and
- $F_{N^\mathrm{D}}(n^\mathrm{D})$ is the cumulative mass function (CMF) of the number, N^D, of downlink subframes in an active small cell BS, which is written as

$$F_{N^D}\left(n^D\right) = \Pr\left[N^D \le n^D\right] = \sum_{i=0}^{n^D} f_{N^D}(i). \tag{11.18}$$

Given such a result, the conditional probability of performing an uplink transmission in the lth subframe can be computed as

$$q_{l,k}^U = 1 - q_{l,\tilde{k}}^D = F_{N^D}(l-1). \tag{11.19}$$

Finally, the unconditional probabilities of performing a downlink and an uplink transmission in the lth subframe, i.e. q_l^D and q_l^U, can be respectively derived by calculating the expected values of the previously derived conditional probabilities, $q_{l,\tilde{k}}^D$ and $q_{l,\tilde{k}}^U$, over all the possible values of the number, $\tilde{k}$, of UEs per active small cell BS as shown in equation (11.16), which concludes our proof. □

11.4.5 Inter-Cell Interlink Interference Probabilities

From Theorem 11.4.1, we can derive the probability of each subframe being a downlink or an uplink one. Building on those, the probabilities of inter-cell inter-link interference can be derived in dynamic TDD systems. To clarify, such probabilities are formally defined as follows:

- The probability of the downlink-to-uplink interference is defined as $\Pr^{D2U} \triangleq \Pr[Z = \text{"D"}|S = \text{"U"}]$, while the probability of the uplink-to-uplink interference is defined as $\Pr^{U2U} \triangleq \Pr[Z = \text{"U"}|S = \text{"U"}]$, where Z and S denote the link directions for the inter-cell interference and the carrier signals, respectively, and the link direction may take the string value, "D," for the downlink case and the string value, "U," for the uplink one. Importantly, $\Pr^{D2U} + \Pr^{U2U} = 1$.
- Similarly, the probability of the uplink -to-downlink interference is defined as $\Pr^{U2D} \triangleq \Pr[Z = \text{"U"}|S = \text{"D"}]$, while the probability of the downlink-to-downlink interference is defined as $\Pr^{D2D} \triangleq \Pr[Z = \text{"D"}|S = \text{"D"}]$, with $\Pr^{U2D} + \Pr^{D2D} = 1$.

Our main results on the probability, $\Pr^{D2U}$, of downlink-to-uplink interference and the probability, $\Pr^{U2D}$, of uplink-to-downlink interference are summarized in Theorem 11.4.2.

THEOREM 11.4.2 *The probability, $\Pr^{D2U}$, of downlink-to-uplink interference and the probability, $\Pr^{U2D}$, of uplink-to-downlink interference can be derived in closed-form expressions as*

$$\begin{cases} \Pr^{D2U} = \dfrac{\sum_{l=1}^{T} q_l^D q_l^U}{\sum_{j=1}^{T} q_j^U} \\[4mm] \Pr^{U2D} = \dfrac{\sum_{l=1}^{T} q_l^U q_l^D}{\sum_{j=1}^{T} q_j^D} \end{cases}, \tag{11.20}$$

where

- q_l^D and q_l^U are obtained from equation (11.16).

Proof   By examining the inter-cell interlink interference in all the lth subframe ($l \in \{1, 2, \ldots, T\}$), one by one, the probability of downlink-to-uplink interference, Pr^{D2U}, can be derived as

$$Pr^{D2U} = Pr[Z = \text{``D''}|S = \text{``U''}]$$

$$= \sum_{l=1}^{T} Pr[(Z = \text{``D''}|L = l)|S = \text{``U''}]$$

$$\times Pr[L = l|S = \text{``U''}]$$

$$\overset{(a)}{=} \sum_{l=1}^{T} q_l^D Pr[L = l|S = \text{``U''}]$$

$$\overset{(b)}{=} \sum_{l=1}^{T} q_l^D \frac{Pr[S = \text{``U''}|L = l]Pr[L = l]}{Pr[S = \text{``U''}]}$$

$$\overset{(c)}{=} \sum_{l=1}^{T} q_l^D \frac{q_l^U \frac{1}{T}}{\frac{1}{T}\sum_{j=1}^{T} q_j^U}, \tag{11.21}$$

where

- step (a) follows from the equality,

$$Pr[(Z = \text{``D''}|L = l)|S = \text{``U''}] = Pr[Z = \text{``D''}|L = l] = q_l^D,$$

due to the independence of the event, $(Z = \text{``D''}|L = l)$, and the event, $(S = \text{``U''})$,
- step (b) follows from Bayes' Theorem and
- step (c) follows from the calculation on the probability of the carrier signal being an uplink one, which can be written as

$$Pr[S = \text{``U''}] = \sum_{j=1}^{T} Pr[S = \text{``U''}|L = j]Pr[L = j]$$

$$= \frac{1}{T}\sum_{j=1}^{T} q_j^U. \tag{11.22}$$

The results for the probability of uplink-to-downlink interference, Pr^{U2D}, can be derived in a similar manner, which concludes our proof. $\square$

Let us also remark at this point that, as discussed at the beginning of Section 11.4, for static TDD, we have that the probabilities, Pr^{D2U} and Pr^{U2D}, amount to 0, i.e. $Pr^{D2U} = Pr^{U2D} = 0$, since all the TDD subframes are synchronized and of the same direction. On the other hand, as shown in Theorem 11.4.2, for dynamic TDD we have that such probabilities, Pr^{D2U} and Pr^{U2D}, may be larger than 0, and can have a major impact on the evaluation of the aggregated downlink and uplink originated inter-cell interference, I_{agg}^D and I_{agg}^U, in equation (11.6).

For completeness, let us also indicate that, in asynchronous dynamic TDD systems, the probabilities, Pr^{D2U} and Pr^{U2D}, equate to $Pr^{D2U} = p^D$ and $Pr^{U2D} = p^U$ [241], respectively, due to the randomness governing the collision process between asynchronous dynamic TDD subframes in neighbouring cells.

11.4.6 Average Downlink, Average Uplink and Average Total Time Resource Utilizations

From Theorem 11.4.1, as well as equations (11.2) and (11.3), we can derive the average downlink, the average uplink and the average total TRUs, κ^D, κ^U and κ, in a dynamic TDD network, which are presented in Lemma (11.4.3).

LEMMA 11.4.3 *In a dynamic TDD system, the average downlink, the average uplink and the average total TRUs, κ^D, κ^U and κ, are given by*

$$\begin{cases} \kappa^D = \frac{1}{T} \sum_{l=1}^{T} \sum_{\tilde{k}=1}^{+\infty} \left(1 - \sum_{i=0}^{l-1} f_{N^D}(i)\right) f_{\tilde{K}}(\tilde{k}) \\ \kappa^U = \frac{1}{T} \sum_{l=1}^{T} \sum_{\tilde{k}=1}^{+\infty} \sum_{i=0}^{l-1} f_{N^D}(i) f_{\tilde{K}}(\tilde{k}) \\ \kappa = 1 \end{cases} \qquad (11.23)$$

where

- *the PMFs, $f_{\tilde{K}}(\tilde{k})$ and $f_{N^D}(i)$, are given by equations (11.9) and (11.14), respectively.*

Proof The proof is straightforward by plugging equation (11.16) into equation (11.2) and subsequently into equation (11.3). Moreover, considering the results of the average downlink and the average uplink TRUs, κ^D and κ^U, in equation (11.23), we can compute the average total TRU, $\kappa = \kappa^D + \kappa^U$, as

$$\kappa = \frac{1}{T} \sum_{j=1}^{T} \sum_{\tilde{k}=1}^{+\infty} f_{\tilde{K}}(\tilde{k}) = \frac{1}{T} \sum_{j=1}^{T} 1 = 1, \qquad (11.24)$$

which concludes the proof. ☐

Importantly, note that the results in Lemma (11.4.3) can be applied to dynamic TDD in both the asynchronous and the synchronous cases, because such TRUs characterize the efficiency of the resource usage in dynamic TDD, no matter what the synchronization assumption is. Lemma (11.4.3) not only quantifies the average MAC layer performance of a dynamic TDD cell, but also shows from a theoretical viewpoint that dynamic TDD can always achieve a full resource utilization, i.e. $\kappa = 1$, provided that there is traffic for it, thanks to the smart adaptation of the downlink and the uplink subframes to the downlink and the uplink data requests.

In the following, and for comparison purposes, we also derive the average down-link, the average uplink and the average total TRUs, κ^D, κ^U and κ, in a static TDD network, which are presented in Theorem 11.4.4.

THEOREM 11.4.4   *In static TDD system, the average downlink, the average uplink and the average total TRUs, κ^D, κ^U and κ, are given by*

$$
\begin{cases}
\kappa^D = \left(1 - \sum\limits_{\tilde{k}=1}^{+\infty} \left(1 - p^D\right)^{\tilde{k}} f_{\tilde{K}}(\tilde{k})\right) \frac{N_0^D}{T} \\[2ex]
\kappa^U = \left(1 - \sum\limits_{\tilde{k}=1}^{+\infty} \left(1 - p^U\right)^{\tilde{k}} f_{\tilde{K}}(\tilde{k})\right) \frac{N_0^U}{T} \\[2ex]
\kappa = \frac{1}{T} \sum\limits_{\tilde{k}=1}^{+\infty} \left[\left(1 - \left(p^U\right)^{\tilde{k}}\right) N_0^D + \left(1 - \left(p^D\right)^{\tilde{k}}\right) N_0^U\right] f_{\tilde{K}}(\tilde{k})
\end{cases}
\tag{11.25}
$$

where

- *the PMF, $f_{\tilde{K}}(\tilde{k})$, is given by equation (11.9) and*
- N_0^D *and* N_0^U *are the number of subframes designated for downlink and uplink use, respectively, in static TDD system, which satisfy, $N_0^D + N_0^U = T$.*

Proof   In a static TDD system, for a given number, $\tilde{k}$, of UEs per active small cell BS, the probabilities that (i) no UE requests any downlink data and (ii) no UE requests any uplink data can be calculated from the PMF, $f_{M^D}(0)$, in equation (11.11) and from the PMF, $f_{M^U}(0)$, in equation (11.12), respectively. In such two cases, static TDD unwisely allocates N_0^D and N_0^U subframes to the downlink and the uplink, respectively, which results in a resource waste.

The probabilities of such a resource waste in the downlink and the uplink can be denoted by w^D and w^U, respectively, and can be calculated as

$$
\begin{cases}
w^D = \sum\limits_{\tilde{k}=1}^{+\infty} f_{M^D}(0) f_{\tilde{K}}(\tilde{k}) = \sum\limits_{\tilde{k}=1}^{+\infty} \left(1 - p^D\right)^{\tilde{k}} f_{\tilde{K}}(\tilde{k}) \\[2ex]
w^U = \sum\limits_{\tilde{k}=1}^{+\infty} f_{M^U}(0) f_{\tilde{K}}(\tilde{k}) = \sum\limits_{\tilde{k}=1}^{+\infty} \left(1 - p^U\right)^{\tilde{k}} f_{\tilde{K}}(\tilde{k})
\end{cases}
\tag{11.26}
$$

Excluding such a resource waste from the number of subframes designated for downlink and uplink use, N_0^D and N_0^U, we can obtain the average downlink, the average uplink and the average total TRUs, κ^D, κ^U and κ, in a static TDD system as

$$
\begin{cases}
\kappa^D = \left(1 - w^D\right) \frac{N_0^D}{T} \\[2ex]
\kappa^U = \left(1 - w^U\right) \frac{N_0^U}{T}
\end{cases}
\tag{11.27}
$$

The proof is completed by plugging equations (11.26), (11.11) and (11.12) into equation (11.27), followed by computing the average total TRU, κ, from equation (11.3).   □

From Lemma (11.4.3) and Theorem 11.4.4, we can further quantify the additional average total TRU gain that dynamic TDD can achieve as

$$\kappa^{\text{ADD}} = \frac{1}{T} \sum_{\tilde{k}=1}^{+\infty} \left[\left(p^{\text{U}} \right)^{\tilde{k}} N_0^{\text{D}} + \left(p^{\text{D}} \right)^{\tilde{k}} N_0^{\text{U}} \right] f_{\tilde{K}}(\tilde{k}), \tag{11.28}$$

where

- κ^{ADD} measures the difference of the average total TRU, κ, in equation (11.23) and that in equation (11.25).

In addition, in the following, we characterize the limit of the additional average total TRU, κ^{ADD}, in Lemma (11.4.5).

LEMMA 11.4.5   *When the small cell BS density tends to infinity, i.e. $\lambda \to +\infty$, the limit of the additional average total TRU, κ^{ADD}, is given by*

$$\lim_{\lambda \to +\infty} \kappa^{\text{ADD}} = \frac{p^{\text{D}} N_0^{\text{U}}}{T} + \frac{p^{\text{U}} N_0^{\text{D}}}{T}. \tag{11.29}$$

Proof   From equation (11.9), we have the following equality, $\lim_{\lambda \to +\infty} \Pr[\tilde{K} = 1] = 1$. Thus, using Lemma (11.4.3), we can draw the following conclusions about dynamic TDD:

$$\begin{cases} \lim_{\lambda \to +\infty} \kappa^{\text{D}} = 1 - f_{N^{\text{D}}}(0) = 1 - p^{\text{U}} = p^{\text{D}} \\ \lim_{\lambda \to +\infty} \kappa^{\text{U}} = 1 - f_{N^{\text{U}}}(0) = 1 - p^{\text{D}} = p^{\text{U}} \end{cases} \tag{11.30}$$

Similarly, based on the same equality, $\lim_{\lambda \to +\infty} \Pr[\tilde{K} = 1] = 1$, and using Theorem 11.4.4, we can draw the following conclusion about static TDD:

$$\begin{cases} \lim_{\lambda \to +\infty} \kappa^{\text{D}} = \left(1 - \left(1 - p^{\text{D}} \right) \right) \frac{N_0^{\text{D}}}{T} = \frac{p^{\text{D}} N_0^{\text{D}}}{T} \\ \lim_{\lambda \to +\infty} \kappa^{\text{U}} = \left(1 - \left(1 - p^{\text{U}} \right) \right) \frac{N_0^{\text{U}}}{T} = \frac{p^{\text{U}} N_0^{\text{U}}}{T} \end{cases} \tag{11.31}$$

The proof is completed by comparing the limits in equation (11.30) with those in equation (11.31).   □

Let us highlight that in equation (11.29) of Lemma (11.4.5), the first and the second terms represent the contributions from the downlink and the uplink, respectively.

11.5   Discussion

In this section, we use numerical results from static system-level simulations to evaluate the accuracy of the theoretical performance analyses presented in this chapter.

11.5.1 Case Study

To evaluate the performance of a dynamic TDD ultra-dense network, the same 3GPP case studies as in Chapter 8 for the downlink and in Chapter 10 for the uplink have been used to allow an apple-to-apple comparison. For readers interested in a more detailed description of these 3GPP case studies, please refer to Section 8.5.1 and Section 10.4.1, respectively.

As a summary, let us highlight that the path loss model recommended by [153] is adopted, which is comprised of two slopes, i.e. $N = 2$, and for each slope, $n \in \{1, 2\}$, we have that

- $A_n^{B2U, L} = 10^{-10.38}, \alpha_n^{B2U, L} = 2.09, A_n^{B2U, NL} = 10^{-14.54}, \alpha_n^{B2U, NL} = 3.75,$
- $A_n^{B2B, L} = 10^{-9.84}, \alpha_n^{B2B, L} = 2, A_n^{B2B, NL} = 10^{-16.94}, \alpha_n^{B2B, NL} = 4,$
- $A_n^{U2U, L} = 10^{-9.85}, \alpha_n^{U2U, L} = 2, A_n^{U2U, NL} = 10^{-17.58}$ and $\alpha_n^{U2U, NL} = 4.$

The following LoS probability function is used for both the BS-to-UE and the BS-to-BS links, $\Pr^{B2U, L}(d)$ and $\Pr^{B2B, L}(d)$, which is defined as a two-piece exponential function:

$$\Pr^{B2U, L}(d) = \Pr^{B2B, L}(d) = \begin{cases} 1 - 5\exp(-R_1/d), & 0 < d \leq d_1 \\ 5\exp(-d/R_2), & d > d_1 \end{cases}, \quad (11.32)$$

where

- $R_1 = 156\,\text{m},$
- $R_2 = 30\,\text{m}$ and
- $d_1 = \frac{R_1}{\ln 10}.$

The following – more simple – LoS probability function is instead utilized for the UE-to-UE links, $\Pr^{U2U, L}(d)$:

$$\Pr^{U2U, L}(d) = \begin{cases} 1, & 0 < d \leq 50\,\text{m} \\ 0, & d > 50\,\text{m} \end{cases}. \quad (11.33)$$

Moreover, and also according to [153], note that transmit power values are set to

- $P^D = 24\,\text{dBm},$
- $P^U = 23\,\text{dBm},$
- $P_N^D = -95\,\text{dBm}$ (including a noise figure of 9 dB at the UE) and
- $P_N^U = -91\,\text{dBm}$ (including a noise figure of 13 dB at the BS).

The UE density, ρ, is set to $300\,\text{UEs/km}^2$ with a fitting parameter, q, equal to 3.5, while the antenna height difference, L, takes the value of 8.5 m, where the small cell BS antenna height and the UE antenna height are assumed to be 10 m and 1.5 m, respectively. We also assume that the total number of subframes, T, in a TDD frame is 10, and that the probability, p^D, of a UE requesting downlink data in a subframe is $\frac{2}{3}$. Thus, for static TDD, this results in $N_0^D = 7$ downlink subframes and $N_0^U = 3$ uplink subframes, as per equation (11.25), which achieves the best match between the numbers, N_0^D and N_0^U, of downlink and uplink subframes and the probabilities, p^D and

p^U, of a UE requesting downlink or uplink data in a subframe, according to equation (11.13).

11.5.2   Validation of the Results on the Subframe-Dependent Time Resource Utilization

In Figure 11.5, analytical and simulation results with respect to the subframe-dependent downlink TRU, q_l^D, are presented. For brevity, we do not show the results of the subframe-dependent uplink TRU, q_l^U, because of its duality with the variable, q_l^D, shown in equation (11.19).

From this figure, we can observe that

- the analytical results of the subframe-dependent downlink TRU, q_l^D, match well with the simulation ones. More specifically, the maximum difference between the analytical and the simulation results is shown to be less than 0.3%. It should also be noted that the curves shown in Figure 11.5 are not smooth. This is because of the non-continuous mapping in equation (11.13), and not due to an insufficient number of simulation experiments.
- The subframe-dependent downlink TRU, q_l^D, monotonically decreases with the subframe number, l, because of the considered downlink-before-uplink TDD structure shown in Figure 11.1. In other words, if scheduled by the small cell BSs, the downlink transmissions are to be conducted in the first few subframes of the TDD frame.

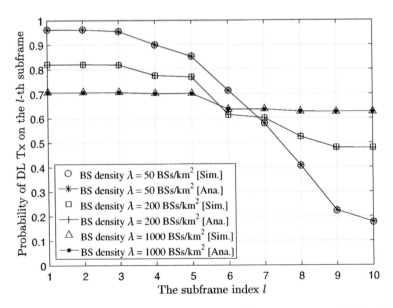

Figure 11.5 Subframe-dependent downlink TRU, q_l^D ($\rho = 300$ UEs/km^2, $p^D = \frac{2}{3}$, $T = 10$, various λ).

- When the small cell BS density, λ, is relatively small compared with the UE density, ρ, e.g. $\lambda = 50\,\text{BSs/km}^2$, note that q_1^D approaches 1 and q_{10}^D is close to 0. This is because:

  □ the number, $\tilde{k}$, of UEs per active small cell BS tends to be relatively large, e.g. when $\lambda = 50\,\text{BSs/km}^2$, we have that $\tilde{k} = 4\,\text{UEs/BS}$, as shown in Figure 11.2.
  □ As a result, the number, m^D, of downlink data requests is non-zero for most cases, e.g. when $\tilde{k} = 4$ and $p^D = \frac{2}{3}$, we have that $m^D = 3$, as shown in Figure 11.3, and
  □ the number, n^D, of downlink subframes also tends to be relatively large to accommodate for the mentioned number, m^D, of downlink data requests, e.g. when $\tilde{k} = 4$, $p^D = \frac{2}{3}$ and $T = 10$, we have that $n^D = \{3, 5, 8\}$, as shown in Figure 11.4.
  □ Due to the downlink-before-uplink TDD structure shown in Figure 11.1, the first subframe has a very high probability to be a downlink one, i.e. q_1^D is almost 1, while the last subframe is mostly likely an uplink one, i.e. q_{10}^D is close to 0.

- When the small cell BS density, λ, is relatively large compared with the UE density, ρ, e.g. $\lambda = 1000\,\text{BSs/km}^2$, note that the subframe-dependent downlink TRU, q_l^D, is almost a constant for all values of the subframe number, l, and that such a constant roughly equals to the probability, p^D, of a UE requesting downlink data, i.e. $p^D = \frac{2}{3}$. This is because:

  □ more than 80% of the active small cell BSs will serve $\tilde{k} = 1$ UE when the small cell BS, λ, is $1000\,\text{BSs/km}^2$, as shown in Figure 11.2.
  □ As a result, when an active small cell BS serves $\tilde{k} = 1$ UE, both the number, m^D, of downlink data requests and the number, n^D, of downlink subframes show a high fluctuation, as shown in Figures 11.3 and 11.4, respectively, due to the strong traffic dynamics.
  □ Thus, when the active small cell BSs serves $\tilde{k} = 1$ UE, in most cases, all the subframes are dynamically used as either downlink or uplink ones, the probability of which solely depends on the probabilities, p^D and p^U, of such a UE requesting downlink or uplink data in a subframe. Since $p^D = \frac{2}{3}$, we have that $q_l^D \approx \frac{2}{3}$.

- It is easy to see that the case where small cell BS density, λ, is $200\,\text{BSs/km}^2$ plays in the middle of the above presented extreme cases, i.e. $\lambda = 50\,\text{BSs/km}^2$ and $\lambda = 1000\,\text{BSs/km}^2$.

11.5.3   Validation of the Results on the Probabilities of Inter-Cell Interlink Interference

In Figure 11.6, analytical and simulation results with respect to the probability, Pr^{D2U}, of downlink-to-uplink interference and the probability, Pr^{U2D}, of uplink-to-downlink interference are presented.

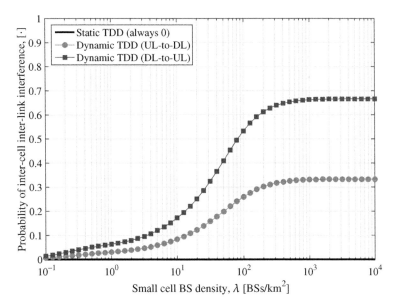

Figure 11.6 Probability of inter-cell interlink interference ($\rho = 300$ UEs/km^2, $p^{\mathrm{D}} = \frac{2}{3}$ and $T = 10$).

From this figure, we can observe that

- the analytical and simulation results match well, which validates the accuracy of the presented analysis.
- For static TDD, both the probability, $\mathrm{Pr}^{\mathrm{D2U}}$, of downlink-to-uplink interference and the probability, $\mathrm{Pr}^{\mathrm{U2D}}$, of uplink-to-downlink interference are 0, as discussed in Section 11.4.5.
- For dynamic TDD, as the small cell BS density, λ, increases, both the probability, $\mathrm{Pr}^{\mathrm{D2U}}$, of downlink-to-uplink interference and the probability, $\mathrm{Pr}^{\mathrm{U2D}}$, of uplink-to-downlink interference gradually increase and converge to the probabilities, p^{D} and p^{U}, of a UE requesting downlink or uplink data in a subframe, respectively. This is because,
  - ☐ when the small cell BS, λ, increases, the number, $\tilde{k}$, of UEs in an active small cell BS decreases, and thus both the numbers, m^{D} and m^{U}, of downlink and uplink data requests and the numbers, n^{D} and n^{U}, of downlink and uplink subframes show a high fluctuation.
  - ☐ Thus, when the small cell BS, λ, is high enough to reach the limit of one UE per active small cell BS, all the subframes will be used as either downlink or uplink ones, the probability of which solely depends on the probabilities, p^{D} and p^{U}, of such a UE requesting downlink or uplink data in a subframe.

11.5.4 Validation of the Results on the Average Time Resource Utilization

In Figure 11.7, analytical and simulation results with respect to the average downlink TRU, κ^{D}, and the average uplink TRU, κ^{U}, are presented.

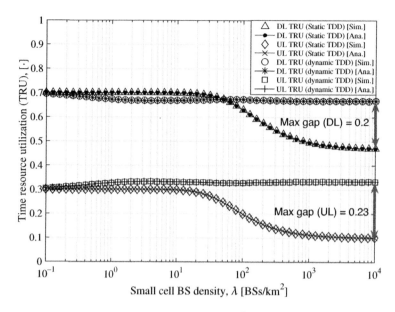

Figure 11.7 Average downlink and uplink TRU, κ^D and κ^U ($\rho = 300$ UEs/km^2, $p^D = \frac{2}{3}$, $T = 10$, various λ).

From this figure, we can observe that

- the analytical and simulation results match well, which validates the accuracy of the presented analysis.
- For static TDD, the total average TRU, κ, is shown to be always smaller than one in ultra-dense small cell networks, e.g. $\lambda > 10$ BSs/km^2, which verifies Theorem 11.4.4, and shows the main drawback of this technology. In more detail, the average downlink TRU, κ^D, when using static TDD starts from 0.7, and then decreases, as the small cell BS, λ, increases. Given the number, $T = 10$, of subframes in a TDD frame and the number, $N_0^D = 7$, of downlink subframes in the statistically optimal static TDD configuration, static TDD maintains $N_0^U = 3$ uplink subframes, even if there is no uplink data request in a small cell BS, leading to a subframe waste. Also note that the total average TRU, κ, is much less than 1 when the small cell BS, λ, is large, e.g. $\lambda = 10^4$ BSs/km^2, showing the inefficiency of static TDD in dense small cell networks.
- For dynamic TDD, the total average TRU, κ, is shown to be always equal to 1, which verifies Lemma 11.4.3, and shows the main advantage of this technology. In more detail, the average downlink TRU, κ^D, when using dynamic TDD starts from 0.7, and converges to $p^D = \frac{2}{3}$, as the small cell BS, λ, increases. This is because when the small cell BS, λ, is high enough to reach the limit of one UE per active small cell BS, all the subframes will be used as downlink ones with a probability of p^D.

- According to Lemma 11.4.5, the additional average total TRU, κ^{ADD}, achievable by dynamic TDD over static TDD, i.e. $\lim_{\lambda \to +\infty} \kappa^{ADD}$, should be composed of two parts. One part, $\frac{p^D N_0^U}{T}$, contributed by the downlink and another one, $\frac{p^U N_0^D}{T}$, contributed by the uplink, which in this case are 0.2 and 0.23, respectively. These values are validated in Figure 11.7, and thus the limit, $\lim_{\lambda \to +\infty} \kappa^{ADD}$, can be estimated to be 0.43 (+75.4%).

11.5.5   The Coverage Probability Performance

In Figures 11.8 and 11.9, simulation results with respect to the downlink coverage probability and the uplink coverage probability are presented, respectively. It is important to note that the analytical results on the probability, Pr^{D2U}, of downlink-to-uplink interference and the probability, Pr^{U2D}, of uplink-to-downlink interference are used for the simulation of the aggregated downlink and uplink originated inter-cell interference, I_{agg}^D and I_{agg}^U, discussed in equation (11.6). Moreover, since the uplink SINR is vulnerable to the downlink-to-uplink interference, in Figure 11.9 the effectiveness of both full and partial IC technologies are investigated [91], which remove all and the top three downlink-to-uplink interfering signals based on the small cell BS-to-small cell BS path loss, respectively.

From Figures 11.8 and 11.9, we can importantly observe that

- dynamic TDD without IC results in a poor uplink coverage probability with respect to static TDD due to the mentioned strong downlink-to-uplink interference.

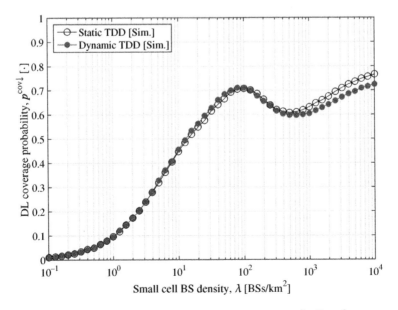

Figure 11.8 Downlink coverage probability ($\rho = 300\,\mathrm{UEs/km^2}$, $p^D = \frac{2}{3}$ and $T = 10$).

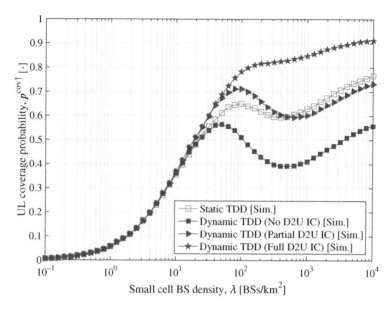

Figure 11.9 Uplink coverage probability ($\rho = 300$ UEs/km^2, $p^D = \frac{2}{3}$ and $T = 10$).

- With the use of full IC and partial IC, the uplink coverage probability of dynamic TDD can be significantly improved.

As in Chapters 3–10, to visualize the fundamental behavior of the downlink and uplink coverage probabilities in a more intuitive manner, Figure 11.10 shows the downlink and uplink coverage probability heat maps for the selected UE density, i.e. 300 UEs/km^2 and three different small cell BS densities, i.e. 50, 250 and 2500 BSs/km^2. These heat maps are computed using NetVisual, which is able to capture not only the mean, but also the standard deviation of the coverage probability as introduced in Section 2.4.

From Figure 11.10, we can see that:

- Comparing Figures 11.10a and 11.10c, we can see that the downlink coverage probability of static TDD is comparable to that of dynamic TDD, mainly because of the comparable transmit powers in the downlink (24 dBm) and in the uplink (23 dBm). Thus, there is not much of a change when swapping the inter-cell interference from a small cell BS to a UE.
- In Figure 11.10b, when considering the ultra-dense network with a small cell BS density, λ, around 2500 BSs/km^2, the uplink coverage probability heat map of the dynamic TDD network is significantly dark, showing that the uplink of such a type of network is particularly vulnerable to the downlink-to-uplink inter-cell interference.
- A potential solution to mitigate the large downlink-to-uplink inter-cell interference is full and partial IC, as mentioned earlier. In Figure 11.10e, where we assume that each small cell BS can remove the top three interfering signals

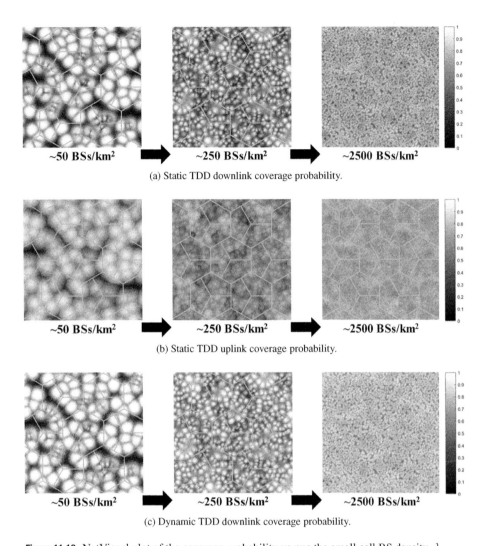

(a) Static TDD downlink coverage probability.

(b) Static TDD uplink coverage probability.

(c) Dynamic TDD downlink coverage probability.

Figure 11.10 NetVisual plot of the coverage probability versus the small cell BS density, λ.

from other neighbouring small cell BSs, the uplink coverage probability heat map of the dynamic TDD network shows a considerable performance improvement with respect to that in Figure 11.10d. Although not shown in this figure, performance is further improved with full IC, as shown in Figure 11.9.

11.5.6   The Area Spectral Efficiency Performance

From Figures 11.8, 11.9 and 11.10, we should not conclude that dynamic TDD without IC exhibits no performance gain with respect to static TDD due to the detrimental downlink-to-uplink interference. This is because the coverage probability is a metric mostly driven by the UE SINRs, and does not capture the effect of the amount of

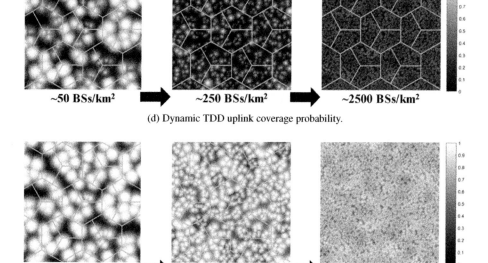

(d) Dynamic TDD uplink coverage probability.

(e) Dynamic TDD uplink coverage probability with partial IC.

Figure 11.10 NetVisual plot of the coverage probability versus the small cell BS density, λ. (cont.)

time-frequency resources allocated by the network to such UEs. To have a complete view we need to investigate the ASE defined in equation (11.7), which characterizes the MAC layer TRU.

In Figures 11.11 and 11.12, simulation results with respect to the downlink ASE and the uplink ASE in a logarithmic and a linear scale are presented, respectively.

From Figures 11.11 and 11.12, we can observe that

- dynamic TDD can achieve a larger total ASE – downlink plus uplink ASE – than static TDD, even without IC, mainly due to the dynamic adaptation of the downlink and the uplink subframes to the downlink and the uplink data requests. For example, under the utilized system model, and when the small cell BS, λ, is 10^4 BSs/km², the total ASE of static TDD and that of dynamic TDD without IC are 387.1 and 587.1 bps/Hz/km² (+51.67%), respectively.
- If the downlink-to-uplink interference is further addressed by means of IC, for the same small cell BS, $\lambda = 10^4$ BSs/km², the ASE of dynamic TDD can be further improved to 675.0 bps/Hz/km² (+74.37%/+14.97%) with partial IC and 953.9 bps/Hz/km² (+146.42%/+62.48%) with full IC. Note that the first and the second percentage improvements are with respect to static TDD and dynamic TDD without IC, respectively, where the latter shows the importance of dealing with the downlink-to-uplink interference.

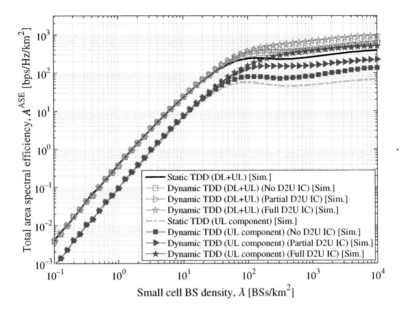

Figure 11.11 ASE (logarithmic scale).

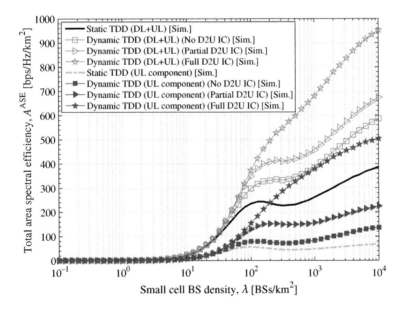

Figure 11.12 ASE (linear scale).

11.6   Conclusions

In this chapter, the MAC layer performance of a synchronous dynamic TDD network in terms of TRU has been investigated as a function of the network densification. Dynamic TDD has been shown to have an increasingly higher average total TRU than static TDD of up to 75.4% in ultra-dense networks. This shows the importance of adapting the downlink and uplink network resources to match the incoming downlink and uplink traffic load. The studies presented have also shown that the probabilities of inter-cell interlink interference vary with the small cell BS density, λ, and the UE density, ρ. Besides, the results have shown the importance of dealing with the downlink-to-uplink interference by means of, for example, full or partial IC to make the most of dynamic TDD, especially in ultra-dense networks.

References

[1] E. S. Grosvenor and M. Wesson, *Alexander Graham Bell: The Life and Times of the Man Who Invented the Telephone*. New York: Harry N. Abrams, 1997.

[2] CISCO, "Cisco visual networking index: Global mobile data traffic forecast update (2017–2022)," Feb. 2019. https://s3.amazonaws.com/media.mediapost.com/uploads/CiscoForecast.pdf

[3] T. Berners-Lee, *Weaving the Web: The Original Design and Ultimate Destiny of the World Wide Web*. New York: Harper Business, 2000.

[4] I. McCulloh, H. Armstrong and A. Johnson, *Social Network Analysis with Applications*. Hoboken, NJ: John Wiley & Sons Ltd., 2013.

[5] M. K. Weldon, *The Future X Network: A Bell Labs Perspective*. Boca Raton, FL: CRC Press, 2015.

[6] U. Black, *Voice over IP*, 1st ed. Upper Saddle River, NJ: Prentice Hall, 1999.

[7] C. Poynton, *Digital Video and HD: Algorithms and Interfaces*, 2nd ed. New York: Elsevier, 2012.

[8] D. López-Pérez, M. Ding, H. Claussen and A. Jafari, "Towards 1 Gbps/UE in cellular systems: Understanding ultra-dense small cell deployments," *IEEE Communications Surveys Tutorials*, vol. 17, no. 4, pp. 2078–2101, Jun. 2015.

[9] C. E. Shannon, "Communication in the presence of noise," *Proceedings of the Institute of Radio Engineers*, vol. 37, no. 1, pp. 10–21, Jan. 1949.

[10] R. W. Heath and A. Lozano, *Foundations of MIMO Communication*. Cambridge: Cambridge University Press, 2018.

[11] W. Webb, *Wireless Communications: The Future*. Chichester: John Wiley & Sons Ltd., 2007.

[12] H. Holma, D. A. Toskala and T. Nakamura, *5G Technology : 3GPP New Radio*. Chichester: John Wiley & Sons Ltd., 2020.

[13] E. Dahlman, S. Parkvall and J. Skold, *5G NR: The Next Generation Wireless Access Technology*. Cambridge, MA: Academic Press, 2018.

[14] Amdocs, "Game changing economics for small cell deployment," White Paper, Oct. 2014. https://docplayer.net/19134752-Game-changing-economics-for-small-cell-deployment-amdocs-oss-white-paper-october-2013.html

[15] Nokia, "Indoor deployment strategies," White Paper, Jun. 2014. www.scribd.com/document/249226439/Nokia-Indoor-Deployment-Strategies

[16] A. C. Stocker, "Enhanced intercell interference coordination challenges in heterogeneous networks," *IEEE Transactions on Vehicular Technology*, vol. 33, no. 4, pp. 269–275, 1984.

[17] R. Iyer, J. Parker and P. Sood, "Intelligent networking for digital cellular systems and the wireless world," in *IEEE Global Telecommunications Conference (GLOBECOM)*, vol. 1, pp. 475–479, Dec. 1990.

[18] L. T. W. Ho, *"Self-organising algorithms for fourth generation wireless networks and its analysis using complexity metrics,"* Ph.D. Thesis, Queen Mary College, University of London, Jun. 2003.

[19] H. Claussen, L. T. W. Ho, H. R. Karimi, F. J. Mullany and L. G. Samuel, "I, base station: Cognisant robots and future wireless access networks," in *Proceedings 3rd IEEE Consumer Communications and Networking Conference (CCNC)*, Las Vegas, NV, pp. 595–599, Jan. 2006.

[20] H. Claussen, L. T. W. Ho and L. G. Samuel, "An overview of the femtocell concept," *Bell Labs Technical Journal*, vol. 15, no. 3, pp. 137–147, Dec. 2008.

[21] L. T. W. Ho and H. Claussen, "Effects of user-deployed, co-channel femtocells on the call drop probability in a residential scenario," in *IEEE International Symposium on Personal, Indoor and Mobile Radio Communications (PIMRC)*, Athens, Greece, Sep. 2007.

[22] H. Claussen, L. T. W. Ho and L. G. Samuel, "Financial analysis of a pico-cellular home network deployment," in *IEEE International Conference on Communications (ICC)*, Glasgow, UK, pp. 5604–5609, Jun. 2007.

[23] H. Claussen, "Performance of macro- and co-channel femtocells in a hierarchical cell structure," in *IEEE International Symposium on Personal, Indoor and Mobile Radio Communications (PIMRC)*, Athens, Greece, Sep. 2007.

[24] H. Claussen, L. T. W. Ho and F. Pivit, "Effects of joint macrocell and residential picocell deployment on the network energy efficiency," in *IEEE International Symposium on Personal, Indoor and Mobile Radio Communications (PIMRC)*, Cannes, France, Sep. 2008.

[25] H. Claussen, L. T. W. Ho and L. G. Samuel, "Self-optimization of coverage for femtocell deployments," in *Proceedings Wireless Telecommunications Symposium (WTS)*, Los Angeles, CA, pp. 278–285, Apr. 2008.

[26] H. Claussen, "Co-channel operation of macro- and femtocells in a hierarchical cell structure," *International Journal of Wireless Information Networks*, vol. 15, no. 3, pp. 137–147, Dec. 2008.

[27] H. Claussen, L. T. W. Ho and F. Pivit, "Leveraging advances in mobile broadband technology to improve environmental sustainability," *Telecommunications Journal of Australia*, vol. 59, no. 1, pp. 4.1–4.18, Feb. 2009.

[28] H. Claussen and F. Pivit, "Femtocell coverage optimization using switched multi-element antennas," in *IEEE International Conference on Communications (ICC)*, Dresden, Germany, Jun. 2009.

[29] H. Claussen and D. Calin, "Macrocell offloading benefits in joint macro- and femtocell deployments," in *IEEE International Symposium on Personal, Indoor and Mobile Radio Communications (PIMRC)*, Tokyo, Japan, pp. 350–354, Sep. 2009.

[30] H. Claussen, L. T. W. Ho and F. Pivit, "Self-optimization of femtocell coverage to minimize the increase in core network mobility signalling," *Bell Labs Technical Journal*, vol. 14, no. 2, pp. 155–184, Aug. 2009.

[31] D. López-Pérez, A. Valcarce, G. de la Roche and J. Zhang, "Access methods to WiMAX femtocells: A downlink system-level case study," in *11th IEEE Singapore International Conference on Communication Systems*, pp. 1657–1662, Nov. 2008.

[32] D. López-Pérez, G. de la Roche, A. Valcarce, A. Juttner and J. Zhang, "Interference avoidance and dynamic frequency planning for WiMAX femtocells networks," in *11th IEEE Singapore International Conference on Communication Systems*, pp. 1579–1584, Nov. 2008.

[33] D. López-Pérez, A. Valcarce, G. de la Roche and J. Zhang, "OFDMA femtocells: A roadmap on interference avoidance," *IEEE Communications Magazine*, vol. 47, no. 9, pp. 41–48, Oct. 2009.

[34] D. López-Pérez, A. Ladanyi, A. Jüttner and J. Zhang, "OFDMA femtocells: A self-organizing approach for frequency assignment," in *IEEE International Symposium on Personal, Indoor and Mobile Radio Communications (PIMRC)*, Tokyo, Japan, Sep. 2009.

[35] G. D. L. Roche, A. Valcarce, D. López-Pérez and J. Zhang, "Access control mechanisms for femtocells," *IEEE Communications Magazine*, vol. 48, no. 1, pp. 33–39, Jan. 2010.

[36] D. López-Pérez, X. Chu, A. V. Vasilakos and H. Claussen, "Power minimization based resource allocation for interference mitigation in OFDMA femtocell networks," *IEEE Journal on Selected Areas in Communications*, vol. 32, no. 2, pp. 333–344, Feb. 2014.

[37] V. Chandrasekhar and J. G. Andrews, "Spectrum allocation in tiered cellular networks," *IEEE Transactions on Communications*, vol. 57, no. 10, pp. 3059–3068, Oct. 2009.

[38] V. Chandrasekhar and J. G. Andrews, "Uplink capacity and interference avoidance for two-tier femtocell networks," *IEEE Transactions on Wireless Communications*, vol. 8, no. 7, pp. 3498–3509, Jul. 2009.

[39] V. Chandrasekhar, J. G. Andrews, T. Muharemovic, Z. Shen and A. Gatherer, "Power control in two-tier femtocell networks," *IEEE Transactions on Wireless Communications*, vol. 8, no. 8, pp. 4316–4328, Aug. 2009.

[40] D. López-Pérez, I. Guvenc, G. de la Roche, et al., "Enhanced intercell interference coordination challenges in heterogeneous networks," *IEEE Wireless Communications*, vol. 18, no. 3, pp. 22–30, Jun. 2011.

[41] D. López-Pérez, I. Guvenc and X. Chu, "Mobility management challenges in 3GPP heterogeneous networks," *IEEE Communications Magazine*, vol. 50, no. 12, pp. 70–78, Dec. 2012.

[42] D. López-Pérez, X. Chu and I. Guvenc, "On the expanded region of picocells in heterogeneous networks," *IEEE Journal of Selected Topics in Signal Processing*, vol. 6, no. 3, pp. 281–294, Mar. 2012.

[43] BeFEMTO. (2016) Broadband evolved femto networks. [Online]. Available: www.ict-befemto.eu/

[44] IEEE. (2016) IEEE Xplore Digital Library. [Online]. Available: http://ieeexplore.ieee.org/

[45] Small Cell Forum press release. (2011, Jun.) 3G femtocells now outnumber conventional 3G basestations globally. [Online]. Available: www.smallcellforum.org/press-releases/3g-femtocells-now-outnumber-conventional-3g-basestations-globally/

[46] Small Cell Forum. (2015, Jun.) Market status statistics June 2015 – Mobile Experts. [Online]. Available: http://scf.io/en/documents/050_-_Market_status_report_June_2015_-_Mobile_Experts.php

[47] S. Reedy. (2013, Sep.) Multimode small cells get stalled in labs. [Online]. Available: www.lightreading.com/mobile/small-cells/multimode-small-cells-get-stalled-in-labs/d/d-id/703334

[48] Small Cell Forum. (2018, Dec.) Small cells market status report. [Online]. Available: https://scf.io/en/documents/050_-_Small_cells_market_status_report_December_2018.php

[49] S. Hamalainen, H. Sanneck and C. Sartori, *LTE Self-Organising Networks (SON): Network Management Automation for Operational Efficiency*. Chichester: John Wiley & Sons Ltd., 2011.

[50] M. Haenggi, J. G. Andrews, F. Baccelli, O. Dousse and M. Franceschetti, "Stochastic geometry and random graphs for the analysis and design of wireless networks," *IEEE Journal on Selected Areas in Communications*, vol. 27, no. 7, pp. 1029–1046, Sep. 2009.

[51] F. Baccelli and B. Blaszczyszyn, "Stochastic geometry and wireless networks: Volume I theory," *Foundation and Trend R in Networking*, vol. 3, no. 3–4, pp. 249–449, 2009.

[52] M. Haenggi, *Stochastic Geometry for Wireless Networks*. New York: Cambridge University Press, 2012.

[53] H. ElSawy, E. Hossain and M. Haenggi, "Stochastic geometry for modeling, analysis, and design of multi-tier and cognitive cellular wireless networks: A survey," *IEEE Communications Surveys Tutorials*, vol. 15, no. 3, pp. 996–1019, Third quarter 2013.

[54] S. Mukherjee, *Analytical Modeling of Heterogeneous Cellular Networks*. New York: Cambridge University Press, 2014.

[55] N. Deng, W. Zhou and M. Haenggi, "The Ginibre point process as a model for wireless networks with repulsion," *IEEE Transactions on Wireless Communications*, vol. 14, no. 1, pp. 107–121, Jan. 2015.

[56] M. Haenggi, "The meta distribution of the SIR in Poisson bipolar and cellular networks," *IEEE Transactions on Wireless Communications*, vol. 15, no. 4, pp. 2577–2589, April 2016.

[57] M. Haenggi, "The local delay in Poisson networks," *IEEE Transactions on Information Theory*, vol. 59, no. 3, pp. 1788–1802, Mar. 2013.

[58] T. D. Novlan, H. S. Dhillon and J. G. Andrews, "Analytical modeling of uplink cellular networks," *IEEE Transactions on Wireless Communications*, vol. 12, no. 6, pp. 2669–2679, June 2013.

[59] M. D. Renzo, W. Lu and P. Guan, "The intensity matching approach: A tractable stochastic geometry approximation to system-level analysis of cellular networks," *IEEE Transactions on Wireless Communications*, vol. 15, no. 9, pp. 5963–5983, Sep. 2016.

[60] J. Andrews, F. Baccelli and R. Ganti, "A tractable approach to coverage and rate in cellular networks," *IEEE Transactions on Communications*, vol. 59, no. 11, pp. 3122–3134, Nov. 2011.

[61] H. S. Dhillon, R. K. Ganti, F. Baccelli and J. G. Andrews, "Modeling and analysis of K-tier downlink heterogeneous cellular networks," *IEEE Journal on Selected Areas in Communications*, vol. 30, no. 3, pp. 550–560, Apr. 2012.

[62] H. S. Dhillon, M. Kountouris and J. G. Andrews, "Downlink MIMO hetnets: Modeling, ordering results and performance analysis," *IEEE Transactions on Wireless Communications*, vol. 12, no. 10, pp. 5208–5222, Oct. 2013.

[63] H. S. Dhillon, R. K. Ganti and J. G. Andrews, "Load-aware modeling and analysis of heterogeneous cellular networks," *IEEE Transactions on Wireless Communications*, vol. 12, no. 4, pp. 1666–1677, Apr. 2013.

[64] Y. J. Chun, S. L. Cotton, H. S. Dhillon, A. Ghrayeb and M. O. Hasna, "A stochastic geometric analysis of device-to-device communications operating over generalized fading channels," *IEEE Transactions on Wireless Communications*, vol. 16, no. 7, pp. 4151–4165, July 2017.

[65] D. Malak, M. Al-Shalash and J. G. Andrews, "Spatially correlated content caching for device-to-device communications," *IEEE Transactions on Wireless Communications*, vol. 17, no. 1, pp. 56–70, Jan. 2018.

[66] N. Kouzayha, Z. Dawy, J. G. Andrews and H. ElSawy, "Joint downlink/uplink RF wake-up solution for IoT over cellular networks," *IEEE Transactions on Wireless Communications*, vol. 17, no. 3, pp. 1574–1588, March 2018.

[67] V. V. Chetlur and H. S. Dhillon, "Downlink coverage analysis for a finite 3-D wireless network of unmanned aerial vehicles," *IEEE Transactions on Communications*, vol. 65, no. 10, pp. 4543–4558, Oct. 2017.

[68] T. Bai, A. Alkhateeb and R. W. Heath, "Coverage and capacity of millimeter-wave cellular networks," *IEEE Communications Magazine*, vol. 52, no. 9, pp. 70–77, Sep. 2014.

[69] T. Bai and R. W. Heath, "Coverage and rate analysis for millimeter-wave cellular networks," *IEEE Transactions on Wireless Communications*, vol. 14, no. 2, pp. 1100–1114, Feb. 2015.

[70] A. K. Gupta, J. G. Andrews and R. W. Heath, "Macrodiversity in cellular networks with random blockages," *IEEE Transactions on Wireless Communications*, vol. 17, no. 2, pp. 996–1010, Feb. 2018.

[71] A. Thornburg and R. W. Heath, "Ergodic rate of millimeter wave ad hoc networks," *IEEE Transactions on Wireless Communications*, vol. 17, no. 2, pp. 914–926, Feb. 2018.

[72] Y. Zhu, L. Wang, K. K. Wong and R. W. Heath, "Secure communications in millimeter wave Ad Hoc networks," *IEEE Transactions on Wireless Communications*, vol. 16, no. 5, pp. 3205–3217, May 2017.

[73] R. Jurdi, A. K. Gupta, J. G. Andrews and R. W. Heath, "Modeling infrastructure sharing in mmWave networks with shared spectrum licenses," *IEEE Transactions on Cognitive Communications and Networking*, vol. 4, no. 2, pp. 328–343, Jun. 2018.

[74] L. Wang, K. K. Wong, R. W. Heath and J. Yuan, "Wireless powered dense cellular networks: How many small cells do we need?" *IEEE Journal on Selected Areas in Communications*, vol. 35, no. 9, pp. 2010–2024, Sep. 2017.

[75] G. Nigam, P. Minero and M. Haenggi, "Coordinated multipoint joint transmission in heterogeneous networks," *IEEE Transactions on Communications*, vol. 62, no. 11, pp. 4134–4146, Nov. 2014.

[76] H. Sun, M. Sheng, M. Wildemeersch, T. Q. S. Quek and J. Li, "Traffic adaptation and energy efficiency for small cell networks with dynamic TDD," *IEEE Journal on Selected Areas in Communications*, vol. 34, no. 12, pp. 3234–3251, Dec. 2016.

[77] Y. S. Soh, T. Q. S. Quek, M. Kountouris and H. Shin, "Energy efficient heterogeneous cellular networks," *IEEE Journal on Selected Areas in Communications*, vol. 31, no. 5, pp. 840–850, May 2013.

[78] G. de la Roche, A. Valcarce, D. López-Pérez and J. Zhang, "Access control mechanisms for femtocells," *IEEE Communications Magazine*, vol. 48, no. 1, pp. 33–39, Jan. 2010.

[79] M. Ding, P. Wang, D. López-Pérez, G. Mao and Z. Lin, "Performance impact of LoS and NLoS transmissions in dense cellular networks," *IEEE Transactions on Wireless Communications*, vol. 15, no. 3, pp. 2365–2380, Mar. 2016.

[80] Qualcomm, "1000x: More smallcells. Hyper-dense small cell deployments," Jun. 2014. www.qualcomm.com/media/documents/files/1000x-more-small-cells.pdf

[81] X. Zhang and J. Andrews, "Downlink cellular network analysis with multi-slope path loss models," *IEEE Transactions on Communications*, vol. 63, no. 5, pp. 1881–1894, May 2015.

[82] M. Ding and D. López-Pérez, "Performance impact of base station antenna heights in dense cellular networks," *IEEE Transactions on Wireless Communications*, vol. 16, no. 12, pp. 8147–8161, Dec. 2017.

[83] J. Liu, M. Sheng, L. Liu and J. Li, "How dense is ultra-dense for wireless networks: From far- to near-field communications," *arXiv:1606.04749 [cs.IT]*, Jun. 2016.

[84] V. Coskun, K. Ok and B. Ozdenizci, *Near Field Communication (NFC): From Theory to Practice*. Chichester: John Wiley & Sons Ltd., Dec. 2011.

[85] A. AlAmmouri, J. G. Andrews and F. Baccelli, "SINR and throughput of dense cellular networks with stretched exponential path loss," *IEEE Transactions on Wireless Communications*, vol. 17, no. 2, pp. 1147–1160, Feb. 2018.

[86] M. Franceschetti, J. Bruck and L. J. Schulman, "A random walk model of wave propagation," *IEEE Transactions on Antennas and Propagation*, vol. 52, no. 5, pp. 1304–1317, May 2004.

[87] 3GPP, "TR 36.842: Study on small cell enhancements for E-UTRA and E-UTRAN, higher layer aspects," Dec. 2013. https://portal.3gpp.org/desktopmodules/Specifications/SpecificationDetails.aspx?specificationId=2543

[88] M. Ding, D. López-Pérez, G. Mao and Z. Lin, "Performance impact of idle mode capability on dense small cell networks," *IEEE Transactions on Vehicular Technology*, vol. 66, no. 11, pp. 10 446–10 460, Nov. 2017.

[89] M. Ding, D. López-Pérez, A. H. Jafari, G. Mao and Z. Lin, "Ultra-dense networks: A new look at the proportional fair scheduler," in *IEEE Global Telecommunications Conference (GLOBECOM)*, pp. 1–7, Dec. 2017.

[90] Y. Chen, M. Ding, D. López-Pérez, et al., "Ultra-dense network: A holistic analysis of multi-piece path loss, antenna heights, finite users and BS idle modes," *IEEE Transactions on Mobile Computing*, vol. 20, no. 4, pp. 1702–1713, Apr. 2021

[91] M. Ding, D. López-Pérez, R. Xue, A. Vasilakos and W. Chen, "On dynamic time-division-duplex transmissions for small-cell networks," *IEEE Transactions on Vehicular Technology*, vol. 65, no. 11, pp. 8933–8951, Nov. 2016.

[92] T. Ding, M. Ding, G. Mao, Z. Lin, A. Y. Zomaya and D. López-Pérez, "Performance analysis of dense small cell networks with dynamic TDD," *IEEE Transactions on Vehicular Technology*, vol. 67, no. 10, pp. 9816–9830, Oct. 2018.

[93] A. Goldsmith, *Wireless Communications*. Cambridge: Cambridge University Press, 2012.

[94] Y. Chen, M. Ding and D. López-Pérez, "Performance of ultra-dense networks with a generalized multipath fading," *IEEE Wireless Communications Letters*, vol. 8, no. 5, pp. 1419–1422, Oct. 2019.

[95] D. López-Pérez and M. Ding, "Toward ultradense small cell networks: A brief history on the theoretical analysis of dense wireless networks," *Wiley Encyclopedia of Electrical and Electronics Engineering*, May 2019. https://doi.org/10.1002/047134608X.W8392

[96] A. H. Jafari, M. Ding and D. López-Pérez, "Performance analysis of dense small cell networks with line of sight and non-line of sight transmissions under Rician fading." in T. Q. Duong, X. Chu and H. A. Suraweera (eds.), *Ultra-Dense Networks for 5G and Beyond: Modelling, Analysis, and Applications*, Chichester John Wiley & Sons Ltd., pp. 41–64, Apr. 2019.

[97] J. Yang, M. Ding, G. Mao, Z. Lin and X. Ge, "Analysis of underlaid d2d-enhanced cellular networks: Interference management and proportional fair scheduler," *IEEE Access*, vol. 7, pp. 35 755–35 768, Mar. 2019.

[98] J. Yang, M. Ding, G. Mao, et al., "Optimal base station antenna downtilt in downlink cellular networks," *IEEE Transactions on Wireless Communications*, vol. 18, no. 3, pp. 1779–1791, Mar. 2019.

[99] C. Ma, M. Ding, D. López-Pérez, et al., "Performance analysis of the idle mode capability in a dense heterogeneous cellular network," *IEEE Transactions on Communications*, vol. 66, no. 9, pp. 3959–3973, Sep. 2018.

[100] M. Ding, D. López-Pérez, H. Claussen and M. A. Kaafar, "On the fundamental characteristics of ultra-dense small cell networks," *IEEE Network*, vol. 32, no. 3, pp. 92–100, May 2018.

[101] B. Yang, G. Mao, X. Ge, M. Ding and X. Yang, "On the energy-efficient deployment for ultra-dense heterogeneous networks with NLoS and LoS transmissions," *IEEE Transactions on Green Communications and Networking*, vol. 2, no. 2, pp. 369–384, Jun. 2018.

[102] M. Ding, D. López-Pérez, G. Mao and Z. Lin, "Ultra-dense networks: Is there a limit to spatial spectrum reuse?" in *IEEE International Conference on Communications (ICC)*, pp. 1–6, May 2018.

[103] B. Yang, G. Mao, M. Ding, X. Ge and X. Tao, "Dense small cell networks: From noise-limited to dense interference-limited," *IEEE Transactions on Vehicular Technology*, vol. 67, no. 5, pp. 4262–4277, May 2018.

[104] X. Yao, M. Ding, D. López-Pérez, et al., "Performance analysis of uplink massive MIMO networks with a finite user density," in *IEEE Wireless Communications and Networking Conference (WCNC)*, pp. 1–6, Apr. 2018.

[105] M. Ding and D. López-Pérez, "Promises and caveats of uplink IoT ultra-dense networks," in *2018 IEEE Wireless Communications and Networking Conference (WCNC)*, pp. 1–6, Apr. 2018.

[106] C. Ma, M. Ding, H. Chen, et al., "On the performance of multi-tier heterogeneous cellular networks with idle mode capability," in *IEEE Wireless Communications and Networking Conference (WCNC)*, pp. 1–6, Apr. 2018.

[107] A. H. Jafari, D. López-Pérez, M. Ding and J. Zhang, "Performance analysis of dense small cell networks with practical antenna heights under Rician fading," *IEEE Access*, vol. 6, pp. 9960–9974, Oct. 2018.

[108] X. Yao, M. Ding, D. López-Pérez, Z. Lin and G. Mao, "What is the optimal network deployment for a fixed density of antennas?" in *IEEE Global Telecommunications Conference (GLOBECOM)*, pp. 1–6, Dec. 2017.

[109] M. Ding, D. López-Pérez, G. Mao and Z. Lin, "What is the true value of dynamic TDD?: A mac layer perspective," in *IEEE Global Telecommunications Conference (GLOBECOM)*, pp. 1–7, Dec. 2017.

[110] M. Ding and D. López-Pérez, "Performance impact of base station antenna heights in dense cellular networks," *IEEE Transactions on Wireless Communications*, vol. 16, no. 12, pp. 8147–8161, Dec. 2017.

[111] M. Ding and D. López-Pérez, "On the performance of practical ultra-dense networks: The major and minor factors," *The IEEE Workshop on Spatial Stochastic Models for Wireless Networks (SpaSWiN) 2017*, pp. 1–8, May 2017.

[112] B. Yang, M. Ding, G. Mao and X. Ge, "Performance analysis of dense small cell networks with generalized fading," in *IEEE International Conference on Communications (ICC)*, pp. 1–7, May 2017.

[113] T. Ding, M. Ding, G. Mao, et al., "Uplink performance analysis of dense cellular networks with LoS and NLoS transmissions," *IEEE Transactions on Wireless Communications*, vol. 16, no. 4, pp. 2601–2613, Apr. 2017.

[114] M. Ding and D. López-Pérez, "Please lower small cell antenna heights in 5G," in *IEEE Global Telecommunications Conference (GLOBECOM)*, pp. 1–6, Dec. 2016.

[115] M. Ding, D. López-Pérez, G. Mao and Z. Lin, "Study on the idle mode capability with LoS and NLoS transmissions," in *IEEE Global Telecommunications Conference (GLOBECOM)*, pp. 1–6, Dec. 2016.

[116] J. Wang, X. Chu, M. Ding and D. López-Pérez, "On the performance of multi-tier heterogeneous networks under LoS and NLoS transmissions," in *IEEE Global Telecommunications Conference (GLOBECOM)*, pp. 1–6, Dec. 2016.

[117] T. Ding, M. Ding, G. Mao, Z. Lin and D. López-Pérez, "Uplink performance analysis of dense cellular networks with LoS and NLoS transmissions," in *IEEE International Conference on Communications (ICC)*, pp. 1–6, May 2016.

[118] M. Ding, D. López-Pérez, G. Mao, P. Wang and Z. Lin, "Will the area spectral efficiency monotonically grow as small cells go dense?" in *IEEE Global Telecommunications Conference (GLOBECOM)*, San Diego, CA, pp. 1–7, Dec. 2015.

[119] A. H. Jafari, D. López-Pérez, M. Ding and J. Zhang, "Study on scheduling techniques for ultra dense small cell networks," in *IEEE Vehicular Technology Conference (VTC)*, pp. 1–6, Sep. 2015.

[120] A. Fotouhi, H. Qiang, M. Ding, et al., "Survey on UAV cellular communications: Practical aspects, standardization advancements, regulation, and security challenges," *IEEE Communications Surveys Tutorials*, vol. 21, no. 4, pp. 3417–3442, Fourth-quarter 2019.

[121] Z. Meng, Y. Chen, M. Ding and D. López-Pérez, "A new look at UAV channel modeling: A long tail of los probability," in *IEEE International Symposium on Personal, Indoor and Mobile Radio Communications (PIMRC)*, pp. 1–6, Sep. 2019.

[122] D. López-Pérez, M. Ding, H. Li, et al., "On the downlink performance of UAV communications in dense cellular networks," in *IEEE Global Telecommunications Conference (GLOBECOM)*, pp. 1–7, Dec. 2018.

[123] Z. Yin, J. Li, M. Ding, F. Song and D. López-Pérez, "Uplink performance analysis of base station antenna heights in dense cellular networks," in *IEEE Global Telecommunications Conference (GLOBECOM)*, pp. 1–7, Dec. 2018.

[124] H. Li, M. Ding, D. López-Pérez, et al., "Performance analysis of the access link of drone base station networks with LoS/NLoS transmissions," *Springer INISCOM2018*, pp. 1–7, Aug. 2018.

[125] C. Liu, M. Ding, C. Ma, et al., "Performance analysis for practical unmanned aerial vehicle networks with LoS/NLoS transmissions," in *IEEE International Conference on Communications (ICC)*, pp. 1–6, May 2018.

[126] Y. Chen, M. Ding, D. López-Pérez, et al., "Dynamic reuse of unlicensed spectrum: An inter-working of LTE and WiFi," *IEEE Wireless Communications*, vol. 24, no. 5, pp. 52–59, Oct. 2017.

[127] D. López-Pérez, J. Ling, B. H. Kim, et al., "LWIP and Wi-Fi Boost flow control," in *IEEE Wireless Communications and Networking Conference (WCNC)*, pp. 1–6, Mar. 2017.

[128] D. López-Pérez, D. Laselva, E. Wallmeier, et al., "Long term evolution-wireless local area network aggregation flow control," *IEEE Access*, vol. 4, pp. 9860–9869, Jan. 2016.

[129] Y. Chen, M. Ding, D. López-Pérez, Z. Lin and G. Mao, "A space-time analysis of LTE and Wi-Fi inter-working," *IEEE Journal on Selected Areas in Communications*, vol. 34, no. 11, pp. 2981–2998, Nov. 2016.

[130] D. López-Pérez, J. Ling, B. H. Kim, et al., "Boosted WiFi through LTE small cells: The solution for an all-wireless enterprise," in *IEEE International Symposium on Personal, Indoor and Mobile Radio Communications (PIMRC)*, pp. 1–6, Sep. 2016.

[131] F. Song, J. Li, M. Ding, et al., "Probabilistic caching for small-cell networks with terrestrial and aerial users," *IEEE Transactions on Vehicular Technology*, vol. 68, no. 9, pp. 9162–9177, Sep. 2019.

[132] P. Cheng, C. Ma, M. Ding, et al., "Localized small cell caching: A machine learning approach based on rating data," *IEEE Transactions on Communications*, vol. 67, no. 2, pp. 1663–1676, Feb. 2019.

[133] C. Ma, M. Ding, H. Chen, et al., "Socially aware caching strategy in device-to-device communication networks," *IEEE Transactions on Vehicular Technology*, vol. 67, no. 5, pp. 4615–4629, May 2018.

[134] Y. Chen, M. Ding, J. Li, et al., "Probabilistic small-cell caching: Performance analysis and optimization," *IEEE Transactions on Vehicular Technology*, vol. 66, no. 5, pp. 4341–4354, May 2017.

[135] J. Li, Y. Chen, M. Ding, et al., "A small-cell caching system in mobile cellular networks with LoS and NLoS channels," *IEEE Access*, vol. 5, pp. 1296–1305, Mar. 2017.

[136] C. Ma, M. Ding, H. Chen, et al., "Socially aware distributed caching in device-to-device communication networks," in *IEEE Global Telecommunications Conference (GLOBECOM)*, pp. 1–6, Dec. 2016.

[137] M. Ding, D. López-Pérez, G. Mao, Z. Lin and S. K. Das, "DNA-GA: A tractable approach for performance analysis of uplink cellular networks," *IEEE Transactions on Communications*, vol. 66, no. 1, pp. 355–369, Jan. 2018.

[138] M. Ding, D. López-Pérez, G. Mao and Z. Lin, "DNA-GA: A new approach of network performance analysis," in *IEEE International Conference on Communications (ICC)*, pp. 1–7, May 2016.

[139] M. Ding, D. López-Pérez, G. Mao and Z. Lin, "Microscopic analysis of the uplink interference in FDMA small cell networks," *IEEE Trans. on Wireless Communications*, vol. 15, no. 6, pp. 4277–4291, Jun. 2016.

[140] M. Ding, D. López-Pérez, G. Mao and Z. Lin, "Approximation of uplink inter-cell interference in FDMA small cell networks," in *IEEE Global Telecommunications Conference (GLOBECOM)*, pp. 1–7, Dec. 2015.

[141] M. Ding, D. López-Pérez, A. V. Vasilakos and W. Chen, "Analysis on the SINR performance of dynamic TDD in homogeneous small cell networks," *2014 IEEE Global Communications Conference*, pp. 1552–1558, Dec. 2014.

[142] M. Ding, D. López-Pérez, A. V. Vasilakos and W. Chen, "Dynamic TDD transmissions in homogeneous small cell networks," in *IEEE International Conference on Communications (ICC)*, pp. 616–621, Jun. 2014.

[143] M. Ding, D. López-Pérez, R. Xue, A. V. Vasilakos and W. Chen, "Small cell dynamic TDD transmissions in heterogeneous networks," in *IEEE International Conference on Communications (ICC)*, pp. 4881–4887, Jun. 2014.

[144] J. Wang, X. Chu, M. Ding and D. López-Pérez, "The effect of LoS and NLoS transmissions on base station clustering in dense small-cell networks," in *IEEE Vehicular Technology Conference (VTC)*, pp. 1–6, Sep. 2019.

[145] M. Ding and H. Luo, *Multi-Point Cooperative Communication Systems: Theory and Applications*. Berlin/Heidelberg: Springer, 2013.

[146] H. Claussen, D. Lopez-Perez, L. Ho, R. Razavi and S. Kucera, *Small Cell Networks: Deployment, Management, and Optimization*. Hoboken, NJ: Wiley-IEEE Press, 2018.

[147] 3GPP, "TR 25.814: Physical layer aspects for evolved Universal Terrestrial Radio Access (UTRA)," Oct. 2006. https://portal.3gpp.org/desktopmodules/Specifications/SpecificationDetails.aspx?specificationId=1247

[148] Cisco, "Antenna patterns and their meaning," White Paper, Aug. 2007. https://www.industrialnetworking.com/pdf/Antenna-Patterns.pdf

[149] 3GPP, "TR 38.901: Study on channel model for frequencies from 0.5 to 100 GHz," Jan. 2020. https://portal.3gpp.org/desktopmodules/Specifications/SpecificationDetails.a

[150] X. Li, R. W. Heath, Jr., K. Linehan and R. Butler, "Impact of metro cell antenna pattern and downtilt in heterogeneous networks," *arXiv:1502.05782 [cs.IT]*, Feb. 2015. [Online]. Available: http://arxiv.org/abs/1502.05782

[151] X. Chu, D. López-Pérez, F. Gunnarsson and Y. Yang, *Heterogeneous Cellular Networks: Theory, Simulation and Deployment*. Cambridge: Cambridge University Press, 2003.

[152] J. S. Seybold, *Introduction to RF Propagation*. Hoboken, NJ: John Wiley & Sons Ltd., 2005.

[153] 3GPP, "TR 36.828: Further enhancements to LTE Time Division Duplex for Downlink-Uplink interference management and traffic adaptation," Jun. 2012. https://portal.3gpp.org/desktopmodules/Specifications/SpecificationDetails.aspx?specificationId=2507

[154] A. Goldsmith, *Wireless Communications*. Cambridge: Cambridge University Press, 2005.

[155] F. Adachi and T. Tjhung, "Tapped delay line model for band-limited multipath channel in DS-CDMA mobile radio," *Electronics Letters*, vol. 37, no. 5, pp. 318–319, Mar. 2001.

[156] I. Gradshteyn and I. Ryzhik, *Table of Integrals, Series, and Products, 7th ed.* Cambridge, MA: Academic Press, 2007.

[157] 3GPP, "TR 25.996: Spatial channel model for Multiple Input Multiple Output (MIMO) simulations," Jun. 2018. https://portal.3gpp.org/desktopmodules/Specifications/SpecificationDetails.aspx?specificationId=1382

[158] H. Claussen, D. López-Pérez, L. Ho, R. Razavi and S. Kucera, *Small Cell Networks: Deployment, Management, and Optimization*, 1st ed. Hoboken, NJ: Wiley-IEEE Press, 2018.

[159] I. Slivnyak, "Some properties of stationary flows of homogeneous random events," *Theory Probability*, vol. 7, pp. 336–341, 1962.

[160] M. D. Renzo, W. Lu and P. Guan, "The intensity matching approach: A tractable stochastic geometry approximation to system-level analysis of cellular networks," *IEEE Transactions on Wireless Communications*, vol. 15, no. 9, pp. 5963–5983, Sep. 2016.

[161] M. Rupp, S. Schwarz and M. Taranetz, *The Vienna LTE-Advanced Simulators: Up and Downlink, Link and System Level Simulation*, 1st ed. Berlin/Heidelberg: Springer, 2016.

[162] S. Sesia, I. Toufik and M. Baker, *LTE - The UMTS Long Term Evolution: From Theory to Practice*, 2nd ed. Hoboken, NJ: John Wiley & Sons Ltd., 2011.

[163] S. Ahmadi, *5G NR: Architecture, Technology, Implementation, and Operation of 3GPP New Radio Standards*, 1st ed. Hoboken, NJ: Academic Press, Jun. 2019.

[164] P. Frenger, S. Parkvall and E. Dahlman, "Performance comparison of HARQ with Chase combining and incremental redundancy for HSDPA," in *IEEE Vehicular Technology Conference (VTC)*, pp. 1829–1833, Oct. 2001.

[165] J. Gozalvez and J. Dunlop, "Link level modelling techniques for analysing the configuration of link adaptation algorithms in mobile radio networks," in *European Wireless*, pp. 1–6, Feb. 2004.

[166] G. Monghal, K. I. Pedersen, I. Z. Kovacs and P. E. Mogensen, "QoS oriented time and frequency domain packet schedulers for the UTRAN long term evolution," in *IEEE Vehicular Technology Conference (VTC)*, pp. 2532–2536, May 2008.

[167] N. Kolehmainen, J. Puttonen, P. Kela, et al., "Channel quality indication reporting schemes for UTRAN long term evolution downlink," in *IEEE Vehicular Technology Conference (VTC)*, pp. 2522–2526, May 2008.

[168] K. I. Pedersen, T. E. Kolding, F. Frederiksen, et al., "An overview of downlink radio resource management for UTRAN long-term evolution," *IEEE Communications Magazine*, vol. 47, no. 7, pp. 86–93, Jul. 2009.

[169] S. D. Lembo, "Modeling BLER performance of punctured turbo codes," Ph.D. Thesis, School of Electrical Engineering, Aalto University, May 2011.

[170] C. B. Chae, I. Hwang, R. W. Heath and V. Tarokh, "Interference aware-coordinated beamforming in a multi-cell system," *IEEE Transactions on Wireless Communications*, vol. 11, no. 10, pp. 3692–3703, Oct. 2012.

[171] K. S. Gilhousen, I. Jacobs, R. Padovani, et al., "On the capacity of a cellular CDMA system," *IEEE Transactions on Vehicular Technology*, vol. 40, no. 2, pp. 303–312, May 1991.

[172] A. J. Viterbi, A. M. Viterbi and E. Zehavi, "Other-cell interference in cellular power-controlled CDMA," *IEEE Transactions on Communications*, vol. 42, no. 2/3/4, pp. 1501–1504, Feb.–Apr 1994.

[173] D. Gesbert, S. Hanly, H. Huang, et al., "Multi-cell MIMO cooperative networks: A new look at interference," *IEEE Journal on Selected Areas in Communications*, vol. 28, no. 9, pp. 1380–1408, Dec. 2010.

[174] A. D. Wyner, "Shannon-theoretic approach to a Gaussian cellular multi-access channel," *IEEE Transactions on Information Theory*, vol. 40, no. 6, pp. 1713–1727, Nov. 1994.

[175] J. Xu, J. Zhang and J. G. Andrews, "On the accuracy of the Wyner Model in cellular networks," *IEEE Transactions on Wireless Communications*, vol. 10, no. 9, pp. 3098–3109, Jul. 2011.

[176] O. Somekh, B. M. Zaidel and S. Shamai, "Sum rate characterization of joint multiple cell-site processing," *IEEE Transactions on Information Theory*, vol. 53, no. 12, pp. 4473–4497, Dec. 2007.

[177] S. Jing, D. N. C. Tse, J. Hou, et al., "Multi-cell downlink capacity with coordinated processing," *EURASIP Journal on Wireless Communications and Networking*, vol. 2008, pp. 1–19, Apr. 2008.

[178] O. Simeone, O. Somekh, H. V. Poor and S. Shamai, "Local base station cooperation via finite-capacity links for the uplink of linear cellular networks," *IEEE Transactions on Information Theory*, vol. 55, no. 1, pp. 190–204, Jan. 2009.

[179] T. S. Rappaport, *Wireless Communications: Principles and Practice*, 2nd ed. Hoboken, NJ: Prentice-Hall, 2002.

[180] D. Stoyan, W. Kendall and J. Mecke, *Stochastic Geometry and Its Applications*, 2nd ed. Hoboken, NJ: John Wiley & Sons Ltd., 1996.

[181] D. Daley and D. V. Jones, *An Introduction to the Theory of Point Processes. Volume I: Elementary Theory and Methods*, 2nd ed. New York: Springer, 2003.

[182] D. Daley and D. V. Jones, *An Introduction to the Theory of Point Processes. Volume II: General Theory and Structure*, 2nd ed. Berlin/Heidelberg: Springer, 2008.

[183] R. G. Bartle and D. R. Sherbert, *Introduction to Real Analysis*, 4th ed. Hoboken, NJ: John Wiley & Sons Ltd., 2010.

[184] A. S. Kechris, *Classical Descriptive Set Theory*. Berlin/Heidelberg: Springer-Verlag, 1995.

[185] D. W. Stroock, *Probability Theory: An Analytic View*, 2nd ed. Cambridge: Cambridge University Press, 2012.

[186] A. M. Bruckner, J. B. Bruckner and B. S. Thomson, *Real Analysis*, 2nd ed. Scotts Valley, CA: CreateSpace Independent Publishing Platform, 2008.

[187] G. Last and M. Penrose, *Lectures on the Poisson Process*, 1st ed. Cambridge: Cambridge University Press, 2017.

[188] N. Campbell, "The study of discontinuous phenomena," *Mathematical Proceedings of the Cambridge Philosophy Society*, vol. 15, pp. 117–136, 1909.

[189] M. Dacey, "Two-dimensional random point patterns: A review and an interpretation," *Papers of the Regional Sciency Association*, vol. 13, no. 1, pp. 41–55, 1964.

[190] P. Hertz, "Uber den geigerseitigen durchschnittlichen Abstand von Punkten, die mit bekannter mittlerer Dichte im Raume angeordnet sind," *Methematiche Annalen*, vol. 67, pp. 387–398, 1909.

[191] S. Chandrasekhar, "Stochastic processes in physics and chemistry," *Review of Modern Physics*, vol. 15, pp. 1–89, 1943.

[192] J. Skellam, "Random dispersal in theoretical populations," *Biometrika*, vol. 38, pp. 196–218, 1951.

[193] M. Moroshita, "Estimation of population density by spacing methods," *Memoirs of the Faculty of Science, Kyushi University*, vol. 1, 187–197, 1954.

[194] H. Thompson, "Distribution of distance to n-th neighbour in a population of randomly distributed individuals," *Ecology*, vol. 37, no. 2, pp. 391–394, Apr. 1956.

[195] D. Moltchanov, "Distance distributions in random networks," *Ad-Hoc Networks*, vol. 10, no. 6, pp. 1146–1166, Aug. 2012.

[196] Small Cell Forum, "Small cell siting challenges and recommendations," Small Cell Forum Release 10.0 - Document 195.10.01, Aug. 2018. www.5gamericas.org/wp-content/uploads/2019/07/Small_Cell_Siting_Challenges__Recommendations_White paper_final.pdf

[197] F. Baccelli and S. Zuyev, "Stochastic geometry models of mobile communication networks," in J. H. Dshalalow (ed.), *Frontiers in Queueing: Models and Applications in Science and Engineering*. Boca Raton, FL: CRC Press, pp. 227–243, 1996.

[198] F. Baccelli, M. Klein, M. Lebourges and S. Zuyev, "Stochastic geometry and architecture of communication networks," *Journal of Telecommunication Systems*, vol. 7, no. 1, pp. 209–227, Jun. 1997.

[199] T. X. Brown, "Cellular performance bounds via shotgun cellular systems," *IEEE Journal on Selected Areas in Communications*, vol. 18, no. 11, pp. 2443–2455, Nov. 2000.

[200] A. Al-Hourani, R. J. Evans and K. Sithamparanathan, "Nearest neighbour distance distribution in hard-core point processes," *arXiv:1606.03695 [cs.IT]*, Jun. 2016.

[201] C. Choi, J. O. Woo and J. G. Andrews, "Modeling a spatially correlated cellular network with strong repulsion," *arXiv:1701.02261 [cs.IT]*, Jan. 2017.

[202] M. J. Nawrocki, M. Dohler and A. H. Aghvami, *Understanding UMTS Radio Network Modelling, Planning and Automated Optimisation: Theory and Practice*, 1st ed. Hoboken, NJ: John Wiley & Sons Ltd., 2006.

[203] J. Laiho, A. Wacker and T. Novosad, *Radio Network Planning and Optimisation for UMTS*, 2nd ed. Hoboken, NJ: John Wiley & Sons Ltd., 2006.

[204] Ajay R. Mishra, *Fundamentals of Network Planning and Optimisation 2G/3G/4G: Evolution to 5G*, 2nd ed. Hoboken, NJ: John Wiley & Sons Ltd., 2018.

[205] X. Zhang and J. Andrews, "Downlink cellular network analysis with multi-slope path loss models," *IEEE Transactions on Communications*, vol. 63, no. 5, pp. 1881–1894, May 2015.

[206] T. Bai and R. Heath, "Coverage and rate analysis for millimeter-wave cellular networks," *IEEE Transactions on Wireless Communications*, vol. 14, no. 2, pp. 1100–1114, Feb. 2015.

[207] C. Galiotto, N. K. Pratas, N. Marchetti and L. Doyle, "A stochastic geometry framework for LOS/NLOS propagation in dense small cell networks," *arXiv:1412.5065 [cs.IT]*, Jun. 2015. [Online]. Available: http://arxiv.org/abs/1412.5065

[208] M. Ding, P. Wang, D. López-Pérez, G. Mao and Z. Lin, "Performance impact of LoS and NLoS transmissions in dense cellular networks," *IEEE Transactions on Wireless Communications*, vol. 15, no. 3, pp. 2365–2380, Mar. 2016.

[209] R. L. Burden and J. D. Faires, *Numerical Analysis*, 2nd ed. Boston, MA: PWS Publishers, 1985.

[210] R. Pettijohn, "There's nothing small about small cell deployments," *ICT Solutions and Education*, Jul. 2019. https://isemag.com/2019/07/theres-nothing-small-about-small-cell-deployments/

[211] Small Cell Forum, "Deployment issues for urban small cells," Small Cell Forum Release 7.0 - Document 096.07.01, Jun. 2014. https://scf.io/en/documents/096_-_Deployment_issues_for_urban_small_cells.php

[212] G. Fischer, F. Pivit and W. Wiesbeck, "EISL, the pendant to EIRP: A measure for the receive performance of base stations at the air interface," in *2002 32nd European Microwave Conference*, pp. 1–4 Sep. 2002.

[213] H. Holma and A. Toskala, *WCDMA for UMTS: Radio Access for Third Generation Mobile Communications*, 3rd ed. Hoboken, NJ: John Wiley & Sons Ltd., 2002.

[214] H. Holma and A. Toskala, *LTE for UMTS - OFDMA and SC-FDMA Based Radio Access*. Hoboken, NJ: John Wiley & Sons Ltd., 2009.

[215] I. Ashraf, L. Ho and H. Claussen, "Improving energy efficiency of femtocell base stations via user activity detection," in *IEEE Wireless Communications and Networking Conference (WCNC)*, Sydney, Australia, pp. 1–5 Apr. 2010.

[216] E. Dahlman, S. Parkvall and J. Skold, *4G, LTE-Advanced Pro and The Road to 5G*, 3rd ed. Cambridge, MA: Academic Press, 2016.

[217] S. Lee and K. Huang, "Coverage and economy of cellular networks with many base stations," *IEEE Communications Letters*, vol. 16, no. 7, pp. 1038–1040, Jul. 2012.

[218] Z. Luo, M. Ding and H. Luo, "Dynamic small cell on/off scheduling using Stackelberg game," *IEEE Communications Letters*, vol. 18, no. 9, pp. 1615–1618, Sep. 2014.

[219] C. Li, J. Zhang and K. Letaief, "Throughput and energy efficiency analysis of small cell networks with multi-antenna base stations," *IEEE Transactions on Wireless Communications*, vol. 13, no. 5, pp. 2505–2517, May 2014.

[220] T. Zhang, J. Zhao, L. An and D. Liu, "Energy efficiency of base station deployment in ultra dense HetNets: A stochastic geometry analysis," *IEEE Wireless Communications Letters*, vol. 5, no. 2, pp. 184–187, Apr. 2016.

[221] J. G. Proakis, *Digital Communications*, 4th ed. New York: McGraw-Hill, 2000.

[222] A. Pokhariyal, K. I. Pedersen, G. Monghal, et al., "HARQ aware frequency domain packet scheduler with different degrees of fairness for the UTRAN long term evolution," in *IEEE Vehicular Technology Conference (VTC)*, pp. 2761–2765, Apr. 2007.

[223] T. Chapman, E. Larsson, P. von Wrycza, et al., *HSPA Evolution: The Fundamentals for Mobile Broadband*. Cambridge, MA: Academic Press, 2014.

[224] E. Dahlman, S. Parkvall and J. Skold, *4G: LTE/LTE-Advanced for Mobile Broadband*. Cambridge, MA: Academic Press, 2013.

[225] J. G. Choi and S. Bahk, "Cell-throughput analysis of the proportional fair scheduler in the single-cell environment," *IEEE Transactions on Vehicular Technology*, vol. 56, no. 2, pp. 766–778, Mar. 2007.

[226] G. Miao, J. Zander, K. W. Sung and S. B. Slimane, *Fundamentals of Mobile Data Networks*, 1st ed. Cambridge: CreateSpace Independent Publishing Platform, 2016.

[227] E. Liu and K. K. Leung, "Expected throughput of the proportional fair scheduling over rayleigh fading channels," *IEEE Communications Letters*, vol. 14, no. 6, pp. 515–517, Jun. 2010.

[228] J. Wu, N. B. Mehta, A. F. Molisch and J. Zhang, "Unified spectral efficiency analysis of cellular systems with channel-aware schedulers," *IEEE Transactions on Communications*, vol. 59, no. 12, pp. 3463–3474, Dec. 2011.

[229] F. Liu, J. Riihijarvi and M. Petrova, "Robust data rate estimation with stochastic SINR modeling in multi-interference OFDMA networks," in *IEEE International Conference on Sensing, Communication, and Networking (SECON)*, pp. 211–219, Jun. 2015.

[230] H. A. David and H. N. Nagaraja, *Order Statistics*, 3rd ed. Hoboken, NJ: John Wiley & Sons Ltd., 2003.

[231] M. Ding, D. López-Pérez, Y. Chen, et al., "UDN: A holistic analysis of multi-piece path loss, antenna heights, finite users and BS idle modes," *IEEE Transactions on Mobile Computing*, vol. 20, no. x, pp. 1, Apr. 2021.

[232] S. Boyd and L. Vandenberghe, *Convex Optimization*. Cambridge: Cambridge University Press, 2004.

[233] Ericsson, "Uplink and slow time-to-content: Extract from the Ericsson mobility report," White Paper, Nov. 2016. www.ericsson.com/en/reports-and-papers/mobility-report/articles/uplink-speed-and-slow-time-to-content

[234] ITU-R, "Minimum requirements related to technical performance for IMT-2020 radio interface(s)," Report ITU-R M.2410, Nov. 2017. www.itu.int/pub/R-REP-M.2410

[235] A. Ghosh, J. Zhang, J. G. Andrews and R. Muhamed, *Fundamentals of LTE*. Hoboken, NJ: Prentice Hall, 2010.

[236] T. Ding, M. Ding, G. Mao, et al., "Uplink performance analysis of dense cellular networks with los and nlos transmissions," *arXiv:1609.07837 [cs.IT]*, vol. abs/1609.07837, Sep. 2016. [Online]. Available: http://arxiv.org/abs/1609.07837

[237] M. Haenggi, "User point processes in cellular networks," *IEEE Wireless Communications Letters*, vol. 6, no. 2, pp. 258–261, Apr. 2017.

[238] 3GPP, "TR 36.814: Further advancements for E-UTRA physical layer aspects," Mar. 2010. https://portal.3gpp.org/desktopmodules/Specifications/SpecificationDetails.aspx?specificationId=2493

[239] H. Sun, M. Wildemeersch, M. Sheng and T. Q. S. Quek, "D2D enhanced heterogeneous cellular networks with dynamic TDD," *IEEE Transactions on Wireless Communications*, vol. 14, no. 8, pp. 4204–4218, Aug. 2015.

[240] B. Yu, L. Yang, H. Ishii and S. Mukherjee, "Dynamic TDD support in macrocell- assisted small cell architecture," *IEEE Journal on Selected Areas in Communications*, vol. 33, no. 6, pp. 1201–1213, Jun. 2015.

[241] A. K. Gupta, M. N. Kulkarni, E. Visotsky, et al., "Rate analysis and feasibility of dynamic TDD in 5G cellular systems," in *IEEE International Conference on Communications (ICC)*, pp. 1–6 May 2016.

[242] S. Goyal, C. Galiotto, N. Marchetti and S. Panwar, "Throughput and coverage for a mixed full and half duplex small cell network," in *IEEE International Conference on Communications (ICC)*, pp. 1–7 May 2016.

Index